Chaos

Steven Strogatz, Ph.D.

THE
GREAT
COURSES

PUBLISHED BY:

THE GREAT COURSES
Corporate Headquarters
4840 Westfields Boulevard, Suite 500
Chantilly, Virginia 20151-2299
Phone: 1-800-832-2412
Fax: 703-378-3819
www.thegreatcourses.com

Steven Strogatz, Ph.D.

Jacob Gould Schurman Professor
of Applied Mathematics
Professor of Theoretical and Applied Mechanics
Cornell University

Professor Steven Strogatz is the Jacob Gould Schurman Professor of Applied Mathematics and Professor of Theoretical and Applied Mechanics at Cornell University. After receiving his B.A. in Mathematics summa cum laude from Princeton University in 1980, Professor Strogatz won a Marshall Scholarship to Trinity College, Cambridge. He did his doctoral work in Applied Mathematics at Harvard University, where he received his Ph.D. in 1986; afterward, he was awarded a National Science Foundation Postdoctoral Fellowship. In 1989, he joined the faculty of the Department of Mathematics at MIT, where his research was nationally recognized with a Presidential Young Investigator Award from the White House in 1990. Professor Strogatz began teaching at Cornell University in 1994. In 2004, he was appointed as an external faculty member of the Santa Fe Institute.

Professor Strogatz has done seminal research in chaos and complexity theory with applications to physics and biology. A review in *Nature* described him as one of the "most creative biomathematicians of the past few decades." A theme in his work is the hidden mathematics of everyday life; he and his students have explored the geometry of DNA, the rhythms of human sleep, the synchronous flashing of fireflies, the decline and death of languages, and the mathematics behind "six degrees of separation" in social networks. In particular, his 1998 *Nature* paper (with his former student Duncan Watts) titled "Collective dynamics of 'small-world' networks" is the most highly cited article about networks in the past decade across all scientific disciplines. Professor Strogatz's research has been featured in many mass media outlets, including *The New York Times*, *Newsweek*, CBS News, *U.S. News and World Report*, *The New Yorker*, *Discover*, *Nature*, *Science*, *Scientific American*, *Die Zeit*, and *The Daily Telegraph*.

Professor Strogatz has been lauded for his exceptional abilities as a teacher and communicator. In 1991, he was honored with the E. M. Baker Memorial Award for Excellence in Undergraduate Teaching, MIT's only institute-wide teaching prize selected and awarded solely by students. He also has won several teaching awards from Cornell University's College of Engineering, including the 2006 Tau Beta Pi Excellence in Teaching Award, given to a faculty member selected by engineering students for exemplary teaching. At the national level, Professor Strogatz received the 2007 Communications Award—a lifetime achievement award for the communication of mathematics to the general public—from the Joint Policy Board for Mathematics, which represents the four major American mathematical societies. He has delivered dozens of public lectures nationwide and has made many appearances on radio and television, including National Public Radio, BBC, Discovery Channel, and C-SPAN.

His books include *Nonlinear Dynamics and Chaos* (Perseus, 1994)—the most widely used textbook on chaos theory—and *Sync: The Emerging Science of Spontaneous Order* (Hyperion, 2003), which is aimed at nonscientists and was chosen as a Best Book of 2003 by *Discover*. ∎

Acknowledgments

Thanks to the many friends and colleagues who helped me prepare this course by providing advice, information, visual material, or just plain encouragement. In particular, thanks to Alan Alda, Wolfgang Beyer, Joe Burns, Victor Donnay, Irv Epstein, Michael Frame, David Harrison, Eric Heller, Leon Iasemidis, Martin Lo, Benoit Mandelbrot, Frank Moon, Lou Pecora, Heinz-Otto Peitgen, Richard Rand, Dan Rockmore, Shane Ross, Richard Taylor, and Steve Wheatcraft.

Hubert Hohn—computer artist, friend, and collaborator—was indispensable. Hu created many of the computer simulations and animations shown in this course. Thanks for your brilliant craftsmanship, generosity, and never-ending good cheer.

The team at The Teaching Company was insightful, professional, meticulously attentive to detail, yet light-hearted and fun. Thank you to Joan Burton, Tony Hidenrick, Zach Rhoades, Lisa Robertson, and Jay Tate. All of you were delightful to work with.

Above all, I thank my wife, Carole, and our two young daughters, Leah and Joanna. I love you all so much, and I especially appreciated your patience and support during all those months when I was distracted by that chaotic guest in our house. ■

Table of Contents

Table of Contents

Table of Contents

Chaos

Scope:

Newton's laws describe the motion of nearly everything under the Sun and give physics great power in its ability to relate cause and effect. Yet the path of a ball in a pinball machine seems to defy human control—it seems chaotic. What is the science of chaos? How can something be chaotic yet follow deterministic laws? This course investigates how apparent disorder arises from extreme sensitivity to initial conditions.

One of the goals of the course is to explain the main ideas of chaos clearly and honestly, in a way that any interested layperson can understand. The material is inherently accessible because it is rooted in our daily experience; the phenomena we will be discussing are familiar, not esoteric. Furthermore, the subject is marvelously broad and interdisciplinary, intersecting nearly every field of human knowledge and endeavor from astronomy and zoology to the arts, the humanities, and business. No doubt that sounds far-fetched, but it really is not because chaos is the science of how things change—and everything changes. The systems we will consider are the dynamical systems all around us, from the variability of the weather and animal populations to earthquakes, Internet traffic, and the wobbles of London's Millennium Bridge. Chaos also sheds new light on the dynamical systems inside us; think of the spontaneous pulsing of our hearts or the electrical firing of our brain cells.

Another crucial feature of chaos adds to its accessibility: No advanced math is needed to grasp its key concepts. Instead, pictures turn out to be more powerful than symbols and formulas. Although we will discuss many mathematical ideas, we will always visualize them and will not manipulate x's and y's. On the few occasions when it helps to resort to algebra, the math will be explained in both words and pictures.

After an introductory lecture explaining what makes the science of chaos so revolutionary, the course begins with the story of how chaos was glimpsed and then lost for nearly a century before reemerging explosively in the 1970s and 1980s. This is a tale of twists and turns—in many ways like a detective story. We devote the first half of the course to it because the intellectual ride is thrilling and fast paced, with fascinating ideas coming from every direction. We will meet Isaac Newton, the champion of order who gave us a picture of the universe so orderly that it seemed to outlaw chaos once and for all. Two hundred years later, a chink in Newton's armor appeared and gave humanity its first glimpse of mathematical chaos. When it was revealed to Henri Poincaré, however, he recoiled from it; the picture he uncovered was horrifying in both its implications and its appearance. We will see why Poincaré's discovery was overshadowed and forgotten for 70 years before being rediscovered in even starker form by Ed Lorenz, a meteorologist at MIT. His work gave us the concept of the butterfly effect, which incredibly had no impact until the 1970s and 1980s, when the tidal wave of chaos theory finally hit.

Along the way, we will learn the core ideas of chaos. By developing these ideas in their historical context, we can feel like we are sitting at the elbows of the masters, following them on their false starts and sudden flashes of insight. Not only is this an enjoyable way to learn the fundamentals of the subject, it also helps us appreciate just what an intellectual feat it was to create the science of chaos in the first place.

The third quarter of the course focuses on fractals: infinitely intricate shapes in which the smallest parts resemble the whole. We will learn why fractals so often appear hand in hand with chaos and what distinguishes them from ordinary, Euclidean shapes. We then will examine their utility in describing a wide variety of erratic phenomena and processes, ranging from the shapes of coastlines and the gyrations of the stock market to the hidden structure of Jackson Pollock's drip paintings.

The final section of the course takes us to the frontiers of chaos research today. We will see how chaos is being used for practical purposes in encryption and space mission design. Then we will look at the medical implications of chaos and the tantalizing but mysterious linkages among chaos, quantum mechanics, and number theory.

The course concludes with a vision for where the science of chaos is headed next, a path that is clear but daunting. Chaos theorists were among the earliest scientists to focus on nonlinear systems in which the whole is more (or less) than the sum of the parts. Many of the major unsolved problems of science today have precisely this character. From cancer to consciousness, these sorts of problems involve thickets of interlocking feedback loops in which everything affects everything else and the simple logic of cause and effect breaks down. Solving such problems will require new ways of thinking, and these problems are going to challenge scientists for decades to come. ■

The Chaos Revolution
Lecture 1

There's a genuine reason for the excitement about chaos theory. People sense, correctly, that there is something revolutionary happening in science. We're starting to find our way through a no-man's-land, a territory that science had long dreaded and feared to explore. This is the realm of chaos—the erratic side of nature, the discontinuous, the jagged, the irregular and unpredictable.

Welcome to chaos! I realize that sounds silly, and it's meant to. I want to approach this course in a playful spirit, to help you enjoy learning about the amazing subject of chaos theory. By now, everyone has heard of chaos theory. It's one of those rare branches of science to have become a pop sensation. James Gleick's bestselling book *Chaos: Making a New Science* (Viking, 1987) was translated into more than 20 languages and sold more than half a million copies. Meanwhile, multicolored images of fractals seemed to be appearing everywhere in the late 1980s—on T-shirts, posters, and screensavers. The superstar of fractals was the Mandelbrot set, in which dazzling structure reveals itself endlessly as you magnify small parts of the image.

Still, it's remarkable that the public became so fascinated by chaos. Why this curiosity about a seemingly arcane branch of math and physics? Some cynics dismissed the public interest in chaos as no more than fascination with pretty pictures and a catchy name. But I believe there's a genuine reason for the excitement. Science, especially physics, has always been obsessed with order, patterns—the lawful side of nature. But in every field, chaos has always been there, an irritating reminder of what we don't know and don't understand. It was there in ecology, in the fluctuations of wildlife populations. It was there in cardiology, in the arrhythmic quivering of a human heart in the moments before death. And it was there in fluid mechanics, in the turbulent motion of the sea and atmosphere. Scientists regarded all these as vexing puzzles, or worse, as pathologies, monsters.

But starting about 30 years ago, a few scientists began finding strange, unexpected connections between different forms of chaos. Geologists noticed surprising patterns in the frequency of earthquakes. The same patterns appeared in the variability of human heart rates and in bursts of traffic on the Internet. The rules of chaos were turning out to be universal—independent of the stuff behaving chaotically—the same for electronic circuits, lasers, chemical reactions, or nerve cells. It was as if disorder was a thing in itself. It didn't matter what was behaving chaotically; the *process* of becoming chaotic was turning out to be lawful—but the laws were like nothing science had ever seen before.

This is a course about the new science of chaos. By the time we're through, you'll never look at the world the same way again. But what *is* chaos, exactly? It's a paradoxical state, a kind of unpredictable behavior in a system governed by deterministic laws.

We can illustrate it by playing with a double pendulum. The pendulum swings back and forth according to Newton's laws. Indeed, Newton's laws describe the motion of nearly everything under the Sun. For almost 350 years they've given physics great power to relate cause and effect. So you'd think we should be able to predict how this double pendulum will behave, and, until about 30 years ago, that's what everyone thought. It was even a textbook exercise to analyze its motion. I remember solving it in my sophomore physics class.

But under certain conditions (that our professor never asked us to consider), it swings crazily. It seems "chaotic." Worse than that, it's unpredictable. How many times will it whirl over the top before stopping? That's the paradox: How can something be chaotic and unpredictable yet follow deterministic laws?

As unlikely as it seems, the analysis of simple chaotic systems like this one have revealed new principles about how the world works—principles so powerful and far reaching that they have reshaped nearly every branch of science. By the end of the course, you'll appreciate why some have called chaos the third great revolution of 20th-century physics, along with relativity and quantum mechanics.

But is it really so revolutionary? Arguably, yes. Chaos breaks many rules of conventional science. First, it asks unusual questions. Unlike relativity and quantum physics, which originated in the study of phenomena that are extraordinarily tiny, fast, or huge, chaos theory addresses questions at the everyday scale of ordinary life—questions that many scientists thought were already answered, or intractable, or unworthy of study (like the irregular dripping of a leaky faucet). Second, chaos takes the focus off the laws of nature and shifts it to their consequences. Third, it relies on the computer and uses it as a laboratory, not just a number cruncher. Additionally, chaos emphasizes holism, not reductionism (but not fuzzy, soft-headed holism—rather, a holism grounded in rigorous science and mathematics). Chaos is radically interdisciplinary in an era of specialization. Experts in fluid mechanics exchange ideas with cardiologists; physicists work with economists. Also, the picture of the world that chaos offers is topsy-turvy. Simple systems show complex behavior and vice versa. Broader academic understanding of chaos and its implications for a range of fields came suddenly. There were no courses in chaos when I was a graduate student; now every research university offers one. For all these reasons, it seems fair to speak of the "chaos revolution."

The goal of this course is to explain the main ideas of chaos clearly and honestly, in a way that any interested layperson can understand. The material is inherently accessible because it is rooted in our daily experience. Chaos is a science of the curiosities all around us: the puffy shape of clouds, the notorious unpredictability of the weather. Furthermore, no advanced math is needed to grasp the key concepts of the subject. Pictures turn out to be much more powerful here than symbols and formulas. So although we'll be discussing many mathematical ideas, we'll always be visualizing them, not manipulating x's and y's. When we do use algebra, it'll always be explained with words and pictures too.

Above all, chaos theory is marvelously broad. Its study intersects nearly every field of human knowledge and endeavor, from astronomy to zoology, to the arts, humanities, and finance. No doubt that sounds far-fetched, but it really isn't, because chaos theory is the science of how things change—and everything changes.

There are at least two good ways to approach the study of chaos theory. We could focus on the purely mathematical side of chaos, which is very beautiful and deep. But my own preference is to take a much broader approach, because I think that best captures the true spirit of the field. So we're going to look at chaos from every conceivable angle—from the standpoint of science and mathematics, of course, but also from historical, philosophical, artistic, and practical perspectives.

[The] study [of chaos theory] intersects nearly every field of human knowledge and endeavor. … No doubt that sounds far-fetched, but it really isn't, because chaos theory is the science of how things change—and everything changes.

The course begins with the story of how chaos was glimpsed in the late 1800s and then lost again for nearly a century before reemerging explosively in the 1970s and '80s. This is a tale of twists and turns, in many ways like a detective story. It's a story of blunders and breakthroughs, rejected papers and serendipitous discoveries—a very human story. We'll devote the first half of the course to it, because the intellectual ride is thrilling and fast paced, with fascinating ideas coming from every direction. We'll meet Isaac Newton, the champion of order, who gave us a picture of the universe so orderly that it seemed to outlaw chaos once and for all. Two hundred years later, a chink in Newton's armor appeared, and that gave humanity its first glimpse of mathematical chaos. But when it was revealed to Henri Poincaré, he recoiled from it. The picture he uncovered was horrifying, in its implications as well as its appearance. We'll see why Poincaré's discovery was overshadowed and forgotten for 70 years before being rediscovered in even starker form by Ed Lorenz, a meteorologist at MIT, in work that gave us the concept of the butterfly effect. And yet, incredibly, it too had no impact until the 1970s and '80s, when the tidal wave of chaos theory finally hit.

By developing the core ideas of chaos in historical context, we can feel like we're sitting at the elbows of the masters, following them on their false starts and sudden flashes of insight. This first part of our journey will take us from planetary astronomy to weather prediction to population biology to condensed-matter physics—a path almost as chaotic as the problems we're studying! And with the help of computer animations and some beautiful pictures and ideas, we'll make the mathematical foundations of chaos theory come to life.

The third quarter of the course focuses on fractals—shapes or processes whose structure repeats ad infinitum, such that their tiniest parts resemble the original whole. We'll learn why fractals are so intimately related to chaos—you could think of them as the footprints of chaos, the remnants that chaos leaves behind.

Scientists use fractal dimensions to characterize the roughness of highly irregular objects, such as coastlines.

And we'll see what distinguishes fractals from ordinary, Euclidean shapes. Then we examine their utility in describing a wide variety of erratic phenomena and processes, such as the jagged coastline of Norway, the gyrations of the stock market, and the hidden structure of Jackson Pollock's drip paintings.

The final section of the course takes us to the frontiers of chaos research today. We'll see how chaos is being used for practical purposes, in encryption and in space mission design. Next we'll look at the medical implications of chaos for epilepsy and cardiac arrhythmias. And then we'll discuss the tantalizing but mysterious linkages among chaos, quantum mechanics, and number theory.

The course concludes with a vision for where the science of chaos is headed next. We'll see why the new ways of thinking developed by chaos theorists may hold the key to unlocking the greatest problems facing science today, from cancer to the mystery of consciousness. ■

Essential Reading

Gleick, *Chaos*, prologue.

Schroeder, *Fractals, Chaos, Power Laws*, chap. 1.

Stewart, *Does God Play Dice?* prologue.

Supplementary Reading

Strogatz, *Nonlinear Dynamics and Chaos*, chap. 1.

Questions to Consider

1. What are the characteristics of a genuine scientific revolution, as opposed to an advance merely hyped as such? What do you think were the greatest revolutions in the history of science?

2. Is "chaos" just another word for "chance" or "randomness"? What exactly do we mean by "chance"?

The Chaos Revolution
Lecture 1—Transcript

Welcome to chaos. I realize that sounds a little bit silly, and actually I mean it that way because I want to have some fun in this course, and I hope that you'll have fun, too. By having fun, I think we can learn better about this amazing new subject of chaos theory. By now, everybody has heard of chaos theory. It's one of those rare branches of science that has become a pop sensation. For instance, in 1987, there was a book that appeared called *Chaos: Making a New Science*, by James Gleick, and that book has now been translated into more than 20 languages. It sold more than half a million copies, and it was on *The New York Times* best seller list for I don't know how many weeks—maybe 10, 20 weeks. That's sort of amazing in itself.

Furthermore, you may have seen dazzling, multicolored images called *fractals*. Do you remember them? You kind of couldn't get away from them in the 1980s. They seemed to be everywhere. They were on T-shirts; they were on coffee mugs, screen savers. The superstar of fractals at that time was one called the *Mandelbrot set*. Here's what it looks like. Does that look familiar? It's an incredible structure. It sort of looks to some people like some kind of a bug or maybe a gingerbread man on its side, and what's curious about it is that it seems to have lots of little copies of itself all around it. You see these little Mandelbrot sets attached to its body and head. Also, notice the lightning bolts shooting off. It's kind of gorgeous. It doesn't look like a normal piece of art, but there's something even more spectacular about it, and this was a game that became a craze. People had just gotten home computers in the 1980s, and the craze was to burrow into the Mandelbrot set and watch what new structures were revealed.

I'm going to do that with you right now by showing you a zoom going deeper and deeper into the Mandelbrot set by sort of magnifying the picture as if we were looking at a microscope, and then I'll stop at a certain point because something surprising will happen. So let me go to my zoom here. All right, fasten your seatbelt. Zooming. Let's pause to appreciate what we see there. We've zoomed in on some of those little bugs, and now you start to see curious structures on top of the bugs that look almost like seahorses or something. Keep zooming, zooming in on one of those seahorse tails. Pretty.

There's even more structure there of a different kind but sort of reminiscent of the original. Keep your eye on something at the center of the picture. I want to zoom in on that. Maybe [you] can sense there's a little black dot in there. Let's zoom in for a closer look at that. Incredible. It's a copy of the Mandelbrot set itself, now magnified by a factor of 1000. In other words, this structure contains tiny, tiny copies of itself. We can keep going down to factors of a million. I'll just let this zoom continuously so you can see this proliferation of wild structures emerging.

The public became fascinated by things like this, the beauty of fractals. And that was, in itself, intriguing. What was the public doing suddenly playing math games on their computers? But that is what was happening. Why this curiosity about a seemingly arcane branch of math and physics? Some cynics dismiss it as just pretty pictures, like this. Others say, "Well, chaos, it sounds very impressive. It's a cool name, and it's a catchy name. That's why people got interested." I think the cynics are wrong. I think they're missing something. Yes, it has a catchy name, and yes, the pictures are beautiful, but there's a genuine reason for the excitement about chaos theory. People sense, correctly, that there is something revolutionary happening in science. We're starting to find our way through a no-man's-land, a territory that science had long dreaded and feared to explore. This is the realm of chaos—the erratic side of nature, the discontinuous, the jagged, the irregular and unpredictable.

Science has long been obsessed with the opposite side of things—the orderly, the regular, the predictable, patterns, the laws of nature, the lawful side of nature—especially in physics. That's what we like, and we've devoted ourselves to that for hundreds of years. But in every field, chaos has been there, lurking in the wings, a kind of irritating reminder of what we don't know and don't understand even still. It was there in ecology, in the unpredictable, wild fluctuations of wildlife populations. It was there in cardiology, in the arrhythmic quivering of a human heart in the moments before death. And it was there in the turbulent sea and in the atmosphere in fluid mechanics.

Turbulence is an especially annoying problem to scientists, a very frustrating problem, because we've known what turbulence looks like for 500 years already, ever since Leonardo da Vinci captured its distinctive pattern of swirls within swirls—Leonardo's study of flowing water from the early 1500s. See the bigger swirls, little swirls inside? That's what turbulence looks like. But sometimes, in ocean waves, turbulence can take a more menacing form. That famous picture by Hokusai, *The Great Wave*—look at the splitting of the wave into little droplets and the poor fishermen being buffeted by it, with Mount Fuji looking tiny in the background compared to this monster wave.

Turbulence is doubly frustrating because not only do we know what it looks like, but we know what math governs it. We know the equations for turbulence, and we've known them for 150 years. So according to traditional science, we should be basically done. We've got the equations. Why can't we understand turbulence? But we can't. It still lies beyond our scientific reach.

Scientists have regarded these kinds of things, the fluctuations of wildlife, turbulence, and so on, arrhythmias, as frustrating puzzles, yes—but sometimes they've come to scorn them as pathologies, monsters, things that you just don't want to deal with. But starting about 30 years ago, a few mavericks in every field started to say, "Now, wait a second. We've got to come to grips with chaos."

When they did, they started to find strange and unexpected connections between different forms of chaos and different disciplines. Geologists, for instance, noticed patterns in the frequency of earthquakes. The same patterns appeared in the variability of human heart rates or, later, in Internet traffic. There are some images of Internet traffic compared to heart rate variability, and what you can see in both is that if you look at what is happening over a short time scale and blow it up and then blow it up again, sort of like when we zoomed in on the Mandelbrot set, you see structures that look reminiscent of the whole. That's true for Internet traffic or for the heart, and they look reminiscent in the same way. They follow the same mathematical laws.

In fact, the rules of chaos, which were just starting to be discovered, were turning out to be universal. Amazing: universal. Independent of the stuff that was behaving chaotically. The laws were turning out to be the same for chaos in electronic circuits, in lasers, chemical reactions, or nerve cells. It was as if disorder was a thing in itself. It didn't really matter so much what was behaving disorderly. It was disorder that you wanted to study, the *process* of going chaotic. Those processes, those laws of chaos, turned out to be as lawful as anything when we studied order, except that they were laws of a very strange and unfamiliar kind. They were like nothing that science had ever seen before.

So this is a story about this new science of chaos, and by the time we're through, you'll never look at the world the same way again.

But what *is* chaos exactly? I've been using the word a lot. Let me try to be a little more precise about what scientists mean by chaos today. It's a paradoxical state. It sounds self-contradictory. On the one hand, it's a kind of unpredictable behavior, like the name suggests. But on the other hand, it occurs in a system that obeys deterministic laws. Deterministic unpredictability—that's chaos. Now, what does *deterministic* mean? That means that the future is determined by the present. There are laws of nature that tell us, given the current conditions, in the laws, we know what's going to happen next. It's not a matter of probability or randomness. It's not like two different things could happen next with some probability. There is only one possible future given the present in a deterministic system.

We can illustrate a system like this that is on the one hand deterministic and on the other unpredictable, by playing with this device that I've got here right next to me. It's called a "double pendulum," and you see that it has, in addition to the usual sort of pendulum, a second leg here so it can swing and do fairly calm things like that. In fact, this pendulum swings back and forth according to Newton's laws of motion. No doubt you've heard of them and maybe even studied them in physics classes. Newton's laws of motion are the paradigm of classical physics. They describe the motion of nearly everything under the Sun. For 350 years, Newton's laws have given physics tremendous power to relate cause and effect.

So you would think, looking at this thing, it's pretty simple. We should be able to predict how this double pendulum will behave, and in fact, that's pretty much what everyone did think until about 30 years ago. It was even a textbook exercise, to calculate the motion of a double pendulum. I remember doing it in my sophomore physics class. This was a standard exercise, and the teacher gave us pretty specific instructions. The instructions were: Solve for the motion of the two arms of the pendulum under the assumption that the angles are small. Now, small angles means it's basically pointing straight down, or if you pull it off to the side, not by very much. Don't start doing this, okay? Those are big angles. You're supposed to consider small angles. So we did that, and we calculated this. We could explain that. In fact, our formulas worked perfectly. It was a hard calculation, by the way. If you've ever done the problem of the double pendulum, it's not trivial. It takes some pretty good knowledge of physics to do it. You have to learn about something called "normal modes of oscillation," and it's tricky, but you can do it. Anyone can do it with enough training.

So that was what double pendulums used to mean to physicists. But under certain conditions, the conditions that the professor told us not to think about—and I'm not sure why, actually, because I don't know if the professor knew that something bad and different would happen or whether it was just custom, tradition. You just don't do the large-angle case. I don't know. But in any case, we never did consider large angles. Well, what's so bad about them? What happens if we do? Do you have intuition about what this pendulum will do if I start it like this? Let me let it go, but make a mental note. How good is your intuition about what you're about to see? So think about it first. All right, here it goes.

Now, that's a little bit surprising, isn't it? That is, it seemed to turn over a few times more than you might have thought. The motion was a bit crazier and wilder than you might have imagined. Let's look at it again. You might think it's done, and yet it just went one more time. Now, that motion wasn't exactly as it was the first time. In other words, there's a kind of unpredictability to this motion. How many times will the bottom leg whirl over the top before it stops? That's the kind of thing that's paradoxical here—that this system, totally governed by Newton's laws, can be unpredictable as well as deterministic.

As unlikely as it seems, the analysis of simple chaotic systems like this one has revealed new principles about how the world works—principles so powerful and so far reaching that they have reshaped nearly every branch of science. By the end of the course, you will appreciate why some people have called chaos the third great revolution in 20th-century physics, along with relativity and quantum mechanics.

"Seriously? Is it really so revolutionary?" you might be asking. Fair enough. Arguably, yes. Let me try to give you a series of arguments why I think it is justified to say that chaos is a revolution. Those other revolutions are not to be diminished. They're incredible and important, but chaos is also a pretty serious revolution. Here's why:

First of all, it breaks a lot of the rules of conventional science. It asks unusual questions. Unlike relativity or quantum mechanics, which originated in the study of phenomena that were very fast, close to the speed of light; very small inside the atom; or enormous, at the scale of the whole universe, where general relativity, Einstein's theory of gravity, applies—instead of going to these extreme regimes, chaos theory addresses questions at the everyday scale of ordinary human life, questions that scientists thought, "These are questions we already understand the answers. We know Newton's laws." Or questions that seemed intractable, "Turbulence, all right, we know we can't solve turbulence. Let's put that aside." Or questions that just seemed pointless and worthless, not worthy of study, like that faucet in your kitchen sink that goes drip, … , drip, drip, drip. It's got a strange leak. It's not a regular pattern of drips. It's sort of a chaotic pattern of drips. Is there anything worth studying there? It turns out yes, that by studying leaky faucets, chaos theorists later found patterns that also turned out to apply in things like the irregular rhythms of heart cells undergoing arrhythmia. So chaos looked at things that seemed mundane and found fantastic new mysteries and new discoveries in them.

A second point: Chaos takes the focus off the laws of nature and shifts it onto their consequences. Let me explain what I mean by that. The game in theoretical physics ever since Newton has been "find the new laws of nature." Find fundamental laws, from Newton's laws, through the laws of electricity and magnetism, laws of relativity, quantum mechanics, today's string theory—this is a great enterprise, the fundamental laws, the search for the basic foundational patterns in nature. Yes, a worthy enterprise, but there's another game that can be played, which is: Having found the laws, what are their implications? What are their consequences? We know the laws of fluid mechanics, say; can we deduce the structure of turbulence from them? That's a different question. It was not a glamorous question for a long time. People thought that was a matter of technique. That's not creative. I've got the laws;

I'm done. Now, I'll leave it to you, sort of like, "Give this to the workers to deal with that part." That's wrong. There are a lot of great problems in figuring out the consequences of the laws. In fact, knowing the laws only takes you a very small part of the way toward the goal of science, which is understanding nature. We have to be able to solve the laws, understand what they're telling us, not just find them. Chaos theory is a lot about that, about finding clever ways of inferring consequences.

A third radical departure from traditional science is the way that chaos uses the computer. Traditionally, the computer was viewed as a glorified calculator. It's just something that you use to crunch numbers, and that's it. It's a tool. But in the hands of chaos theorists, computers became an intuition amplifier, a way of expanding your mind, of seeing things that were inconceivable. It became an arena for doing thought experiments and for exploring the consequences of these laws that I just mentioned, by letting the computer grind forward instant by instant to work out, to elaborate, the predictions following from those laws.

Fourth: Chaos emphasizes a different side of nature than usual. There's a strategy that's different about chaos theory; that is, a time-honored strategy in science—all branches of science—is what we call "reductionism." Take a problem that's hard, reduce it by breaking it into smaller parts, try to analyze each part separately, and then that should be easier. If you can solve what's happening in each of the parts, put them back together, and you have solved your original problem. That works extremely well and has been a foundational strategy for all of modern science.

But there are times when reductionism is not sufficient. There are some problems, which by their very nature don't really permit reductionism— in this sense: that if you break them into their parts, you lose something essential. There are systems in which the interaction between the parts creates something where the whole is greater than the sum, that is, where there are cooperative effects taking place or some kind of interference. In systems like that, we have to find a way to do scientific holism, to go from the parts to the whole. "Holism" sort of sounds fuzzy. Maybe it sounds to you like this is New Age thinking or something. I'm not talking about that. I'm talking about rigorous science, provable mathematics, but about trying to put the parts back together to understand the whole. That's what chaos theory has aimed to do and to a large extent is succeeding at doing.

A fifth point is that chaos is radically interdisciplinary in an era of specialization. That is, just like we reduce problems by chopping them into smaller parts, science itself has become chopped into smaller parts, where different disciplines put up walls between themselves and adjacent disciplines, and everyone works in their own little corner of the universe and makes a lot of progress by specializing. But chaos theorists weren't like that. They said, "Hey, I can learn from you. The chaos you're seeing in ecology might teach me something about the chaos I'm seeing in fluid mechanics." So you had this weird, unconventional situation: a physicist talking to economists, mathematicians working with artists. It's actually very refreshing and fun, and it has been enormously fruitful.

Another thing is that chaos paints a topsy-turvy picture of the world. That is, we used to think that simple laws led to simple consequences, and if you saw complex behavior in the world, it was because there were complex causes, like maybe lots of components of the system that you hadn't taken account of or external influences making things appear complex. But chaos has this topsy-turvy view in which we say, "No, simple systems can show complex behavior. You don't have to look for complex causes." Even simple things, like this double pendulum, can act chaotic and complicated. Conversely, things that you would think would be enormously complex sometimes can show strikingly simple patterns. They sort of organize themselves.

A final point about chaos and why I think it's revolutionary is that it seems to qualify as a revolution in the sense that the change in thinking came suddenly. When I was a graduate student interested in this topic in the early 1980s, there were no courses in chaos and essentially no textbooks. There was no way to get information about this exciting area and all the discoveries that were being made at that time. All that has changed, and now, only 25 years later or so, every major university has a course in chaos. So for all these reasons, I think it's fair to speak of a "chaos revolution."

Let's talk about our course now. The goal of this course is to explain the main ideas of chaos as clearly and honestly as I can. No hype, just giving it to you straight, but in a way that any interested layperson can understand. I don't expect a lot of background on your part. You don't need it. The material is inherently accessible. It's delightfully accessible because it's rooted in our daily experience. Chaos is a science of the everyday world, of clouds, coastlines, the unpredictability of the weather. Furthermore, no advanced math is needed, at least not to grasp the key concepts of the subject, which is really what we will be doing here. I'm not going to teach you how to solve differential equations. That's not our goal. Our goal is to understand the core concepts of chaos and its main accomplishments. For understanding these core concepts, pictures turn out to be much more powerful than formulas, and we'll be relying on these pictures. [These will be] much more important than the symbols you're used to in algebra class, for instance.

Although we'll be discussing many mathematical ideas, some of them very deep ideas, we will always be visualizing them. We will not be manipulating those dreaded x's and y's. On the other hand, when we do use algebra, and sometimes it will come to our rescue even, I will always explain it with pictures and with words, too.

Above all, chaos theory is marvelously interdisciplinary. Its study intersects virtually every branch of human endeavor, every field of human knowledge from astronomy to zoology, the arts, commerce, humanities, all of it. Seriously. No doubt that sounds far-fetched, but it's not, because chaos is really the science of how things change—and everything changes.

There are at least two good ways to approach the teaching of chaos. On the one hand, we could focus on the purely mathematical side of chaos. It's a beautiful, deep, marvelous story, and I would enjoy doing it that way. I'm not so sure you would enjoy learning it that way. You might, but my own preference is to take a different approach, a much broader approach than focusing on pure math because I think that best captures the true spirit of this field. So we're going to be looking at chaos from every conceivable angle—from the standpoint of science and math, of course, but also from a historical perspective, philosophical, artistic, and practical.

The course begins with the story of how chaos was first glimpsed in the late 1800s and then lost again for nearly a century before reemerging explosively in the 1970s and 1980s. It's a tale of twists and turns. In many ways, it's really like a detective story. It's a story of blunders, breakthroughs, the heartbreak of rejected scientific papers, and the thrill of serendipitous discoveries. It's a very human story, and I'm going to be telling it that way. We will devote the whole first half of the course to it, about 12 lectures, because the intellectual ride is thrilling and fast paced, with fascinating ideas coming at us from every direction.

We'll meet Isaac Newton, the great champion of order, who gave us a picture of the universe so regular, so orderly, that it seemed to outlaw chaos once and for all. But then, 200 years later, we'll see a chink appearing in Newton's armor, a chink that gave humanity its first glimpse of mathematical chaos. But when it was revealed to the person of Henri Poincaré, he recoiled from it. The picture he uncovered was horrifying. Here's what it looked like. That was Poincaré's nightmare, and it's nightmarish in its implications, as well as its appearance. We'll see why Poincaré's discovery was overshadowed. Surprisingly, given that this is such an enormous discovery, you would think that the bells would be ringing. No, [all was] quiet. It was forgotten, overshadowed for nearly 70 years before being rediscovered in an even starker form by Ed Lorenz, a meteorologist at MIT, in work that gave us the concept of the butterfly effect. And yet, incredibly, Lorenz's work, too, had no impact for another 10 years, until the early 1970s, when the tidal wave of chaos theory finally hit.

By developing the core ideas of chaos in this historical context, we can feel like we're sitting at the elbows of the masters, watching them, following them on their false starts and their sudden flashes of insight. The first part of our journey will take us from planetary astronomy, to weather prediction, to population biology, and then to condensed-matter physics, a path almost as chaotic as the subject that we'll be studying. Then, with the help of computer simulations and some beautiful pictures and ideas, we'll make the mathematical underpinnings and foundations of chaos theory come to life.

In the third quarter of the course, we'll be focusing on fractals, shapes or processes whose structure repeats ad infinitum—like the Mandelbrot set— such that their tiniest parts resemble the original whole, as shown in this next image for two types of trees, one real and one imagined. Here is a real tree picture, and I take this region in the red box and blow it up to full size. You see that the blown-up picture sort of looks like a real tree even though it's just a piece of a tree. Take this little piece of the tree and blow it up to full size, and it also still looks very much like a real tree, although you're starting to see that this one doesn't have enough twigs to be a real tree. So the fractal structure is breaking down in a real system, whereas in idealized mathematical fractals, you can blow up a little piece to make the whole and you can just keep zooming in, all the way down infinitely deep.

We'll learn why this strange construction of things that are self-similar— why that's so intimately related to chaos. You could even say that fractals are the footprints of chaos, the remnants that chaos leaves behind. We'll see what distinguishes these kinds of shapes and processes in time from ordinary shapes, Euclidean shapes that you're more accustomed to thinking about: circles, spheres, and cubes.

Then, we'll examine the utility of fractals in describing a wide variety of shapes in nature, erratic phenomena, and all kinds of processes, ranging from the jagged coastline of Norway, with all of its invaginations, fjords, and inlets; to the gyrations of the stock market; to the hidden structure in Jackson Pollock's drip paintings.

The final section of the course takes us to the frontiers of chaos research today. We'll see how chaos is being used for practical purposes, in encrypting secret messages and in space mission design—how to get to the Moon with practically no fuel. Next, we'll look at the medical implications of chaos for understanding and predicting when epileptic seizures might occur, or possibly treating cardiac arrhythmias. And then, we'll discuss the tantalizing but still totally mysterious linkages among chaos, quantum mechanics, and number theory.

The course concludes with a vision—you might even say a messianic vision—for where the science of chaos is headed next. We'll see why the new ways of thinking being developed by chaos theorists may hold the key to unlocking the greatest challenges that science faces today, from cancer to the mystery of consciousness. I'm really looking forward to taking this incredible journey with you. See you next time.

The Clockwork Universe
Lecture 2

To appreciate … what is so revolutionary about [chaos theory], it will help if we can understand more about what it overturned. … Our main goal is to trace the rise of Isaac Newton's clockwork universe. … But before [that], I think it's worth going back even farther to the origin of the idea of chaos itself. It's one of humanity's oldest and deepest philosophical ideas.

We begin by surveying the diverse concepts of chaos in the creation myths of the ancients. From the dawn of humanity, people have struggled to make sense of the world. Much of life is predictable: sunrise, sunset, the changing of the seasons. But much of life is also unpredictable: plagues and famines, floods and wars.

The ancient Greeks summarized the tension between order and disorder with two opposing words: cosmos and chaos. *Cosmos* means order. (When you use cosmetics, you're putting your face in order.) *Chaos* initially meant the chasm, the abyss, the bottomless pit. Later, it came to mean the primeval state before creation—a state of utter disorder. This sense of "chaos" as utter confusion persisted into the modern era and gave rise to the scientific term "gas" (denoting trillions of molecules in frenzied motion).

The ancient Hebrew word for chaos is *tohu va-vohu*. A rare phrase, it appears in Genesis 1:2, where it describes the primal, chaotic state of the universe before God brings order to it. Notice that according to this tradition, divine intent was required to wrest order from chaos.

There is also the suggestion, which becomes explicit in the New Testament, that the primeval chaos was not merely a dark abyss or an empty void; it was an active, malevolent confusion that God needed to overcome to create the universe. For many ancient cultures, that evil side of chaos took the form of a serpent, sea monster, or dragon. In the case of the Babylonians, chaos was personified by the ocean goddess Tiamat, a sea monster who dwells in the dark, watery abyss beneath the Earth.

Against this backdrop of mythology, it was a tremendous breakthrough to conceive of the world as ruled by natural laws that human intelligence could find and comprehend. In Western civilization, this breakthrough came with the Ionian Greeks of the 7th and 6th centuries B.C.E. For example, Thales predicted an eclipse of the Sun in 585 B.C.E. Pythagoras discovered the laws of musical harmony and proved his famous theorem about triangles. And Euclid wrote a textbook on geometry that established a style of rigorous, logical reasoning that served as a model for centuries afterward.

Their rational spirit inspired the great minds whose discoveries in the 1600s launched the Scientific Revolution. Specifically, Galileo discovered the laws of motion on Earth. These include the law of inertia, and regularities in the motion of falling bodies and swinging pendulums. Johannes Kepler deciphered the laws of planetary motion, thus solving an age-old puzzle, and Isaac Newton explained the laws found by Galileo and Kepler, thus unifying the science of heaven and Earth.

Newton's achievement was so stupendous, and so important to the rest of this course, that we need to pause to examine what he accomplished.

Newton's achievement was so stupendous, and so important to the rest of this course, that we need to pause to examine what he accomplished. He used several tools, which he invented. These tools included his three laws of motion and the law of universal gravitation (which played the role of Euclid's axioms), as well as calculus (for deducing how things move, given the forces acting on them).

His explanations had a radically new character. They relied on a calculus-based tool that enabled Newton to predict a system's changing behavior from moment to moment. In effect, Newton had invented a mathematical soothsayer—a device now known as a "differential equation." Armed with his differential equation (his second law of motion) and his law of gravity, he explained and unified the laws that Kepler and Galileo had found before him. Until then, those earlier discoveries were cryptic facts, disconnected and mysterious.

Newton's account of natural phenomena was different from Galileo's and Kepler's. To understand how, we begin by imagining watching a movie of an orbiting planet or a flying arrow. Kepler and Galileo could describe such motions from beginning to end using the empirical regularities they'd discovered. In effect, they could give you the story line for the whole movie. Newton's laws, in contrast, let you calculate the "difference" between successive frames of the movie, to predict how nature changes in the next instant—hence the term "differential equations." He needed to find a way to integrate the differences between infinitely many such frames to yield a continuous movie. Hence, this part of Newton's method is known as "integrating" a differential equation.

With this, Newton felt he had found a secret key to understanding the universe—a secret so precious that he published it only in code. Translated into modern language, Newton's secret is: "It is useful to solve differential equations." Newton gave humanity something brand new and shocking: its first deterministic law of nature. "Deterministic" means that a system's future behavior is predetermined by its current state and the governing laws.

But by banishing disorder, the new worldview became disquieting in its own way, especially in its moral and philosophical implications. Once set into motion by its creator, the universe would run on its own, its fate determined by Newton's laws, with no room for chance or free will. The universe appeared to be a vast impersonal clockwork.

Yet we still see random behavior all around us. How do we reconcile the traditional, intuitive understanding of chaos with the clockwork system of orderly processes that Newton developed? Newton's successors believed that chaos was a mirage. It merely reflected our imperfect human knowledge of the countless components of any real system and the myriad uncontrolled forces acting on them. In the meantime, probability theory was developed as a practical tool for handling systems dominated by such uncertainties.

Today, we view chaos as much more deeply ingrained in the universe. Even systems with perfectly known, perfectly deterministic laws can be unpredictable—and those are precisely the ones we call "chaotic." Thus, Newton's notion of determinism is crucial to the rest of this course, because one of the defining features of a chaotic system is that it obeys deterministic laws. ■

Essential Reading

Peterson, *Newton's Clock*, chaps. 3–5.

Stewart, *Does God Play Dice?* chaps. 1–2.

Supplementary Reading

Kline, *Mathematics in Western Culture*, chaps. 12–15.

Questions to Consider

1. Why did so many ancient cultures believe the world originated in chaos? What other primal states could one imagine?

2. Is free will possible in a deterministic world? In a random world? Or is free will an illusion?

The Clockwork Universe
Lecture 2—Transcript

Welcome back to chaos. In the first lecture, I tried to emphasize that there is something revolutionary about chaos theory, and to appreciate just exactly what is so revolutionary about it, it will help if we can understand more about what it overturned. That's the goal of this lecture. Our main goal is to trace the rise of Isaac Newton's clockwork universe, a universe so regular and so orderly that it would seem to have banished chaos. There's no room for chaos in Newton's clockwork.

That will sort of be the second half of the lecture, but before we get to Newton's clockwork, I think it's worth going back even farther to the origin of the idea of chaos itself. It's one of humanity's oldest and deepest philosophical ideas. So let's begin by surveying the diverse concepts of chaos (and I do mean diverse—you will see that there are all kinds of concepts of what chaos meant over the years) in the creation epics of the ancient Greeks, the Hebrews, the Babylonians, and so on.

From the dawn of humanity, people have always noticed that the world is mysterious. They have been trying to make sense of the world. Much of life is predictable: for instance, the changing of the seasons, night following day, the cycles. Those things are reassuring to us. The soothing beat of your mother's heart, even in the womb. There's a lot of regularity in the world, and those are comforting things. But much of life is also unpredictable: plagues, famines, wars, horrible things that happen to good people. It's this unpredictable side of life that, of course, you associate with chaos, but the ancient Greeks, in thinking about these issues, came up with two words to express the tension between order and chaos, or between order and disorder. *Cosmos* was their word for "order," and we still have that word today. You think of it as meaning the whole universe, but specifically, it means the orderly nature of the universe. And so when you use "cosmetics," the same Greek root, you're putting your face in order.

The other word is *chaos*, for "disorder." Initially, it meant something quite different. It meant the abyss, the chasm, the bottomless pit. Later, chaos took on yet another meaning: the primeval state before creation, a state of utter disorder. For example, in later years, this is how the Roman poet Ovid described chaos in his *Metamorphoses*. He wrote:

> Before the ocean was, or earth, or heaven,
> Nature was all alike, a shapelessness,
> Chaos, so-called, all rude and lumpy matter,
> Nothing but bulk, inert, in whose confusion
> Discordant atoms warred.

I love that. Shapelessness, rude, and lumpy matter. That's chaos, this crazy confusion of atoms jostling around, and that sense of chaos as utter confusion persists into the modern era. We have a word that sounds a lot like chaos that means the same thing down to the atomic scale. It's the word "gas." That's where the word "gas" comes from. It's the same root as "chaos." It's disorder, trillions of molecules in frenzied motion.

Turning now to the Hebrews, the ancient Hebrews had a very mysterious concept of chaos. They used the phrase *tohu va-vohu*. First, I want you to notice the word play in that gorgeous phrase. *Tohu va-vohu*. It has a repetitive sound. We have words like that in English that mean something similar: *hurly-burly*, *higgledy-piggledy*. They all suggest turning. The repetition suggests something swirling, almost like the swirling of turbulent water, which has often been a symbol for chaos, an ancient symbol.

Tohu va-vohu literally, though, has nothing to do with water. Although people are not sure what it meant, it seems to have meant a desert that was so untouched that it was featureless, a desert with no tracks on it, total blandness, uniformity. No one had ever been there. The total absence of any feature of interest. A featureless desert. So it's a rare phrase, which is one reason we find it confusing to translate. It only appears in the Torah in a few places. Actually, every time it appears, it seems to be alluding to the first time that it appears, which is in the second line of Genesis. It describes the beginning of creation, the primal chaotic state of the universe before God brings order to it. Now I'm not going to try to read the Hebrew for you, but let me give you the King James Version, which you might be more familiar with, the translation that we all know: "In the beginning God created the heaven and the earth. And the earth was without form, and void [that's *tohu va-vohu*, "without form, and void"]; and darkness *was* upon the face of the deep." The deep, the abyss, the dark waters. Notice that in this tradition, there is a claim being made about chaos implicitly because after those majestic opening lines, we're told, "And God said, Let there be light: and there was light. … And God [separated] the light from the darkness." That is, it takes the divine hand of God to bring order from chaos. Order does not arise spontaneously. Chaos does not give birth to order in this tradition.

There is also the suggestion, which becomes explicit later, in the New Testament that this primeval chaos was not merely a dark abyss or just pure emptiness, a void. It was much worse than that. It was the embodiment of evil. It was a sort of a harbinger of all the evil to come, an active, malevolent confusion that God needed to overcome to create the universe—a devil essentially. This view of chaos is often embodied in other ancient cultures as some sort of frightening creature, like a sea monster or a dragon or a serpent. In the case of the Babylonians, for instance, chaos was personified by the saltwater ocean goddess Tiamat, who was a sea monster who dwelled in the dark watery abyss beneath the Earth. I certainly don't claim to be an expert on Babylonian creation epics, but my understanding is that in that story, Tiamat gave birth to the whole universe but also to other gods, who got in some kind of horrific battle with her, and chopped her up, and created the universe out of her body parts.

Against this backdrop of mythology, with its dark places and chopped-up sea monsters, it was a tremendous breakthrough for humanity to conceive of a universe that was orderly, ruled by natural laws that were so perfect and simple that human beings could discover them and understand them. This came in Western civilization with the breakthrough brought by the Ionian Greeks of the 7th and 6th centuries B.C.E. The Ionian thinkers sought to discover a basic order in the universe, to discover laws of nature that they could test through predictions. Let me just mention three in this tradition.

Thales, often considered the father of Western science, predicted an eclipse of the Sun. He seems to have traveled to Egypt and learned about astronomy from them, and his prediction in 585 B.C.E. was thought to stop a war between two opposing factions, who were so mystified that he was able to predict this solar eclipse that they thought he must have serious powers and maybe they should listen to him and not fight. We know that it was 585 B.C.E. because we can track it back with modern astronomical methods. There was a solar eclipse that would have been visible in Greece at that time.

Pythagoras, a giant of mathematics and science. You know his name from the Pythagorean theorem of geometry about right triangles, but I think in some ways his more magnificent discovery was the discovery of the laws of musical harmony. He showed that if you took a string of a lyre and another string that was twice as long but held at the same tension and made of the same material, if you pluck those two strings, the longer one would sound a note exactly one octave lower. What was interesting about that is that he was finding mathematical patterns in harmony. You could understand music with math. That was a gorgeous thing because it suggested that number was somehow intrinsic to music, to the pleasure of music. What this did, then, in Pythagoras's vision, was suggest that actually nature was deeply numerical. In fact, as he put it, "all is number," that the Earth and the world is not made of earth, or air, or fire, or water; it was made of number, and that by understanding math, you would understand the secrets of the universe.

Euclid wrote a textbook on geometry that established a style of rigorous logical thinking that held on for centuries afterward, that was a model for all logical thinkers after him.

So you had these three tremendous minds and many other scholars. I don't need to go into Greek science here, but I want to now fast-forward to the second part of our story, the creation of this idea of the clockwork universe, which is in the same spirit as what the Ionian Greeks were trying to do.

Their rational spirit inspired the great scientists and mathematicians who discovered different kinds of laws of nature, specifically laws of motion— how things move on Earth and in the heavens—a series of laws so precise and so profound that they seem to predict the motion of everything that we can see around us, from a falling apple to the Moon in its orbit.

We're going to fast-forward now to Galileo, Kepler, and Isaac Newton, whose discoveries in the 1600s launched the Scientific Revolution. Their great legacy is the culmination of the Ionian dream, the demonstration that the laws of nature were comprehensible to human beings and that the universe is orderly and ruled by mathematics. Most thrilling of all, they taught us that the laws of nature are simple, and beautiful, and comprehensible, accessible to our feeble human minds. For the first time in history, superstition started to recede and the Age of Reason was born.

Let me just quickly review what those geniuses did. Galileo: laws of motion on Earth, the law of inertia. You often hear it said: A body at rest will stay at rest unless acted on by a force. A body moving at a constant velocity will keep moving at a constant velocity unless acted on by a force. He also discovered laws about falling bodies, the parabolic shape of a projectile when thrown through the air or launched from a cannon. He discovered laws about the motion of a swinging pendulum, its period, how long it takes to swing back and forth. These were things that Galileo figured out.

Galileo's great contemporary, Johannes Kepler, looked at, not the Earth, but the solar system, the heavens. Kepler deciphered the basic laws of planetary motion, thus solving an age-old puzzle. Specifically, he looked at the motion of the planets that were known at that time. There were only, I think, five known back then. He found that they obeyed certain gorgeous laws, three of them: One [was]—and this one took him a long time and a nervous breakdown to figure out—that the planets don't move in perfect circles as the Greeks had taught. He wanted to believe that, but the data would not allow it. He figured out that the planets move in an oval shape that was not very well studied at the time, called an "ellipse." He also figured out from the data that the motion around the ellipse was not uniform. The planet didn't just move at a constant speed as it went in this oval path, but it went faster and sort of whipped around the Sun and then slowed down when it got far from the Sun and then whipped around again. He figured out the rule that gave a precise measure of how fast it whipped around and how slow it went far away. He also found a third law, which related the different planets to each other. He could compare the length of time it took for one revolution around the Sun for each of the planets, depending on their distance from the Sun.

These were three phenomenally beautiful mathematical laws right there in the data. [Kepler] was very much a Pythagorean, a believer in this idea that all was number, and so he was, along with Galileo, one of the great predecessors of Newton.

Newton is really the man. We're going to talk about Newton for the rest of the lecture. Some would say he's the greatest genius of all time. It's hard to think who would be superior. It depends on what field you're interested in. Bach, Beethoven, Einstein, Shakespeare, Archimedes—they're right up there, but Newton is with them certainly. In any case, Newton is a towering genius. What do you know about Newton? You're thinking about the apple, right? That's the first thing everyone thinks of. Okay, he did have an apple tree. What's the point of this story about the apple? Not that Newton discovered gravity. People knew that gravity existed. Newton saw an apple drop. That didn't take a genius, to see the apple dropping, but what he did do was make a genius connection. He realized that the force pulling the apple down to the Earth was the same force that was holding the Moon in its orbit around the Earth. That was a brilliant thing, to see the connection between this earthly apple and the celestial Moon, that there was a single operating principle that was governing both.

Maybe that doesn't seem so impressive to you, but a lot of people thought that gravity ended at the edge of the atmosphere, that there was no gravity past the sky, that things would fall from the sky, but beyond that there was no gravity—so this really took some thinking and some imagination, to imagine that gravity could extend all the way out to the Moon. In fact, Newton said everything in the universe pulls on everything else by the force of gravity.

His accomplishments were so stupendous that we need to pause to examine exactly what he did. They will be with us for the rest of the course. He did so much. He was 24 years old when he invented calculus. He discovered the mathematical form of his law of gravity: that the force between the two planets or other bodies pulling on each other got weaker as they got farther apart in a perfectly precise way, that the force dropped off like the inverse square of the distance. That means if I move them 10 times farther apart, the force doesn't go down by 10, it goes down by 100, because 10^2 is 100. So the inverse square law was big.

There were also three laws of motion that he found. You have probably heard about them: the law of inertia that Galileo had found; a famous second law called "$F = ma$," which we'll be talking about a lot, relating force to acceleration; and the third one, action and reaction. But I really want to focus on this second law because that's the important one for our course.

Also, though, another tremendous achievement was the invention of calculus, which he did, too. Calculus was to be used for deducing how things moved according to these force laws that Newton had found. In sort of the way that Euclid had set up axioms for deducing theorems, the force laws were the axioms for Newton, and calculus was his logical tool for deducing the consequences, the corollaries and theorems from those axioms.

Newton's explanations had a radically different character than anything the world had ever seen. They relied on a calculus-based tool that enabled him to predict a system's changing behavior from moment to moment. In effect, Newton had invented a mathematical soothsayer, a crystal ball, a device that we now call a "differential equation." Armed with his differential equation—that's his second law of motion—and his law of gravity, he was able to show that all the different laws that Kepler and Galileo before him had found could be explained in one unified framework. They all followed from Newton's laws. Until then, those earlier discoveries, although fabulous, had seemed like just cryptic, isolated facts. Disconnected and mysterious, true, but with no particular coherence to them. Newton put them all into one framework and showed how they all followed from a few deep principles.

I want to spend some time now trying to explain how Newton used his differential equation to explain those laws and other natural phenomena, and why it was so different from what Galileo and Kepler had done before him. Let me do that by way of an analogy. It's not perfect, but it's very good. We're going to go far with this analogy.

Imagine watching a movie of a planet orbiting the Sun. Imagine that it's sped up so you can see something happening. Or imagine a movie of an arrow flying through the air. Kepler and Galileo had laws for such things that could describe those motions from the beginning to the end. Wonderful! In terms of a movie, they could give you the story line for the whole movie. That was great, but it doesn't always work. That's the problem. Newton realized that this sort of thing that had been achieved earlier was rarely possible—to find a law from beginning to end of a motion. So what Newton did was find a method that will always work every time, at least in principle. But it gives you a different picture than an entire movie. It gives you a frame-by-frame description of the movie. This is a very different concept. What Newton's laws do, they let you calculate—given a frame (think of that as a scene in a movie), Newton would then apply his differential equation to say, What is the logical next scene in the movie an instant later? He could calculate the difference between one frame and the next, which is why we use the [term] "differential equation." It's all about the tiny differences one instant later.

The problem with his method, of course, is that it only allows you to see one instant ahead. It's very myopic. So what you want to do is put the frames together to make the complete movie. That is hard. What you have to do is integrate all this information, integrate all these differences to push forward to make an entire movie, and that's why we use the [terms] "integral calculus," "integration of a differential equation." Those are the terms you may have heard if you've taken calculus, and it's precisely because of this kind of analogy.

What's really remarkable about this is that this idea of Newton's—such a fantastic idea—holds even today in quantum mechanics. It has not been overturned by all the other revolutions in physics. Even in quantum theory, the goal is still to write down a differential equation and integrate it. [Newton] gave us a way of doing theoretical science that has stood the test of time for 350 years now, and there's no sign of it disappearing.

With this discovery, Newton felt he had found the secret to the universe, truly. A secret that he thought was so precious that he published it only in code as an anagram in Latin. If we were to translate this code today—are you ready?—[it] would be: "It is useful to solve differential equations." That's it. That is the secret. So he gave humanity something shocking and brand new, its first deterministic law of nature. By that I mean, "deterministic" refers to the idea that a system's future is completely predetermined by its present and by the laws of nature. What that would mean in terms of the movie, if you rewound the movie to a certain frame and then played it again, the future will play out the same way every time.

That's Newton's view of the universe, and on the one hand, it's sort of appealing because things seem very orderly. You can see why people refer to it as the clockwork, that all you have to do is set the gears in motion, and the clock knows how to tick, tick, tick, and tell time, and it will just run forever, assuming it stays wound up. So this gives a philosophical view of the universe, where God's place would only be to create the universe and the laws, and then God could sort of stand back and things would unfold by themselves, like clockwork.

There's something very disquieting about this universe, too, which is that it totally banished disorder. There didn't seem to be any room not only for disorder but for free will. That has very disturbing moral implications. If you rewound the movie to the moment of truth when you had to make some important decision, you would behave the same way every time according to this view, and so in that sense, how could you say you had any free will? You did what you had to do every time. Given the chance to do it again, you'd do the same thing again. "Judge, I had to kill him. Newton's laws made me do it."

This becomes a really strange view of life, that everything is predetermined, and yet that's what the best science of the era seemed to suggest. That if, although it's only a matter of doing this in principle, not in practice, but still, in principle, if you could say where every atom was in the universe— including the atoms inside all of us, all the planets, all the galaxies, every animal, the atmosphere, the oceans, everything—if you could track the instantaneous position of everything, the future would be set forever. They would all be bouncing off each other, and interacting, and colliding, and doing whatever they need to do, but you would have no free will. It's just an illusion to think that you ever had a choice. So there doesn't seem to be any room, not only for chance or disorder, but for morality either in a world like this, and that bothered people. That's the picture that came out of Newton's clockwork universe.

On the other hand, you can say, "Come on; give me a break. First of all, we feel like we have free will, but more importantly, scientifically, we seem to see all kinds of disorder in the world around us. You said it was banished. Is it really banished? How come I see things that are unpredictable?" So how do we reconcile the traditional, intuitive understanding of chaos with the clockwork system that Newton had developed? Newton's successors puzzled over this, and they weren't really sure, but they thought the answer was something like this: that chaos, yes, does exist. They couldn't deny that, but they thought it was not fundamental, that it was basically a mirage. It merely reflected our imperfect understanding, as human beings, of the countless positions of the particles that we're thinking about in that crazy description I gave you of the whole universe, and the ocean, and the people, and the animals. If you could measure all those particles, if you knew where all of them were, if you could keep track of all the myriad forces acting on them, which no human being could do, but in principle, if you could do it, yes, chaos would disappear if you could do all that. That's what the post-Newtonian scholars thought.

There is a famous story about that. It's probably apocryphal, but I'll tell it anyway. It's a tradition to tell this story. The great French astronomer Laplace was talking to Napoleon. We know that those two did interact. In the story, Napoleon is reading Laplace's *The System of the World*, which came out of this Newtonian thinking, and he said, "I noticed that in your book, you never mentioned God." According to this story, Laplace said, "*Je n'ai pas besoin de cette hypothèse-là*," "I don't have any need for that hypothesis, sir." [In other words,] "I don't need God. There's no need for that in my theory." So that's the view of the clockwork universe.

Meanwhile, just to deal with the practical problem of studying disorderly systems and making predictions, admitting that they exist, probability theory was developed by people like Laplace, and earlier, Pascal and Fermat. It was developed as a practical tool for handling systems that were dominated by such uncertainties in the meantime, until we could figure out how to calculate things better. It was sort of a stopgap measure, very effective but not really fundamental. That was the view.

Today, though, our view is different. There is some debate about this, but at least I would argue that our view is quite a bit different in that we see chaos as much more deeply ingrained in the structure of the universe. That is, in the next few lectures, we'll see that even systems with perfectly known, perfectly deterministic laws—that is, I give you everything that Laplace wants and even all the positions of all the particles, all the forces known, everything known—those can still be unpredictable. That's a shocker, and those are precisely the systems that we're going to call "chaotic."

Let me say that again. Deterministic—in principle, you should be able to calculate everything using the laws—but still unpredictable. That's what we mean by chaos. How can there be such a thing? So we have to understand this more deeply, and in this, Newton's notion of determinism will be crucial to the rest of the course because, as I've just tried to say, that's one of the defining features of a chaotic system, that it has this crazy pairing of two things that seem incompatible—determinism and unpredictability—at the same time.

In the next lecture, we'll see how mankind first stumbled across the hint of this chaos in Newton's clockwork. It's a stunning discovery, and I can't wait to tell you about it, but our time is up. Let's wait, and I'll see you next time. Thanks.

From Clockwork to Chaos
Lecture 3

> Why was [the three-body problem] so intractable? One possibility was that the difficulty was simply technical. … There was a more disturbing possibility, which was … that perhaps there was some kind of deep mathematical obstacle … that prevented the problem from being solved, no matter how clever you were. Even a super-Newton couldn't do it. That turned out to be the case.

We ended Lecture 2 with a cliffhanger: How can chaos survive in Newton's orderly universe? In this lecture and the next, we'll see how chaos squirmed out in the late 1800s by turning the logic of Newton's clockwork on its head. Newton's laws don't forbid chaos; they require it. That was a shocking turnabout—unthinkable to nearly all scientists at the time, including many who still didn't grasp its significance decades later. To appreciate this next chapter in the story of chaos, we need to remind ourselves just how unassailable the Newtonian paradigm seemed at the time.

The Newtonian worldview dominated Western thought for 200 years, from the late 1600s to the late 1800s. Newton and his successors helped create modernity. Their magnificent scientific advances laid the foundations for the creation of cars, skyscrapers, airplanes, electrical power, and everything else that we associate with the Industrial Revolution. Also, Newton was a hero to such Enlightenment philosophers as Voltaire and Locke. His successes in science inspired a sense of confidence that everything in the world, even human affairs, could be understood and tamed by rational thought. The Newtonian spirit is even visible in the language of the Declaration of Independence. Jefferson's phrase "We hold these truths to be self-evident" is straight out of Newton and Euclid.

But by the end of the 1800s, three cracks began to appear in the foundations of determinism. One led to Einstein's relativity; another, to quantum mechanics; and the third, to chaos. Relativity came about because of inconsistencies between Newton's laws of motion and the new science of electricity and magnetism. Einstein resolved them by overhauling our intuitive (and Newtonian) notions of space and time. Quantum mechanics grew out of an attempt to understand the strange phenomena seen in experiments on light and atomic radiation, which defied explanation in Newtonian terms.

Chaos, on the other hand, grew directly out of the same classical problems that had given Newton his greatest successes: the motion of the planets in their orbits. How ironic that the very topic that had triumphantly ushered in the clockwork universe would later play a role in its demise!

How ironic that the very topic that had triumphantly ushered in the clockwork universe [the motion of the planets in their orbits] would later play a role in its demise!

So what happened? How did we get from clockwork to chaos? First, let's recall what inspired the idea of the clockwork universe in the first place. In a beautiful, satisfying calculation, Newton derived the laws of planetary motion from two deeper principles: his law of gravity and his second law of motion (his soothsayer differential equation). The key assumption is that only the Sun pulls on the planet. All other gravitational pulls (from other planets, moons, asteroids, etc.) are ignored because they are much smaller in comparison. Thus, the calculation is known as the solution of the "two-body problem" (planet and Sun). The solution took tremendous mathematical wizardry because of the nonlinear character of the inverse square law of gravity, which makes the resulting differential equation tough to solve. Newton found a trick—a mathematical transformation—that converted his problem to a much simpler, linear one that he could solve. For decades afterward, mathematicians tried to find similar tricks to solve other differential equations, and often they could.

The logical next step was to tackle the three-body problem (Sun and two planets; or Sun, planet, and Moon). The task was made more urgent by improved astronomical observations, which revealed that the planets didn't move in perfect ellipses. The deviations were thought to arise from the gravitational effects of other planets, which Newton had neglected.

Over the subsequent decades, many great mathematicians tried to solve the three-body problem, but no one could. Meanwhile, an approximate method known as "perturbation theory" was invented. It produced some spectacular successes, such as predicting the existence of Neptune (then undetected but later observed in 1846, exactly where the mathematicians said it would be).

Given these successes, it was natural to want more—to solve the three-body problem completely, with no approximations. But the efforts of all the best minds kept failing, just as they had for 200 years. Even when the problem was idealized and simplified drastically, still no one could solve it. One simplification assumed the third body was a tiny speck of dust—pulled on by the other two planets, but exerting a negligible effect back on them—but the problem was still impenetrable.

Why was the three-body problem so difficult to solve? One possibility was that the issue was simply technical—after all, Newton needed tremendous ingenuity to solve the two-body problem, and perhaps the three-body problem was much harder still. A more disturbing possibility—one that in fact turned out to be true—was that there was a mathematical obstacle that prevented the problem from being solved. We can see this through a modern computer simulation showing the orbits in the three-body problem, which can be hopelessly complicated.

In fact, the three-body problem contained the seeds of chaos—a phenomenon as yet unimagined. In saying that, we're now using "chaos" in the contemporary scientific sense of unpredictable, random-looking behavior in a system that is nevertheless governed by nonrandom, deterministic laws. In the next lecture we'll see how this new kind of chaos came to light—and why it was just as quickly buried again. It's an odd little story that holds broader lessons about the sociology of science—and the psychology of scientists. ∎

Essential Reading

Peterson, *Newton's Clock*, chaps. 4–6.

Stewart, *Does God Play Dice?* chap. 2.

Supplementary Reading

Cohen, *Science and the Founding Fathers*, chap. 2.

Kline, *Mathematics in Western Culture*, chaps. 16–18, 21.

Questions to Consider

1. Pick any mundane piece of technology (such as a toaster, car, computer, or airplane) and explain how it relies on something that Newton discovered. As far as possible, summarize the chain of intermediate discoveries and ideas that connect the object to Newton's work.

2. Besides the Declaration of Independence, where else did Newton's work have an impact outside of science, say in the arts, humanities, philosophy, or politics?

3. Knowing what we know today about relativity, quantum mechanics, and chaos, what are some of the most serious flaws in the Newtonian view of the universe?

From Clockwork to Chaos
Lecture 3—Transcript

Welcome back to chaos. We ended Lecture Two on a cliffhanger: How can chaos survive in Newton's orderly universe? In this lecture and the next, we'll see how chaos squirmed out of this straightjacket by turning the logic of Newton's clockwork on its head. Newton's laws don't forbid chaos; they actually require it. This was a shocking turnabout, unthinkable really, to all scientists at the time, including many scientists who still didn't grasp its significance decades later.

To appreciate this next chapter in the ongoing story of chaos, we'll need to remind ourselves just how unassailable the Newtonian paradigm seemed at the time. You have to appreciate that the Newtonian worldview dominated Western thought for 200 years, from the late 1600s to the late 1800s. Newton and his successors helped to create the modern world. They created so many things in science—for instance, the science of fluid mechanics.

In fluid mechanics, the point of view that we take is to treat a fluid, that is, like a body of water or the atmosphere, as made up of countless numbers of tiny particles. This was even before atomic theory was established, but Newton and his successors thought, Let's treat the atmosphere or the sea as a continuum of infinitely many particles and then assume that those particles interact with each other according to Newton's laws of motion already worked out. So understanding the motion of this continuum, this fluid, led to the science of aerodynamics, of course, which gave us airplane flight.

In a similar spirit, they looked at a continuum of, again, infinitely many particles, but now rather than sliding past each other as they would do in a fluid, these particles tugged on each other elastically, corresponding to what you would see in a solid. That gave us the science of elasticity theory and structural mechanics, which allowed, in our own era, for the design of skyscrapers. It improved the building of bridges and so on.

But that wasn't all of it. We've got fluids now under control scientifically, [and] solids. What about gases? The air we're breathing? We could think about a gas, again, as in this Newtonian paradigm, as made up of trillions of molecules. And then they are now able to move freely, but occasionally they bang off of each other in hard collisions. That gave rise to the subject called *kinetic theory* to study the interactions of all the molecules in a gas basically flying free except when they bang into each other. From kinetic theory, again using this Newtonian approach, we got to the subject of thermodynamics, all of which allowed us to understand the workings of steam engines and, later, combustion engines used in cars.

It didn't stop there. This was just the follow-up from Newton and his successors, but there were other people who didn't directly build on Newton's laws but followed the Newtonian spirit. For instance, the rise of the science of electricity and magnetism is so essential to leading to the invention of telegraphs, electric lights, radios, telephones, television. All of that was built on this Newtonian thinking that we need to discover the differential equations that govern electricity and magnetism. That was the Newtonian approach, remember, to science. Find the underlying differential equations with the supreme confidence that that's how nature will be described. Those differential equations did exist for electricity and magnetism, and today they're called *Maxwell's equations* after James Clerk Maxwell, who first really wrote them down. Taken together, these magnificent advances in the Newtonian spirit laid the foundations for everything that we associate with the Industrial Revolution and with the modern world.

Newton was also a hero outside of science. That is, he indirectly influenced subjects like philosophy and politics. He was a hero to the Enlightenment philosophers Voltaire and John Locke. The successes of Newton in science inspired people like this, gave them a sense of confidence that everything in the universe, everything even in human affairs, could be understood and tamed by rational thought. This was the Enlightenment; this was the Age of Reason.

Let me take a little excursion here to show you the influence of this Newtonian approach. You can even see it in our Declaration of Independence. That is, our own system of government has a Newtonian flavor that you may not have realized before, but it's in there. Think about the language of the Declaration, the most ringing phrase of all: "We hold these truths to be self-evident, that all men are created equal." It's wonderful language, but there is a particular phrase there that may strike you, especially if you've studied a bit of Euclidean geometry. What is the phrase? "Self-evident truth." What is a self-evident truth? That is a truth that is undeniable, a truth that is so obvious, it's self-evident. "Self-evident truth" is a phrase that comes to us from Euclid. When Euclid established his approach to geometry, remember how it went? He had certain axioms. These were the self-evident truths. A line is that which has no thickness, a point is that which has no part, and so on. These are things that were laid down as the axioms at the beginning, and Euclid's approach to geometry was to build on axioms using logic to deduce theorems, corollaries, consequences of the axioms.

Newton, of course, was very influenced by Euclid. The approach of Euclid was taken as the model of logical reasoning for all of Western thought for thousands of years after Euclid. Newton himself fell under its sway, and when he wrote his great masterpiece, the *Principia*, he wrote it in a Euclidean framework; that is, there were axioms, the self-evident truths, and the propositions deduced from those self-evident truths. Jefferson was very much in this tradition when he wrote the Declaration.

If this seems like a stretch to you, it's not. We know that Jefferson really admired and loved Newton. We know that from his letters. For example, here's a letter that he wrote to his friend John Adams. This is after—Jefferson was no longer president at this point. It's January 21, 1812. Jefferson has stopped being president for 3 years now, and he says in his letter that he doesn't think about politics much any more, and he writes to Adams, "[I have] given up newspapers in exchange for Tacitus and Thucydides, for Newton and Euclid; and I find myself much the happier." Interesting, isn't it? Who was it that Jefferson was reading for fun? Two great historians and the two great mathematicians Newton and Euclid. It's a little bit creepy, but [Jefferson] was so interested in Newton that he acquired one of his death masks.

There's one other phrase I would like to just point out to you in the Declaration because I find it sort of memorable and a bit puzzling. Let's talk about the opening, the first section of the Declaration, called the introduction. I'll just briefly read it to you. You'll recognize the language. "When in the course of human events, it becomes necessary for one people to dissolve the political bands which have connected them with another … ." Remember what this is about. Jefferson is trying to explain why we have the right to separate ourselves from England.

> When … it becomes necessary for one people to dissolve the political bands which have connected them with another and to assume among the powers of the earth, the separate and equal station to which the Laws of Nature and of Nature's God entitle them, a decent respect to the opinions of mankind requires they should declare the causes which impel them to the separation.

So what he's saying is, given that we have the right to separate ourselves, we should, given our respect, explain why we're going to do it. What are the causes that impel us to the separation?

Think about some of that language, the causes that impel us to separate. That's physics language, the causes that impel something to happen. This is now Jefferson channeling Newton's approach to the world. It's about cause and effect. Causes impel certain things to happen, but the really important phrase here is what gives us the right. He says that we are entitled by the laws of nature and of nature's God. The laws of nature are what gives us the right. That's the point, that the laws of nature as revealed by Newton—he's talking here about Newton's laws and their generalization, that the universe is lawful—that is what gives us our political rights, what gives us our freedom, and so on. So these two opening sentences, the introduction and the preamble, are a testament to the domination of Newtonian thinking in all spheres of human activity at that point. So that's what I want to emphasize before we move on, that Newton's deterministic view of the world had a monumental impact on all spheres of human activity, from science and technology, of course, to philosophy and politics, for centuries afterward.

By the end of the 1800s, three tiny cracks began to appear in the foundations of determinism. One of them led to Einstein's theory of relativity; another, to quantum mechanics; and the third, to chaos.

Relativity: Let's think about what happened in relativity theory, and why did it come about? Because of inconsistencies between Newton's laws of motion on the one hand and Maxwell's laws of electricity and magnetism, the new science of electricity and magnetism. For example, there was the issue of Einstein imagining, What would happen if I were riding alongside a light beam? On the one hand, thinking in a Newtonian way, since I'm moving with the light beam, I wouldn't see it moving. I would just see electric and magnetic fields oscillating in place but not propagating. On the other hand, Maxwell's equations said that's not possible. Light cannot exist as a static pair of undulating waves. It has to propagate "at the speed of light." So there was a serious inconsistency in the theory, and by thinking about that, relativity emerged. So there was that—relativity—because of incongruities in the laws of nature as known at that time, Newton versus Maxwell. The resolution, as Einstein showed, was radical: to overhaul our ideas about time and space, overthrowing Newtonian conceptions of absolute time and absolute space.

Quantum mechanics, then—the second crack in the foundation—grew out of an attempt to understand the strange phenomena that were being revealed in experiments on light and on atomic radiation, which defied explanation in Newtonian terms. The behavior of electrons as they orbited the nucleus didn't seem to make sense in a Newtonian way. According to Newton and Maxwell, as these electrons were orbiting, they should be radiating, they should collapse into the nucleus, but they didn't. Quantum mechanics partly gave us an explanation why, but it was radical in what it said. What happened, you would have electrons jumping from one place to another without passing through the space in between. Quantum mechanics was also radical, maybe even more radical than relativity. It changed our whole conception of determinism; that is, it spoke of a world of probability where nothing was certain. Only probabilities could be stated of events happening.

Alongside those two revolutions, then you have chaos, which has a very different character. Chaos grew directly out of the same classical problems that had given Newton his greatest success, the motion of the planets in their orbits. It seems ironic that the same subject that had triumphantly ushered in the Newtonian clockwork, the problem of planetary motion, would later play a role in its demise.

What happened? How did we get from clockwork to chaos? First, let's recall what inspired the idea of the clockwork universe in the first place. In a beautiful, satisfying calculation, Newton derived the laws of planetary motion from two deeper principles. There was, on the one hand, his law of gravity, universal gravitation, saying that any two particles or masses in the universe pull on each other through gravity with a certain force. That force was proportional to their masses—that is, bigger masses pull harder—and it dropped off like an inverse square of the distance between them. The famous inverse square law of gravity. I'll have more to say about that in a few minutes.

So there was the law of gravity on the one hand. Then, on the other hand, Newton had this mathematical soothsayer that I described in the last lecture, his second law of motion, $F = ma$, which allows him to predict into the future for any system of interest. Remember how that works? The idea of a differential equation, we said, was that we think about the position and velocity of, say, a planet at a given time. And then, Newton's law, combined with the law of gravity, allows us to step forward one instant from there and recalculate the position and velocity, sort of like in a movie moving frame by frame through successive frames to build up a whole picture of an unfolding story.

The key assumption Newton had to make in solving the two-body problem, the Earth going around the Sun, was that only the Sun is pulling on the Earth or, in general, on whatever planet we're talking about, as if only those two bodies existed in the universe—nothing else—just the Sun and the planet. All other gravitational pulls, which his theory said would be there from other planets, from moons, from asteroids, they were all deliberately ignored because they would be much smaller in comparison. Those bodies are much tinier than the Sun—far away, but mainly because of how tiny they were. So the hope was, just ignore those other effects. Just focus on the Sun's effect on the planet.

This calculation, then, of course, is known as the two-body problem, the planet and the Sun. The solution of the problem took tremendous mathematical wizardry. Let's try to understand exactly why it's so tough in terms of this movie analogy. First of all, you've got the planet moving around the Sun. It's at some distance, so we could calculate the force using the inverse square law. That force dictates where the planet will move in the next instant, and as it does—also how its velocity will change—as it does all that, a nanosecond ahead. Actually, I shouldn't even say nanosecond. I'm talking about a mathematical instant. It's a time shorter than any time you can conceive of. It's not even a very well-defined idea. It's kind of an infinitesimal amount of time. So we move forward an infinitesimal amount of time, and now the planet is at some new place with a new speed. You have to recalculate the force, and then you have to recalculate the motion, and recalculate the force. So everything has to constantly be getting recalculated, and somehow as these frames in this amazing movie get calculated, you're going to put them all together to see the whole unfolding story, the whole vision, basically a good, clear, coherent movie of reality.

Let me try to give that to you more precisely for those of you especially who like math. I'm not going to go into details here, but I want you to appreciate that the wizardry of Newton involved a certain ingenious trick. The law of gravity involves an inverse square law. So the distance between these two bodies, the planet and the Sun, drops off like 1 over the distance squared between them. That's what it means. So if we were to say, [we have] a certain distance between the planet and the Sun, and now we make it 10 times larger than it was, the force wouldn't go down by 10. It goes down by 100, because 102 is 100.

The important point here, though, is that this kind of inverse square relationship is what we call "nonlinear." "Linear" would just mean a force proportional to distance. That's what you get with a rubber band or a spring. If I stretch the spring twice as far, it pulls back twice as hard. Or think of pulling a string on a bow in archery. Pull it twice as far, it pulls back twice as hard. That's a linear force, and that's a simple kind of force. That's the simplest possible force. It turns out linear forces give rise to very easy differential equations and also very simple motion. You know that a mass hanging from a spring will just vibrate. It's simple. In fact, high school students can be taught to solve this problem of a mass hanging from a spring. It's called the "simple harmonic oscillator." You may have done it yourself at some point in a physics class. We like the simple harmonic oscillator very much in physics, not because springs and masses are so important, but because this is a case we can solve, and because it governs small vibrations of all kinds of systems.

What Newton did was take his awkward inverse square, nonlinear force, which basically looks like it should be very hard to solve, and he found a trick, a mathematical transformation that allowed him to convert this nonlinear inverse square problem into a much simpler linear problem. That's how he solved it: converting a hard problem to an easier one, taking a nonlinear problem into a linear one. I can't show you the detailed math, of course, and the truth is, we're not really even sure how he did this because he wrote the *Principia* using geometry, suggesting that was the way he argued, but we know that he was in the process of inventing calculus and differential equations, so he may have known enough about those to solve the problem that way.

In any case, we're not sure, $U = \frac{1}{r}$ but here's the gist of how we do it today. You define a new variable. $U = \frac{1}{r}$ would be the language: r is the distance between the Sun and the planet, u is 1 over that distance. That would convert the inverse square to just a square because of the 1-over action, the reciprocal. Then, if I can borrow from John Kennedy, ask not how u changes with time; ask how u changes with respect to angle around the Sun. In other words, we're going to do two transformations. Instead of thinking about distance from the Sun, we look at 1 over the distance, and instead of thinking about how the planet moves in time, we think about how it moves with respect to the angle that the planet makes relative to the Sun as it orbits. This is not obvious that this will be a good thing to do, but if you do, it turns out that the equation relating this new variable u to the angle comes out to be the equation for a simple harmonic oscillator. That's magnificent because now we can solve the problem. We just look up the answer with sines and cosines to a simple harmonic oscillator, and we're done.

When I talked earlier about the double pendulum in the first lecture, why do we only study it for small angles? Because the problem behaves linearly for small angles. That's why. We like the linear case. That's what we're always taught to deal with.

For decades afterward, mathematicians tried to mimic this success, to find similar tricks to solve other differential equations as they arose in different parts of science, and often they could. It was a very happy time in the history of science, especially in physics and astronomy and math, a happy time, full of optimism.

The logical next step, having conquered the two-body problem, then naturally would be to solve the three-body problem, like the Sun with two planets going around it. Or, say, the Earth and the Moon and the Sun. Suppose you wanted to understand the motion of the Moon. You would want those three bodies involved. This task was made more urgent at the time—somewhat after Newton—by improved astronomical observations, which were showing that the planets did not actually move in perfect ellipses. They were a little bit off, and there were deviations, but that wasn't so surprising because we had a feeling we knew where they were coming from. The scientists of the time thought they would probably arise from the gravitational pull of these other planets that we've been neglecting all along, especially big ones, potentially, Jupiter or Saturn.

Over the subsequent decades, many great mathematicians tried to solve the three-body problem in the same way that Newton had solved the two-body problem, but nobody could do it. Nobody could find a trick. No one could find a way to transform the problem into a linear problem. People didn't want to give up, naturally, so they looked for something second best. Instead of solving the problem completely, a good approximation might suffice, and so there was an approximation method developed that is wonderful and has stood the test of time even today, called "perturbation theory." This method doesn't seek a perfect answer; it breaks a problem into its most important contributions, most important aspects, and then secondary aspects, and tertiary, and so on.

So in perturbation theory, you would say the main effect, as we say, the zeroth-order effect, is caused by the action of the Sun on the planet. Then a first-order effect, the next one after that, would be caused by the tugs of these other planets, and so on. So using these perturbation methods, building in successive corrections, mathematicians and astronomers produced spectacular successes. They were able to predict the existence of a new planet, later named Neptune. It was undetected at the time, but based on wobbles in the motion of Uranus, people thought there must be something out there pulling on Uranus. Maybe there's another planet that we haven't seen yet. So using perturbation theory kind of in reverse to ask where would this imagined planet have to be, then using perturbation theory to calculate its effects, where would it have to be to produce the deviations we see?

The mathematicians said to the astronomers, "If you look there, you will find a new planet," and it was observed exactly where it was supposed to be in 1846. This was a marvelous vindication of the whole Newtonian approach and of perturbation theory in particular; [they] could make predictions about nature, predict new planets, and they would be there.

Given these successes, it was natural to be even more optimistic. Don't give up. Let's go back to the three-body problem. Maybe we can just solve it, not just approximate it. Let's solve it. Come on. Be clever. Think of something. But the best minds kept failing, just as they had for 200 years.

Then another idea was suggested. Maybe we should simplify the problem. There's a simpler version of the three-body problem. If we can't solve that, we surely can't solve the harder problem, so let's simplify it as much as we can, like this. Here's one way to simplify it: Take two masses to be very big, two big planets, and make the third body so tiny that you could think of it as a tiny speck of dust, the point being that that speck would be pulled on by these planets, but it wouldn't significantly pull back on them. That is, you could ignore the speck in terms of its action on the planets, and in particular, you could do a really simple thing: Make the two planets just orbit each other in a circle, and put the speck in the same plane as this circle, and then just watch how this pair of rotating, orbiting planets pulls on the speck. That came to be called the "restricted circular three-body problem"—restricted to a plane, circular because the two planets are spinning.

Even that problem turned out to defeat everybody. Nobody could even solve the restricted three-body problem. Why was it so intractable? One possibility was that the difficulty was simply technical. After all, Newton needed tremendous wizardry, as we said, to solve the two-body problem, so nobody was as smart as Newton. Maybe the three-body problem, being a much tougher problem—and these minds are not quite at Newtonian caliber—there's just nobody there who can solve it, that's all, but it's not, in principle, unsolvable. Experience had shown that something like that could be true. Over the 200 years since Newton, it could be really hard to solve differential equations.

There was a more disturbing possibility, which was that the difficulty was not just technical but truly fundamental, and that perhaps there was some kind of deep mathematical obstacle inherent in the problem that prevented the problem from being solved, no matter how clever you were. Even a super-Newton couldn't do it. That turned out to be the case.

Let me try to illustrate that for you by showing you a simulation of what such a speck would do as it orbits these two planets that are themselves orbiting each other. I'm going to show you on a computer, and none of this was available to any of these mathematicians that we're talking about. Of course not. These were people living before 1900, so what I'm showing you is a little bit unfair, but we might as well see what was the difficulty they were facing without realizing it. What you're going to see in the simulation is two planets, and they will look motionless, but keep in mind that they're not really motionless. In fact, as I say, they're orbiting each other on circles. But that would make the simulation hard to look at, and so we'll do the standard trick that's done in physics where you go into the rotating frame of reference. If you think of each of the planets as being a pony on opposite sides of a merry-go-round and they're rotating, let's step onto the merry-go-round. So now we're in the rotating frame, and it just looks like the planets are fixed. What you're going to see, then, is a pair of fixed planets, and then I'll start the speck somewhere, and give it an initial velocity, and let it go, and we'll watch the trajectory of this speck.

Initially, then, I'm going to start the speck on the y-axis, and I'll put the planets on the x-axis. Let me go forward just a certain amount of time—say, 3 time units—and plot the path of the speck. It comes out like this. Using a computer to integrate the differential equation, that is, to sort of play out the movie for us, the speck started here, it went veering over, pulled by this planet, but doesn't collapse, doesn't smash into the planet. It kind of careens away and then does a strange loop, and comes over, and it looks like it might be about to crash into this other planet, but here's what's amazing. If we go forward in time even longer, watch what the ultimate orbit looks like. Instead of going 3 units forward in time, let me go 100 units and watch the path of the speck. This is the total path. I'm not showing it moment by moment, just the end result as if this is the orbit.

Wow! That, you can see, is a pretty complicated-looking motion of the speck, and it's no wonder that the mathematicians of this era couldn't solve for the motion of the speck, couldn't find its position at all times in the sense of finding a formula. In fact, what we now realize is that the three-body problem contained the seeds of chaos, a phenomenon as yet unimagined. In saying that, we're now using "chaos" in the contemporary sense of the word, meaning unpredictable, random-looking behavior in a system that's nevertheless governed by nonrandom, deterministic laws.

In the next lecture, we'll see how this new kind of chaos came to light and, strangely, why it was just as quickly buried again. It's an odd little story that holds broader lessons about the sociology of science and about the psychology of scientists. See you next time.

Chaos Found and Lost Again
Lecture 4

Jules Henri Poincaré was a French polymath, a fantastic mind. ... He made marvelously original contributions across a wide spectrum of math and physics, in some cases inventing whole new fields. ... Poincaré is a founder of topology, and his name lives on in popular imagination today through what's called the *Poincaré conjecture*. ... [But] we'll be focusing on his discovery of chaos.

In the next few lectures, we'll see that the path to the discovery of chaos reads like a detective story, with missed clues, false leads, and wrong turns. The first clue comes with Henri Poincaré's groundbreaking work on the three-body problem. Unfortunately, as with love, the course of true discovery never did run smooth—Poincaré blunders at first, then corrects himself, and finally is aghast to perceive what we now call "chaos" and "fractals" lurking in the three-body problem. The implication is that a system governed by deterministic laws can nevertheless behave unpredictably; chaos creeps into the clockwork.

Poincaré (1854–1912) was also a lucid expositor and wrote for the general public on topics ranging from the philosophy of science to the psychology of mathematical creativity. He came from an influential family; his cousin Raymond was the president of France during World War I. He was extremely visual and intuitive, and always brimming with ideas—it was said he flitted from problem to problem like a bee to flowers. But he was also extremely nearsighted, physically clumsy, and pathetic at drawing—as a student, he scored a perfect zero on the drawing exam for entry into the École Polytechnique and needed a special dispensation to be admitted.

Poincaré's work on the three-body problem was triggered by an international contest among the world's finest mathematicians. In 1885, King Oscar II of Sweden and Norway offered a prize for the solution, with the award to be given in honor of his 60th birthday in 1889. An ambitious 31-year-old professor, Poincaré burrowed into the problem for the next 3 years and submitted his entry a few weeks before the closing date.

The approach Poincaré took to the three-body problem was stunningly creative. He invented a new way of thinking, using pictures instead of formulas. It provides tremendous insight with surprisingly little effort. His pictorial method has become central to chaos theory. We're going to develop and use it throughout this course.

To illustrate Poincaré's method, consider a swinging pendulum. There are three ways we could study its motion. One option would be to conduct an experiment. Alternatively, we could try to solve Newton's differential equation. But this requires arcane math (elliptic functions) rarely taught even in graduate school. Our third option would be to draw a picture of the differential equation. This was Poincaré's approach.

Poincaré invented a new way of thinking, using pictures instead of formulas. It provides tremendous insight with surprisingly little effort. His pictorial method has become central to chaos theory.

Our picture would be abstract. It would depict the pendulum's motion as an imaginary point flying on a "trajectory" through an imaginary world called "state space." Continuing our earlier metaphor (from Lecture 2) in which a differential equation is like a movie, a state is like a single frame in the movie, and state space is a gigantic collection of all possible frames, all possible scenes. A trajectory takes you from one frame to the next—it's one possible scenario.

More precisely, the "state" of a system is defined as all the information we need to determine how the system will change in the next instant. A pendulum's state consists of two numbers: its current angle and velocity. Hence, its state space has two axes: one for angle and one for velocity.

Poincaré's great idea was that state space helps us imagine what the movie will look like before we see it, thus enabling us to predict the real system's behavior. Here's how it works: Newton's law (his differential equation) defines a vector at each point in state space. The vector tells the state where to go next, like the arrows showing storms moving on a weather map, or like those old dance lessons with diagrams showing you where to put your feet next, step by step. Following the arrows amounts to solving the differential equation. The resulting trajectory reveals how the pendulum moves. Even if we'd never seen a pendulum swinging or whirling over the top, the picture would disclose those possibilities to us. Similarly, Poincaré hoped the method would reveal all the motions of three gravitating bodies.

So what happened when Poincaré unleashed his method on the three-body problem? He didn't quite solve it, but he illuminated the problem as never before. His entry to the contest won the king's prize: a gold medal and 2500 crowns (equivalent to about half his annual salary).

Unfortunately Poincaré had blundered at an important point. He caught his error, but his manuscript was already being printed. Poincaré stopped the presses and paid for the print run so far—3585 crowns, more than his whole prize. Over the next few months, he worked feverishly to fix his mistake. What he discovered next would change science forever.

His new analysis revealed chaos and fractals lurking in the three-body problem. Poincaré was horrified. The picture showed a certain pair of curves looping back on themselves, crisscrossing to form an infinitely fine mesh (a kind of fractal shape). Neighboring boxes in the mesh represented different long-term motions. Hence two initially similar states could have wildly different futures. This meant that a deterministic system could be unpredictable. Chaos had infected the clockwork.

There is one more twist in this story: Having been so brilliantly uncovered, this first glimpse of chaos was just as quickly obscured, overshadowed, and forgotten. Why? Poincaré never drew a picture of what he had in mind. Was it because of his ineptitude as a draftsman? Computers did not exist, so he could not use computer graphics to help other people see what he was imagining. Only a small mathematical priesthood could follow his arguments. Poincaré didn't welcome chaos. It was an obstacle that prevented solution of the three-body problem. Also, the zeitgeist was against him. Physics was about to be swept up in two other revolutions—relativity and quantum mechanics—and had little interest in such old-fashioned questions. For all these reasons, Poincaré's discovery of deterministic chaos had little impact for 70 years. ∎

Essential Reading

Peterson, *Newton's Clock*, chap. 7.

Stewart, *Does God Play Dice?* chap. 4.

Supplementary Reading

Diacu and Holmes, *Celestial Encounters*, chap. 1.

Galison, *Einstein's Clocks, Poincaré's Maps*, chap. 2.

Internet Resource

You can play with a simulation of a pendulum and its motion in state space here: http://www.aw-bc.com/ide/idefiles/media/JavaTools/pndulums.html.

Questions to Consider

1. To check if you understand what we mean by the "state" of a system, figure out how many numbers you need to describe the state of the three-body problem, assuming that each of the bodies is a point particle of the same mass. Your choices are 3, 6, 12, 18, or 24. (Hint: What do you need to know about each body, at a given instant, to predict the future of the entire system?)

2. Can you think of other scientific discoveries, like the discovery of chaos, that weren't appreciated until decades or even centuries later? What were the reasons for the slow acceptance in each case?

Chaos Found and Lost Again
Lecture 4—Transcript

Welcome back to chaos. In the last lecture, we saw the first chinks developing in the armor of the clockwork universe. In the next few lectures, we will see that the path leading to the discovery of chaos reads like a detective story, with false clues, missed leads, wrong turns, blind alleys. It's a great story, and I'm really looking forward to telling you about this.

We've seen already the first clue, the hint that there's something amiss in the clockwork universe, which has to do with the three-body problem and the greatest scientists' inability to solve it.

In this lecture, we're going to talk about the best attempt to solve it that comes from the work of Henri Poincaré. We'll look at his groundbreaking work on the three-body problem, and we'll see that, as in love, the course of true discovery never did run smooth. In Poincaré's work, it wasn't smooth sailing at all. In fact, he thought he had solved the problem, but he made a blunder, and then in correcting that blunder, he was aghast to perceive what we now call chaos lurking in the three-body problem. The implication was that a system governed by deterministic laws could nevertheless behave unpredictably. Chaos could creep into the clockwork.

Jules Henri Poincaré was a French polymath, a fantastic mind, really one of the most brilliant people of the past 100 years. He made marvelously original contributions across a wide spectrum of math and physics, in some cases inventing whole new fields, like something called *homology theory* in topology. This is a branch of math that has to do with shapes, and it's a kind of generalized geometry. What happens if you bend a shape or twist it? Poincaré is a founder of topology, and his name lives on in popular imagination today through what's called the *Poincaré conjecture*. You may have read about it in the paper recently. It was solved, and there's supposed to be a bounty of $1 million from the Clay [Mathematics] Institute to the first person to solve the problem, who turns out to be a Russian recluse genius, Grigori Perelman. So Poincaré's name has been in the news in the past few years. But we're not going to be talking about his work in topology. We'll be focusing on his discovery of chaos.

In addition to his work in math and physics, Poincaré was a fantastic expositor. He wrote for the general public on all sorts of things: philosophy of science and even the psychology of mathematical invention. Poincaré came from an influential family. His cousin Raymond Poincaré was the president of France during World War I.

Henri Poincaré was extremely visual; a great, great geometer; and intuitive, not a fussbudget. He wasn't so interested in finicky details. He liked looking at the whole problem in a visual way, and it was said that he was [so] brimming with ideas that he flitted from problem to problem like a bee to a flower. But like many mathematicians, he wasn't perfect. He was extremely nearsighted, a bad athlete, physically clumsy, and most pathetically of all, he couldn't really draw a decent figure, anything—terrible. In fact, as a student, he was so pathetic at drawing that when he took the entrance exam to the École Polytechnique, he got a perfect score of zero. Because he was so far off the chart mathematically, they knew they wanted him, but no one had ever been admitted with a zero. They had to give him a special dispensation to let him in.

Poincaré's work on the three-body problem, which is our topic for this lecture, was triggered by an international contest among the world's greatest mathematicians. This contest was in honor of King Oscar of Sweden and Norway. It was announced in 1885, and it was supposed to be to commemorate the 60th birthday of this king, which was going to take place in 1889. At that time, Poincaré was just a young, 31-year-old professor. He was ambitious. He hadn't really made a name yet. He was known, but not the way he would eventually become known. He wanted to take a whack at this three-body problem that had stumped everybody. His approach to the problem was brilliantly created, as we'll see, and he burrowed into it for 3 years, submitting his entry just a few weeks before the closing date.

The approach that he took was incredibly inventive. He invented a whole new way of thinking, in fact, using pictures instead of formulas. That is, everyone else had approached the three-body problem as an exercise in calculus and differential equations, but Poincaré, being the great visualizer that he was, tried to find a way to visualize the solution. This strategy is simple and appealing. It provides tremendous insight with surprisingly little effort, as we'll see. In fact, his method lives on today. It has become absolutely essential to chaos theory. We are all descendants of Poincaré, all chaos theorists. We'll be developing his method and using it throughout the course.

Let me try to illustrate his method using a swinging pendulum that I've got here. Remember, I used this pendulum in Lecture One. At that time, it was in its double-pendulum mode, but here, I've locked its knee, so it's just an ordinary single pendulum. This, of course, is much simpler than the three-body problem. I'm just doing a pedagogical thing here with you to use the pendulum to get across the idea of Poincaré's method. Now, other than Poincaré's method, how could we study a pendulum? Suppose we wanted to learn rules about its motion, how it moves. Of course, the simplest thing would be [that] we could just do an experiment and watch it. No big surprise there. It moves like a pendulum, tick-tock, back and forth. That's not very satisfying from a theoretical point of view. If we wanted to understand this mathematically, how would we approach it?

Newton told us how. Newton has a recipe. He says, remember, $F = ma$. You tell me the force on the pendulum, F, and I will then tell you, in terms of its mass, m, what its acceleration is. And as the pendulum swings, the force on it changes. What's happening here is that gravity is pulling straight down, and so when the pendulum is like this, the force is pulling perpendicular to the pendulum, and then when it's here, you see, it's pulling along the pendulum, so the force keeps changing. Calculating the motion, then, as we discussed before, amounts to integrating, frame by frame, what $F = ma$ is telling us.

That turns out to be a monstrously difficult problem in differential equations, and very few people have done it. It can be done, but it requires such arcane math that they don't even teach it in graduate school any more. The topic involved, if you tried to solve the differential equation for the pendulum exactly, uses what are called *elliptic functions*. It's not a standard topic any more. They're a generalization of the trig functions that you may have learned about in high school, sine and cosine. These are harder than that, much harder, but they turn out to be involved in the solution of the pendulum. As I say, though, nobody does it that way any more.

The point of view that we're going to take here is to draw a picture of the differential equation. This is the radical new idea of Poincaré. The picture necessarily is abstract. It's going to depict the pendulum as moving through an imaginary world that we'll call "state space," or sometimes I might lapse into calling it "phase space," which is another common term.

Let me give you an analogy to make sense of all this, that is, to introduce all the jargon up front and then say, term by term, what I mean. Poincaré's picture is that the motion of a pendulum corresponds to a trajectory moving through phase space. What am I saying? Before getting into the math of all those words, let me go back to my movie analogy, which I think you'll find helpful. Remember, in Lecture Two, we said that a differential equation is like a movie. Not exactly, though. Let's be a little more precise. A differential equation is like the logic behind the movie. It would tell you how one scene in the movie logically leads to another scene. In terms of that analogy, you could think of a state as being like a single frame in the movie that tells you everything you need to know about what's going to happen next. State space, a new concept, is the collection of all possible states. It's an imaginary space, and if you like, in terms of the movie, it would be all possible frames that you could imagine in a movie of the pendulum, sort of all possible scenes in the movie.

Let me give it to you a little more precisely, coming to the math. More precisely, then, the state of a pendulum is all the information that I would need to tell you for you to be able to determine how the pendulum will move in the next instant. It's the totality of information needed to predict the future. So let's see if we can guess what is the state of a pendulum; that is, suppose I take the pendulum off to the side like this. Certainly, I've given you one piece of information now: this angle from the downward vertical. Is that enough information for you to predict how the pendulum will move? What do you think? I can let it go, and then maybe you could predict that, but in fact that one piece of information is not enough. Do you see why? The reason is I could start here, but I haven't told you how fast I'm going to push the pendulum initially. I might just let it go from rest, in which case, you have *that* motion, or I might give it a push, and you see, it's going to do something very different. So you need to know both. You need to know the initial angle, and you need to know the initial velocity.

Having told you those two pieces of information, initial position or angle and initial velocity, is that enough? It turns out, yes. If I give you the initial position and the initial velocity, Newton's differential equation, $F = ma$, his second law, his soothsayer, tells you everything. That will predict the future for this pendulum. So now we know that a pendulum is described by a pair of numbers. Its state requires two numbers, a position—really, we should think of it as an angle—an angle and a velocity. When we look at state space, which is what we're going to do now on the computer, we're going to need two axes. There will be one axis for showing the angle of the pendulum, and another axis will show its velocity.

Before I get into this computer animation, I have to give a word of heartfelt thanks. This simulation that I'm about to show you, like a lot of simulations—in fact, all the simulations I'll be showing you throughout this course—was developed by a friend of mine named Hubert Hohn. He's a computer artist and a good friend, and he put in a tremendous effort. Without him, this course, or at least the computer animation part of it, wouldn't have happened. So thank you very much, Hu.

Now, in most of his simulations, you'll see two things going on at the same time. At first, it might seem confusing to you, but it's going to be very powerful. What Hu does when he does that is, he's trying to help you—and I like it—he's trying to help you see the connection between two ways of looking at things. One way is usually an obvious or familiar way. It turns out not to be so powerful. The other way, although new and puzzling at first, is the powerful way. And so by showing you both ways of looking at things at the same time, you can connect the familiar with the powerful.

In the simulation I'm going to show you, you will see two panels. One is just an animation of a pendulum swinging back and forth, nothing surprising or mysterious. That's the familiar picture, just like if you were watching a real movie of a pendulum. On the left side of your screen will be state space. That's the new picture, unfamiliar, but extremely revealing once you know how to interpret what you're seeing. All right, so here I am, looking at my two panels, and you see there is a make-believe pendulum hanging down here on the right, and then this black patch is state space.

First, I want to orient you in terms of what axes you're seeing. There are two axes, remember. We said we expect [that] the state would have two variables. One of them is an angle measured from the bottom; that's shown horizontally. The other axis is going to be the instantaneous velocity of the pendulum, up here; that's plotted vertically. Now, here's the interesting thing. By moving around in state space, I can change the state of the pendulum. Let me first show you what would happen if I move in a way that changes velocity without changing the angle. Here's how I can do that. Let me go down here to where it says—notice these are written as if they were compass directions: north, east, south, west, and back to north. Let me start here at east, and if this means anything, what it should mean—you see the pendulum is now pointing due east? That's because I fixed the angle at east. But what's that blue arrow? See, as I'm moving, I'm changing the initial velocity I give to the pendulum, and that's shown schematically by this arrow lengthening or shortening. Here, I'm giving it less and less of a push, and then when I go through this axis, now the initial velocity is in the other direction. Moving vertically corresponds to changing velocity.

What about if I move sideways? That's going to correspond to changing the angle of the pendulum, so watch that. I'll try to move as exactly sideways as I can. Notice the blue arrow is not changing much. All that's really happening is [that] we're changing the angle of the pendulum, because I'm only changing that. I'm not changing the velocity. So those are what my two coordinates mean in state space. What I want you to do by the end of this lecture is try to have a feeling for how, when this imaginary trajectory is moving around in state space, how that's going to correspond to the real motion of the pendulum, or at least this animated version of the pendulum.

Let me start a little simulation here that will show you how the pendulum is going to move. If I start from here and I click my mouse, now you see the pendulum animated. It looks like it's just doing what pendulums do, swinging back and forth. But what's going on in state space? You see a little dot moving around on some kind of orbit. What does that mean? Let's orient ourselves. First of all, notice that every time the dot is in the top half of the plane, that is, in the northern hemisphere, the pendulum is swinging counterclockwise. Every time the dot goes back into the bottom half, the pendulum is swinging clockwise. See that? Also, we can stop this anywhere we like. If I wanted to stop right there, the dot is here; that means that the angle of the pendulum is at its most extreme, which you saw was true. That pendulum is at its most extreme angle, but according to this, we have zero velocity. We're on the axis. And so that should mean that at zero velocity, the pendulum is turning around. Let me click "continue," and you'll see the pendulum turn around. It goes back the other way.

What this closed orbit in state space means is that the state of the system is moving around on a loop. It's repeating itself, which is exactly what pendulums do. They move periodically. They repeat their motion. And so this closed loop in state space is a visual version of the periodic behavior of a pendulum.

We've just shown here the simplest kind of motion of the pendulum, and there are other possibilities, and they're all disclosed by state space. So this is the big idea: The different trajectories in state space correspond to different motions of the pendulum. In effect, they represent different movies. Let me show you what some of those movies could look like. Let me pause, and I'll start from a different initial condition. What that would mean in physical terms would be, maybe I've given an initial push to the pendulum or whatever. For instance, suppose I push it really hard. That means, go up to a very high velocity, like where I am now. Let me give it a hard push and see what kind of motion you will see. See, that's something qualitatively different. Now, this pendulum is going over the top; it's whirling. It doesn't form the same kind of closed orbit in state space that it had before, and not surprisingly, because the motion of the pendulum itself is different. This thing is now whirling over the top like a propeller. There's another kind of qualitatively different motion that we can see. Let me start down here. This is an interesting case. Watch this. I've started near the top; I very slowly fall down; I almost come to rest near the top and then go back again. So that's a sort of in-between motion. That thing is almost trying to whirl, but it doesn't quite make it, doesn't quite get over the top.

You can see we've got three qualitatively different trajectories, and they represent these three very different kinds of motion of the pendulum. Now, here's the key question. This is the big idea, as I was saying, behind Poincaré's strategy: Can we figure out what the trajectories look like in state space, that is, on the left of the screen, without seeing an animation to guide us? In other words, if we put our hand over the pendulum, if we had a blind eye—we don't know what the pendulum is really going to do—will state space tell us the answer? Can we use state space to predict how a real system will behave? If we can predict how the movie will play out from what we see in state space, we then, in a sense, can predict what pendulums will do. So that's Poincaré's really great idea. We're going to use state space to imagine what the movie will look like before we see it or even in cases where we can't see it, like in the three-body problem, and we'll rely on that state-space description to predict behavior of the real system.

Here's how it works. Now, remember, I'm covering up the animation. I'm just looking at state space. How do I know where to go? Let me clear the screen and show you how Newton enters the picture. Remember, all of this is being animated by Newton's law, his differential equation, $F = ma$. Where is it in the picture? Well, in Poincaré's beautiful conception, Newton's law, his differential equation, $F = ma$, determines a vector at each point in state space. In other words, knowing what state we're at now, Newton, remember, tells us how to go ahead by one instant forward in time to where I'll be next, what my new state will be. In effect, Newton, at each point in state space, gives me an arrow saying, "Go there next, and now that you're there, go there next," and so on. Let me draw those arrows for you. You see those arrows on state space? That is the visual version of Newton's law, $F = ma$, for a pendulum. It's a vector field, we would say—a field of vectors.

You've seen vector fields whenever you look at weather maps, and they show storms moving across the country, and you see lines of barometric pressure and so on. Those arrows are playing the same role here. Or what it really reminds me of is those old Arthur Murray dance lessons, where to learn how to take a complicated dance step, there is a diagram with an arrow showing you where to put your foot next. This is what these arrows are doing on state space. They tell the pendulum how to update its state from one moment to the next.

All we have to do is start somewhere and then follow the arrows. Just pretend we're dancing and just follow them. This will show us what the pendulum will do. To do that, then, I'll just click somewhere, and watch how the trajectory just follows the arrows. Here we're using the computer to inch forward one instant at a time, and as long as we follow the arrows, everything is consistent, and we're seeing a motion that would correspond to a motion of a real pendulum. So even if we'd never seen a pendulum swinging or whirling over the top, this picture would disclose those possibilities to us.

Similarly, Poincaré hoped that the method would reveal everything that the three-body problem could do. So what happened, then? We're all primed up; we're ready to unleash this method on the three-body problem. What happens? It turns out it didn't quite solve the three-body problem, but Poincaré's work illuminated the problem as never before. He got tremendous penetration into what was going on in the problem. His entry to the contest won. He won the king's prize, brilliant for the young professor. He was the toast of the town. All the best mathematicians of Europe, in the whole world, were aware of this young man, Poincaré. He won the gold medal and the full prize, which was 2500 crowns, which—hard to know exactly what that was worth, but one of the judges, a much more senior mathematician at that time, was making 7000 crowns a year, so assuming Poincaré is more junior, this is probably like half a year's salary: big money for him. But that wasn't the point. He had solved this tremendous—or he was gaining insight into—this really longstanding, important problem. It sounds good, right?

But here is where the course of true discovery never did run smooth. Poincaré had blundered at an important point. There was something seriously wrong in what he submitted. It's a little bit unclear historically how this was figured out. It seems that one of the judges of the contest asked him a question, and I think it was probably an innocent question along the lines of: "Could you just show me a little more of your reasoning here? Not saying you did anything wrong. I just don't follow this step." And Poincaré realized that there was an error. So he caught his own mistake, but his manuscript was already being printed, and this would have been very embarrassing. So Poincaré asked that the presses be stopped, at great expense to himself. He paid for the entire print run. The cost was 3585 crowns, more than his whole prize. Now, over the next few months, he worked feverishly to figure out what was wrong and to fix the mistake if he could, and what he discovered changed the course of science forever.

His new analysis revealed chaos lurking in the three-body problem. Poincaré was horrified. He wasn't looking for that. Let me try to illustrate what it was that he found. It was a nightmare. This was the picture that he saw. I can't quite give you the details of what I'm showing here. It's a bit too abstract for us right now, but all I want you to do is take away the visual impression of this crazy-looking spaghetti diagram. The important thing is that this picture shows a certain pair of curves, which I'm indicating here in red and blue, and they are looping back on themselves infinitely often and crisscrossing in such a way that they form an infinitely fine mesh, a kind of fractal shape. We'll later see that this is an instance of a fractal. Now, Poincaré never drew this picture, at least not for his readers. We don't know if he may have sketched it for himself, but he certainly saw it very clearly in his head, and he understood the implications of it. He also tried to describe it in words. Let me read you his words. Try to imagine yourself trying to make sense of this that I'm about to read. Remember, no picture. Poincaré did not draw a picture in his published work, ever, of this thing that I'm showing you now.

He wrote, "When one tries to depict the figure formed by these two curves and their infinity of intersections, these intersections form a kind of net, web, or infinitely tight mesh. Neither of the two curves can ever cross itself [so the red curve can't cross itself and neither can the blue], but each must fold back on itself in a very complex way in order to cross the links of the web infinitely many times." No picture. You're supposed to picture this in your head. What is this man talking about? And then he says, "One is struck with the complexity of this figure that I am not even attempting to draw. Nothing can give us a better idea of the complexity of the three-body problem."

So he admits, "I'm not even trying to draw it." The picture is complicated, but so what? What does it mean? What does this picture really mean? The point is that neighboring boxes in the mesh, little boxes where the red and blue curves are crossing each other, correspond to different long-term motions of the three-body problem. What this meant was that two states of the three bodies, initially so close as to be indistinguishable, imperceptibly different, two virtually identical states of the system, could have wildly different futures depending on where they started in the mesh. That was the implication of this picture. In other words, a deterministic system could be unpredictable. He had discovered chaos. Chaos had infected the clockwork.

Now there's one more twist to this story. Having been so brilliantly uncovered, this first glimpse of chaos was just as quickly obscured, overshadowed, and forgotten. Why? Well, we don't know. We can speculate. I think one reason is—we know Poincaré never drew a picture of what he had in mind. Was it maybe because of his ineptitude as a draftsman? It's a pretty complicated figure, I admit. In any case, it didn't help his readers to understand what he was talking about, and computer graphics didn't exist. Of course, computers didn't exist, so he couldn't use computer graphics to help himself or others to see what he was imagining, this nightmarish glimpse of chaos. There were people who could follow his arguments, but they were just a tiny mathematical priesthood. Very, very few people had any idea what this man was talking about.

Furthermore, Poincaré himself didn't welcome this newfound discovery. After all, this was an obstacle that prevented the solution of the three-body problem. He wasn't happy about it, but there it was.

Another reason that this may have never really made the splash that you would expect, the big discovery of chaos, is that the zeitgeist was against this sort of thing. In other words, physics was, at that time—remember, this is now around 1890—soon to be swept up in two tremendous revolutions, relativity and quantum mechanics. And the questions that Poincaré was looking at, the three-body problem, this was really old-fashioned stuff. This is back to Newton, 200 years earlier. Nobody wanted to work on such old-fashioned questions. They seemed desiccated and picked over. In fact, Poincaré himself made seminal contributions to relativity theory, and so his interest was going elsewhere, too.

So for all these reasons, the significance of chaos was not appreciated when it was discovered, and Poincaré's work on the three-body problem had little impact on mainstream science for the next 70 years.

In the next lecture, we'll see what finally brought chaos back on the scene. Thanks. See you next time.

The Return of Chaos
Lecture 5

Without any conceptual framework for thinking about chaos, either because people were unaware of Poincaré, or unable to understand him, or just didn't think that his work applied in their context, researchers … dismissed the chaos that they were observing directly, or they ignored it, or I think what was really happening in many cases is they simply couldn't see what they were seeing.

For the next 70 years after Poincaré, chaos remained a quiet backwater of science, studied only by a small mathematical priesthood. This was an era when science was becoming increasingly specialized and mathematics was starting to look inward. Poincaré's flame was kept alive by George David Birkhoff (1884–1944), the first great mathematician to be trained entirely in America, and by a handful of mathematicians in Europe and the Soviet Union. Engineers, meanwhile, were extending Poincaré's methods to analyze the nonlinear oscillations of vacuum tubes and other newfangled technologies relevant to the development of radio, telephones, and radar. Occasionally these researchers stumbled across their own hints of chaos. But because they had no conceptual framework for thinking about it, they dismissed it or ignored it.

The calm ended in 1960 with a thunderclap, due, appropriately enough, to a man fascinated by storms and weather prediction. In this lecture we recount Edward Lorenz's discovery and analysis of chaos in a simplified model of weather patterns.

First, a bit about the man himself. Although Lorenz is universally regarded as one of the pioneers of chaos theory, he's very modest and unassuming. I had the privilege of getting to know Professor Lorenz when I was his colleague at MIT in the early 1990s. Lorenz was born in 1917 and grew up in West Hartford, Connecticut. As a boy, he had always liked weather, but he liked math even more. He thought he would go into math after he graduated from Dartmouth in 1938. He went on to get a master's degree in 1940 from Harvard, where he studied dynamical systems with Birkhoff. But then World

War II broke out, and he found himself working for the Army Air Corps as a weather forecaster. After the war, he continued to be interested in forecasting and wanted to help develop the mathematical theory behind it.

Lorenz wanted to see how well certain forecasting methods would do by testing them on a computer surrogate for the weather. So he needed to make artificial weather. This was a novel strategy. Lorenz was using the computer as an arena for experiments, not as a calculating machine. He kept the computer in his office, also very unusual (personal computers were 25 years away).

His first version of artificial weather was too simple. It would either settle down to an unrealistic equilibrium state or repeat in perfect periodic cycles. Lorenz realized that he needed to concoct a system that was both deterministic and non-periodic, meaning that it would never repeat itself exactly. Otherwise his artificial weather would be easily predictable and hence worthless as a test of forecasting methods. After struggling, he came up with a deterministic system that never repeated itself. (The missing ingredient had been that he needed to vary the warming effect of the sun from east to west, not just from north to south.)

While studying his artificial weather, Lorenz happened across the "butterfly effect," the extreme sensitivity of a chaotic system to tiny changes in its initial conditions (named for an imaginary butterfly flapping its wings in Brazil, creating tiny air currents that ultimately trigger a tornado in Texas).

The discovery came while Lorenz was rerunning a simulation that he had wanted to project farther into the future. He stopped the computer midway through its previous simulation, typed in the numbers it had printed out, and set the computer running again. Then he went for coffee and came back an hour later (by which time 2 months of artificial weather had passed).

The results were nothing like what he had seen the first time! At first Lorenz suspected a computer malfunction (which was not unusual). On closer examination, he noticed that the old and new simulations agreed at first, but then began to differ in the third decimal place, and then in the place before that. The discrepancy was doubling in size every 4 days or so. This clue revealed the culprit: The initial conditions had not actually been the same in the two simulations. To save space on his printouts, Lorenz had rounded off the original numbers to three digits, but the computer was using six digits internally. The initially tiny round-off errors (in the fourth decimal place) were growing exponentially until they eventually overwhelmed the solution itself. The butterfly had changed the weather!

What are the broader lessons of Lorenz's work? The butterfly effect is the signature of chaos. Understanding it will be crucial to all that follows. It is also probably even more severe for real weather than for Lorenz's artificial weather. Hence, long-range weather prediction will never be possible. Even if we had a perfect model of the atmosphere, the inevitable errors in our measurements of current weather conditions will grow until they make our forecast look silly.

The butterfly effect is the signature of chaos. Understanding it will be crucial to all that follows.

There's also a lesson here about the discovery process: Lorenz was serendipitous, not lucky. By deliberately searching for something (a stringent test of a forecasting method), he was keenly vigilant, which helped him spot something unexpected and far more interesting (the butterfly effect).

Lorenz's research strategy changed everything. He adopted Poincaré's pictorial approach but strengthened it enormously by using a modern computer to simulate his system and graph the results. The computer became more than a number cruncher. It became a telescope for the mind, a way of imagining the inconceivable. Suddenly scientists could *see* the consequences of the laws of motion (even though they still couldn't write formulas for them). Computers provided intuition. In retrospect, this is why chaos theory had to wait until the 1960s. Without the computer to perform millions of calculations in the blink of an eye, scientists couldn't begin to fathom what their equations were trying to tell them. ■

Gleick, *Chaos*, 11–31.

Lorenz, *The Essence of Chaos*, chaps. 3 and 4.

Stewart, *Does God Play Dice?* chap. 7.

Strogatz, *Sync*, chap. 7.

Supplementary Reading

Strogatz, *Nonlinear Dynamics and Chaos*, chap. 9.

Internet Resource

If you want to examine the butterfly effect for yourself in a simulation of another of Lorenz's models (to be discussed in Lectures 7 and 8), try this online Java applet: http://www.aw-bc.com/ide/idefiles/media/JavaTools/lrnzdscv.html.

Questions to Consider

1. Do you believe this butterfly joke should be taken seriously? If a butterfly can trigger a tornado, couldn't it just as well prevent one? If so, what is the likely net effect of butterflies on the weather?

2. What other important discoveries have been made by serendipity?

Lecture 5: The Return of Chaos

The Return of Chaos
Lecture 5—Transcript

Welcome back to chaos. Let me remind you where we are in our detective story. The mystery started with Newton's clockwork universe, which seemed to rule out the possibility of chaos once and for all. Its deterministic character seemed to leave no room for chance, no possibility for chaos. But then came the first clue that something was rotten in the clockwork. Remember what it was. That was the intractability of the three-body problem in astronomy. Hundreds of years go by. No one can solve it.

Next comes what seems like a break in the case. Based on visualization, a new technique, Poincaré, the great French mathematician, thinks he may have a way of solving the three-body problem, and in the course of his work, he thinks at first he's solved it, enters a contest, submits his brilliant solution, wins the prize—except that a question comes up about his method, and he then realizes: Oh, wait a second; I made an error. He catches his error, works feverishly to figure out what's going wrong, and then, in the course of that, it turns out he has not solved the problem, but he has discovered something much greater. He has discovered deterministic chaos, the phenomenon being that simple deterministic systems, which have no element of chance in them, nothing random about them, can look like they're behaving randomly and can act wildly unpredictable. Poincaré is not happy about it. He recoils from the implication, but nobody else is really worried because they don't understand what he did. It's so arcane and so abstract, involving thinking in 18-dimensional space, which Poincaré can do, but nobody else can do. So that's where things sit, and that brings us to this lecture right now.

For the next 70 years after Poincaré, chaos remains a really pretty quiet backwater of science, certainly not the hottest area in town, cultivated only by a small priesthood, really, a mathematical priesthood that's devoted to thinking about what Poincaré did, but that's on the order of maybe 5 or 10 people in the world.

You should understand why. This was an era when science was becoming increasingly specialized, and math itself was starting to look inward. That happens once in a while in the history of science, that there are times when scientists are feeling expansive and talking to each other and other times when they become introspective and concerned with their own subject. For example, in the 1700s and 1800s, in the era of Newton and afterward, people were in the expansive mood. The best mathematicians also did astronomy and physics. But now, [at the] beginning [of] the 20th century, mathematicians really only wanted to think about math itself. Why was that? Through all their interactions with physics and astronomy, they had started to think intuitively, and they got into certain logical paradoxes. It turned out that certain things you would want to do by common sense turned out to have mathematical problems, and there came to be a feeling that we need to be more rigorous. So math, to kind of correct its own foundations, turned inward at that point and focused on its own internal structure.

Physicists, meanwhile, had their own problems. It was the era of atomic physics, and people were worrying about electrons and radiation (only recently discovered), radioactivity, things like that, and so they were in no mood to be fretting about the three-body problem in classical Newtonian physics. That was done as far as everybody was concerned, and the action was somewhere else.

This was, then, an era when there was no great motivation to think about what Poincaré had done. His flame almost went out, but it was kept barely flickering by an American mathematician by the name of George David Birkhoff, who taught at Harvard and was himself a very fine mathematician, especially notable for being the first great mathematician to be trained completely in America. That was unusual, right? The best people went to Europe for their training, but not Birkhoff. He stayed here, he taught here, and he trained a number of other American mathematicians who followed him. Along with Birkhoff, there was also work on dynamical systems, which is what Poincaré's area had come to be called, work being done by a handful of mathematicians in the Soviet Union, Europe, and England.

Curiously, what really kept Poincaré's work going was not these mathematicians, although they contributed quite a bit, but the engineers stepped in. We haven't talked much about engineering in this course, but there were mathematically inclined engineers, let's say, as many engineers are, who were interested in what Poincaré had done, but not with regard to chaos. That was an interesting thing. Poincaré's work was so powerful that it touched on all kinds of complicated problems relevant to differential equations, and these engineers wanted to use his methods to analyze what were called *oscillators*, devices that could produce clean periodic signals. For example, [engineers were] thinking about the electrical oscillations of vacuum tubes, which were only recently developed and were being used. These were newfangled technologies being used in the first radios and telephones and then, later, in radar and television and even lasers. So oscillators became a big topic in electrical engineering, and while studying them, these mathematically minded engineers found it useful to extend some of what Poincaré had done, his visual approach.

In the course of their experiments, what sometimes happened is that the engineers would trip across their own hints of chaos. For example, there was a team of Dutch engineers who were noticing a peculiar hiss in a circuit that they were studying. It didn't seem to them like the usual random noise that sometimes afflicted radio circuits, but they didn't know what it was. They reported it, but they couldn't really make sense of it, and that was that. They left it. In retrospect, it was clear that what was happening is [that] they were hearing the sound of chaos, deterministic chaos, in their circuit, later figured out by two British mathematicians, Mary Cartwright and J. E. Littlewood.

Likewise, a Japanese researcher named Hayashi spent a long time looking at oscillators and had observed chaos dozens of times, but he never reported it. Why not? How do we know he was even thinking about it? Well, because he had a student years later. Now I'm fast forwarding to around 1970 or '80, when chaos is in full flower, and this student reports the first electrical chaos in circuits, and his advisor, the famous Hayashi, tells him, "But I've seen that dozens of times." I'm going back many years into the '60s, and his student, Ueda, says, "But, professor, you've never published it. You never told me about this." And that's right; Hayashi kept it to himself. He didn't know what he had. He may have thought that his circuits were malfunctioning.

The point here is that without any conceptual framework for thinking about chaos, either because people were unaware of Poincaré, or unable to understand him, or just didn't think that his work applied in their context, researchers like these dismissed the chaos that they were observing directly, or they ignored it, or I think what was really happening in many cases is they simply couldn't see what they were seeing. It's a curious thing about human psychology, that if you don't have the right mental framework, you sometimes can't see what's right in front of your face.

Let me now come to the main topic of this lecture, which is the first time when this calm period of 70 years after Poincaré ends. The calm ended with a thunderclap, which seems appropriate because the man who created it was a man obsessed with weather prediction and storms. So in the rest of this lecture, I'll tell you the dramatic story of the work of Ed Lorenz, a meteorologist at MIT, and his rediscovery of chaos in a much more dramatic and stark form than what Poincaré had seen. You really can't miss it now, you would think, after Lorenz's work. Lorenz found this chaos, this full-blown chaos, in a model of weather patterns that he developed.

Before I tell you about his fantastic work, I have to tell you about this man. I had the privilege of getting to know him when I was a young assistant professor working at MIT, and he, as I say, was a meteorologist at MIT, working in the Earth and Planetary Sciences Department. Now, everybody regards Professor Lorenz as one of the great pioneers of chaos theory. He's universally admired and respected, and so it's particularly striking when you meet him that this is an absolutely modest and unassuming man who does not come across as brilliant. He really reminds you of some kind of roadside farmer that you might meet if you were going out to buy vegetables or something. He really seems like a Yankee farmer, like the guy that does the Pepperidge Farm commercials.

Every year, we would go through a certain little ritual that I came to have an affectionate feeling of anticipation for. I was teaching a class on chaos, and I felt I had to invite Professor Lorenz to come talk to my students. We had spent three weeks on the Lorenz equations, which I'll be telling you about in this lecture and the next three lectures to come. So the students [had] all this background in the Lorenz equations, and Professor Lorenz, being about 3 minutes away in the next building, let's have him come to the class. So I would call him up. "Professor Lorenz, would you be willing to come talk to my class?" And then he would say, in his monotone, "Yes, I would be happy to come." So then I'd say, "Fantastic!" But then he would always ask—and we did this, I would say, maybe 6 years in a row—he would say, "What would you like me to talk about?"

So what should you talk about, Professor Lorenz? How about the Lorenz equations? I mean, Einstein, would you like to talk about relativity to the class? How about the Lorenz equations? But then he would always say, "That little model? You really want me to talk about that little thing?" And then he would come to the class and not talk about it. At this point, he was 70 or 80 years old, and he would talk about whatever he was working on at the moment. He was still publishing. Our class never cared because they were so awestruck and so thrilled to be getting a glimpse of this man who created modern chaos theory. In fact, whatever he was doing now was pretty interesting, too. So that "little model" is what I'll be telling you about, and it really did change the way we look at the world.

Lorenz was born in 1917 and grew up in West Hartford, Connecticut. As a boy, he liked weather. He certainly was interested in weather, but he wasn't what you would call a weather bug. What he really loved was math. He went to college at Dartmouth, studied math, thought he would go into math, and in 1938 he got his undergraduate degree in math. Then he went on to get a master's degree at Harvard from Birkhoff. There's an interesting link: Poincaré to Birkhoff to Lorenz. So he studied; he got this master's degree. He was ready to go on to become a mathematician, but it's 1938, and you know what's coming: World War II. World War II broke out; Lorenz went into the Army Air Corps and found himself working as a weather forecaster.

After the war, he continued to be interested in weather forecasting, and he wanted to help develop the mathematical theory behind it, which was, at that point, in a pretty rudimentary state. That is, weather forecasting had been something like black magic and maybe even, to some extent, still has some of that. It's very tough to predict the weather. The best people in meteorology knew that and didn't want to go into forecasting. There were other problems, but forecasting was sort of like reading tea leaves, and people felt this is not something that a really talented person should go into. But Lorenz didn't agree. There were three approaches at the time to the problem of weather forecasting, and let me just outline what they are because they will be relevant for our story.

The first is what you could think of as dynamical forecasting in the spirit of Isaac Newton. That is, we're going to write down the correct differential equations for the atmosphere, for pressure, temperature, humidity, everything. It's enormously complicated, much harder even than the three-body problem. And then by solving those differential equations on computers, we will try to inch the weather forward one instant at a time. And the problem [is] very difficult because not only are there all these variables, but you have to keep track of the weather at every point on the map in three dimensions. So it's really tough. But anyway, that's one approach, and it wasn't tremendously successful, especially with the early computers of those days.

Another approach that was common was what you might think of as pattern matching. That is, you look for weather that reminds you of the weather you're seeing today, and you see what happened the last time you saw weather like this. That's a kind of commonsense way to do it: basing forecasts on previous experience.

A third way, which was kind of a newcomer and was challenging these other two, was statistical weather forecasting, where what you would do is—and this might sound ridiculous, just a mathematical version of tea leaves—but what you would do is: Say you want to calculate the temperature in New York at 4:00, you would write down some formula, like maybe the temperature in New York is one-third the temperature in Cleveland 4 hours earlier, plus one-eighth the temperature in St. Louis 10 hours earlier, plus a certain fudge factor for the humidity in some other place, and you have all these numbers that you could adjust in front of all the possible variables that you thought matter. You adjust them until they give the best fit, and then that's your model. It doesn't have any physics in it. It's just totally fitting curves. Okay, so statistical forecasting is that third way of doing things.

Now, what Lorenz was primarily interested in, the problem he set himself, was: Are these statistical methods any good? Could I test them, not just against real weather? And this was a clever trick he thought of. He thought, "Maybe what I could do is create artificial weather in a computer and use that to test these statistical forecasting methods. I'll make a surrogate for the weather that will be simpler than the weather, and maybe I can use—the advantage is, I can run hundreds of days of weather very fast in the computer. I can see if the statistical methods, treating this as if it were real weather, will predict what really happens in the computer, and so on." This was a really cool, novel strategy because Lorenz was using the computer in a new way. It was now an arena for thought experiments. It's not just a calculating machine, calculating the solutions to differential equations or anything like that. It's a place where you can play games. You can do thought experiments. So this was already an interesting bit of creativity on his part.

Another thing that was unusual was that Lorenz was doing so many of these computer experiments that he felt it was worthwhile to have a computer in his office, and that was totally unusual because this is the early 1960s. People would go to computer centers in some big room with a giant machine. But he had a kind of personal computer of its era called the Royal McBee, and it was very unusual. Personal computers as we know them were 25 years away.

What happened when Lorenz did these computer experiments? The trouble was, it's hard to make a good model of the weather in a computer. His first version of artificial weather was too simple. It would always settle down either to some kind of equilibrium state, where it wasn't changing, which isn't like the real weather, or it would change but in too simple a way. It would always repeat after a while, so it would be periodic, just repeating in perfect cycles. That's not the way the weather works either.

He realized that what he needed was to concoct some kind of a system that he could solve in his computer, so it had to be a deterministic system of differential equations, but it shouldn't be periodic; it should never repeat. Otherwise, the artificial weather would just be too simple, and it wouldn't provide a good test of these forecasting methods that he was trying to study.

He had to struggle for a long time to come up with a deterministic system that would do this, that wouldn't repeat itself, and eventually, he did. What he had forgotten to put into his model was the geographical effect of the heating. Now, everyone knows that it's hotter near the equator than up near the poles, so you would want to have some kind of warming that depends on latitude. Certainly, he knew that. But what he didn't put in at first, and what turned out to be the missing ingredient, was that there is also a heating effect that goes east to west; that is, the heating over the oceans is different from the heating over the continents. That makes a difference, too. When he included that east-to-west, or longitudinal, aspect, as well as the latitude, then he found he could get a system that was deterministic and did not repeat itself, just what he was looking for.

Now, where is the chaos in this? While studying his artificial weather, Lorenz happened across something that we now call the "butterfly effect," which refers to the extreme sensitivity of a chaotic system to tiny, imperceptible changes in its initial conditions. We call it the butterfly effect because there's a sort of image that you have in mind of a hypothetical, imaginary butterfly flapping its wings in Brazil, creating little air currents whose long-term effect is to ultimately trigger a tornado in Texas. That's the idea of the butterfly effect. Lorenz found something like this in his computer simulations, and he certainly wasn't looking for it.

Here's how the discovery came. It's a great story. [Lorenz] was rerunning a simulation that he had done. People used to love coming to his office to watch this artificial weather blowing by. Of course, there were no computer graphics. You were just seeing numbers and charts going up and down, but still. So he was rerunning one of his simulations, and he wanted to project farther into the future than he had [previously], so he ran the simulation up to a certain point and then stopped it. The computer was printing out these numbers. He recorded them, wrote them down, and then used those numbers as a new starting point to then integrate even farther forward into the future. Of course, these were slow computers so he knew it would take a while, and he put in these numbers that had occurred midway through the first simulation, typed them in, and then went down the hall to get a cup of coffee. We know this; he told us this story. It's in his book, too, that Lorenz has written.

So he goes down the hall, gets his cup of coffee, and 2 months of artificial weather have now gone by, by the time he comes back. Great. He comes back, looks at the printout of the extension of the earlier weather, and he finds that it doesn't look anything like the weather he was seeing the first time, which seems really weird because he started it midway through an old simulation. It should have given the same weather up to the end and then carried on from there, but it didn't. So what the heck is going on? Lorenz wasn't really as shocked as we might be because this kind of thing used to happen with his computer a lot. First of all, he suspected it could be a weak vacuum tube. Tubes were blowing out, and it was tough in those days, so it could be that. He looked all through there, and he didn't see any problems with any tubes. It could be bugs, literally. Bugs used to get into computers and cause trouble. But it didn't seem to be that; the computer was clean.

He really was puzzled and started scratching his head over it and thought, "All right, I can't see what the problem is with the machine, but maybe if I can localize the problem and figure out where in my computer printout the deviation first occurred, that might help." He could then give that information to the repairman. He knew that from previous experience, and it would speed up the fixing of the machine. He looks at the two printouts, the old one and the new one, and sees that there is no sudden mistake, that in fact, what he had first thought—that they had disagreed—wasn't really true. They agreed perfectly at first. But then as he was tracing through the numbers, he sees that this one now differs by 1 in the last decimal place. He keeps going down the column and sees that it now differs by 2. Now it's off by 3. Now it's off in the next decimal place. The error is growing as he goes farther into the future, with the discrepancy between the old run and the new run doubling in size about every 4 days. So that was a clue.

Can you guess what the problem was? The clue revealed the culprit. The initial conditions that he used when he started the second run midway through the first, they had not actually been the same in the two simulations. Why not? He tried. He wrote down the numbers, but why not? The reason is that what he was printing out from his computer was three digits, .628, numbers like that. He needed to do that to save space on the printout, so he would print out just three-digit numbers. But in fact, his computer, in its internal arithmetic, was computing with six digits, .628172. Those were just getting lopped off, truncated. What had happened, then, is that by ignoring those truncated terms that he couldn't see in the printout, Lorenz was creating round-off errors in the fourth decimal place, and those were growing exponentially fast and, ultimately, overwhelming and changing the weather itself. That little fourth decimal place was like the butterfly, and that changed the weather.

Let me now show you a modern computer simulation of this effect. Keep in mind that it's much easier for us to do what we're about to do than it was for Lorenz, who was just printing out numbers on a slow computer and so on. We're going to see a real computer graphic of this, and very fast. To be honest, the simulation I'm going to show you is not exactly Lorenz's original weather model. It's his later model, the famous Lorenz equations, which are much simpler than his original weather model. This only uses 3 variables, whereas his used 12, but it's basically the same idea, and it shows the same butterfly effect, and so I feel like I'm not being scandalous using this one.

What you're going to see in this simulation, before I start showing you, just to get yourself oriented, is a standard kind of graph called the *time series*. That is, you'll see a variable, which in this case is going to represent the spin of a certain waterwheel that I'll discuss in the next lecture. So there is some variable, spin as a function of time, and you're going to see it bobbling up and down. Watch the graph. We're going to start the solution from some initial condition, and there it goes. You see a variable that at first is oscillating, and the oscillations appear to be growing. They're getting bigger, bigger, bigger, and then something different happens. Now, it's starting to look like it's not repeating. It's aperiodic. This is the sort of thing Lorenz wanted, remember, weather that wouldn't exactly repeat. So you're seeing some kind of aperiodic signal, which, after all, was what he was after. He wanted to make non-periodic behavior from a deterministic system, and there's an example of such a thing.

Next, you're going to see me start midway through the yellow curve. I'll start midway through this simulation, and whereas I've been keeping six digits of accuracy, I'm going to now only keep three. When I hit the restart button, you'll see a blue curve that at first will track the yellow curve. It will go right on top of it, just like Lorenz's numbers agreed at first. But then you'll see it diverge. You see the blue is right on top of the yellow; now it's gone. Now they have totally different futures. The difference was, as I say, the blue curve was this truncated version in which I kept only three digits,

506, from the number that was 506127 originally. So that's what the butterfly effect looks like: good agreement here and then radical divergence after that.

Back to our story. What are the broader implications of Lorenz's work—that is, we've got this butterfly effect, but so what? The butterfly effect is the signature of chaos. This is the thing that defines chaos practically. It was a deterministic system, which in the long term is unpredictable because any tiny error gets amplified exponentially fast and, in this case, made the artificial weather unpredictable.

The implication of this is that the real weather, which is surely more complicated than what you're seeing in the computer, is also very likely to be impossible to predict into the long term. Now, we have to talk about how long term is *long term*. That will be discussed a bit in the next lecture. For now, that's a quantitative detail. The real point is, even if we had a perfect model of the atmosphere and we knew everything there was to know about the laws governing it, the inevitable errors in our measurements of the current state of the atmosphere would very quickly grow, snowballing until they make any forecast look silly.

There's also an interesting lesson here about the discovery process, which I think holds broader lessons for science than just chaos theory, which is that Lorenz was serendipitous in discovering the butterfly effect. That's not the same as being lucky. He didn't discover it by accident. He discovered it by serendipity. Here's the distinction: Lorenz was not just blundering around. He was deliberately looking for something. He was looking for a non-periodic system that was deterministic to test a forecasting method. What that did was make his mind alert. He's looking for something. He's keenly vigilant. He's ready to see trouble or anything unexpected, and he did. It just turned out to be something he wasn't looking for. It was much more interesting than what he was trying to do. It was chaos. But it was because of his frame of mind—that famous line from Pasteur about chance favor[ing] only the prepared mind—this is the way that you make scientific discoveries, by serendipity.

Also, Lorenz's research strategy changed everything about the way we do modern chaos research. He adopted Poincaré's pictorial approach, as we'll see in the next lecture, but he strengthened it enormously by using a modern computer to simulate his system and graph the results. Instead of advancing the simulated weather frame by frame with formulas—the way Newton would have tried to do, by solving a differential equation; or by picturing state space in his head, the way Poincaré had to, not having computers—Lorenz had the computer grind forward one instant at a time, pushing those differential equations and their solutions forward. So with the computer as a weapon for attacking problems, he could go much farther than Poincaré had, but that's not the big innovation. The big innovation was the way he used the computer. As I say, it was not just a number cruncher. It was a kind of mind amplifier, a telescope for the mind that let him imagine the inconceivable. Suddenly, scientists could see the consequences of the laws of motion, even though they still couldn't write formulas for those solutions. Computers were now giving intuition.

In retrospect, I think this is the answer to this question: Why did chaos theory have to wait until the 1960s to really get going? Without the computer to perform millions of calculations in the blink of an eye, scientists couldn't begin to fathom what their equations were trying to tell them.

In the next lecture, we'll take a step back and think about the butterfly effect from a more philosophical perspective and ask questions like: What does it tell us about the course of our own lives, about destiny, about fate? And how can we make sense of a chaotic world given that we can't really predict?

Stay tuned. See you next time.

Chaos as Disorder—The Butterfly Effect
Lecture 6

"We collectively wish to apologize for having misled the general educated public by spreading ideas about the determinism of systems satisfying Newton's laws of motion that, after 1960, were proved to be incorrect."
—Sir James Lighthill, former Lucasian Chair of Mathematics at the University of Cambridge, in a landmark 1986 apology on behalf of all scientists

Lorenz's butterfly effect has now entered popular consciousness, but some aspects of it are commonly misunderstood. In this lecture we clarify its scientific meaning and philosophical significance. The term derives from a 1972 lecture by Lorenz entitled "Predictability: Does the Flap of a Butterfly's Wings in Brazil Set Off a Tornado in Texas?" Lorenz originally had a seagull in the title, but the session chair at the conference replaced it with the more poetic image of a butterfly.

Since then, the butterfly effect has intrigued authors and filmmakers. It featured prominently in Steven Spielberg's blockbuster movie *Jurassic Park* (1993), based on Michael Crichton's bestselling novel (1990). In the movie, Jeff Goldblum plays chaos theorist Ian Malcolm, dressed in leather and looking like a rock star. He demonstrates the butterfly effect by flirtatiously holding Laura Dern's hand and placing a drop of water on the back of it, then asking her to predict which way it will roll off. The unpredictability of the droplet's trajectory is a harbinger of the unpredictability of Jurassic Park, an amusement park with a supposedly fail-safe population of genetically engineered dinosaurs that can never reproduce. The butterfly effect also plays a central role in Tom Stoppard's play *Arcadia* (1993) and in the movies *Sliding Doors* (1998) and *The Butterfly Effect* (2004).

But didn't everyone already know that little things sometimes make a big difference? In fact, they did. Lorenz himself mentions two literary precursors. The first is a 1941 novel called *Storm*, which Lorenz's sister gave him for Christmas when she learned he was going to study meteorology. In the novel, a meteorologist remembers that his professor had remarked that a man sneezing in China can set off a blizzard in New York. The second is the Ray Bradbury time-travel story "A Sound of Thunder" (1952), in which the death of a prehistoric butterfly changes the outcome of a present-day presidential election. Indeed, the idea of the butterfly effect is at least as old as John Gower's familiar verse "For Want of a Nail," published in 1390.

The surprise was that simple systems could also suffer from the butterfly effect. It was always clear that we couldn't predict wars or the course of our own lives. There are too many complexities, with millions of unaccounted variables obeying unknown laws, or no laws at all. But for simple systems like double pendulums, or the three-body problem, or Lorenz's model of artificial weather, we know all the variables and the laws determining their behavior. Yet we still can't predict what they'll do in the long run.

Poincaré certainly realized this. He expressed it clearly in his 1914 book, *Science and Method*. Still, for nearly all other scientists, the recognition of this kind of unpredictability was late in coming and philosophically hard to swallow. It led Sir James Lighthill (who held the same chair at Cambridge that Newton once occupied) to issue a remarkable collective apology. In a 1986 article called "The Recently Recognized Failure of Predictability in Newtonian Dynamics," he expressed regret on behalf of all scientists "for having misled the general educated public by spreading ideas about the determinism of systems satisfying Newton's laws of motion that, after 1960, were proved to be incorrect."

The butterfly effect simply doesn't occur in the kinds of systems most scientists had been studying since Newton. It's not there in systems that relax to equilibrium or … oscillate in regular cycles.

Why were scientists so slow to recognize the butterfly effect? First, the idea of it was philosophically distasteful, and almost threatening to science itself. Science depends on being able to draw a conceptual box around a system of interest, with the confidence that you can neglect tiny effects coming from outside the system. If you had to think about what's happening on Mars just to study the motion of a pendulum, science (and life itself) would be impossible.

Also, the butterfly effect simply doesn't occur in the kinds of systems most scientists had been studying since Newton. It's not there in systems that relax to equilibrium or that oscillate in regular cycles. In those cases, small disturbances stay small, or if they grow, they do so very slowly. A pair of metronomes illustrates this. If you start them in sync and then disturb one, the difference between them stays roughly the same thereafter and doesn't snowball. The tides, the return of Halley's Comet, the timing of eclipses— all of these are strongly periodic and hence predictable, because tiny disturbances do not mushroom into major forecasting errors.

The butterfly effect only afflicts systems that are both deterministic *and* non-periodic. One example is Lorenz's artificial weather model. Another is the double pendulum toy we've played with previously.

The butterfly effect does not imply that chaotic systems are unpredictable. They in fact are predictable in the short term because of their deterministic character. But they become unpredictable after a certain amount of time, called the *horizon of predictability*. It's the time required for tiny errors to double in size. For a chaotic electrical circuit, the horizon is something like a thousandth of a second. For the double pendulum, it's a few tenths of a second. For the weather, it's unknown but seems to be roughly a week or two, and for the entire solar system, it's about 5 million years (as determined by very careful computer simulations).

It's because the horizon is so long for the solar system that the motions of the planets seem utterly predictable to us today; and on the time scales of a human life, or even of the whole history of astronomy, they *are* predictable. When we calculate planetary motions hundreds of years into the past or the future, our predictions are reliable. But any claims about the positions of the planets 4 billion years ago, at the dawn of life on Earth, would be meaningless.

The bottom line is that you can never predict much longer than the predictability horizon, no matter how good your instruments become. The exponential growth of errors in a chaotic system overwhelms even the most meticulous observations. ∎

Essential Reading

Gleick, *Chaos*, 11–31.

Lorenz, *The Essence of Chaos*, chap. 1, app. 1.

Strogatz, *Sync*, chap. 7.

Supplementary Reading

Crichton, *Jurassic Park*.

Lighthill, "The Recently Recognized Failure."

Poincaré, *Science and Method*, chap. IV.

Stoppard, *Arcadia*.

Questions to Consider

1. If you haven't already, read *Jurassic Park* and *Arcadia*. What do you think of Crichton's and Stoppard's accounts of chaos theory? In each case, does the discussion of chaos add to the drama? Are their treatments scientifically accurate? Illuminating?

2. In what ways has the butterfly effect changed the course of your own life?

Chaos as Disorder—The Butterfly Effect
Lecture 6—Transcript

Welcome back to chaos. In the last lecture, we saw how the meteorologist Ed Lorenz happened upon the butterfly effect serendipitously in his simulations of artificial weather. The butterfly effect has now moved beyond science and entered popular consciousness, but some aspects of it are commonly misunderstood. In this lecture, I think it's worth pausing from our hard-core science discussion to discuss the philosophical significance of the butterfly effect to try to clarify what it really means and what it doesn't mean.

Probably the right place to start is by asking, Where does the term come from? It's a felicitous phrase that sticks in the mind. Who created it? You might be surprised to learn that we don't really know. It seems obvious that it must be Lorenz, and there's good evidence for that, but Lorenz himself is not sure. In his own writings, he says he's not quite sure where the term came from. Most likely, though, it derives from the title of a 1972 lecture that he gave at a big meeting of the American Association for the Advancement of Science. The lecture was titled "Predictability: Does the Flap of a Butterfly's Wings in Brazil Set Off a Tornado in Texas?" That would seem to be the source. But Lorenz did not actually write that title. He originally used a seagull in the title. He liked the image of a seagull flapping its wings. However, the session chairman, without telling Lorenz, changed the title to a butterfly, which I think was the right choice, a much more poetic image. It's a dramatic thought that something so delicate as a butterfly could change the weather by flapping its wings.

That choice created a sensation, at least in the world of butterflies. Take a look at this. Here are two butterflies. One is reading the obituary section in the *Butterfly Times* to the other and says, "Oh, he had a short but interesting life. For instance, did you know he was once responsible for a tornado in Texas?" We haven't pinned down which butterfly that is, but it's a fun image, whether it's completely convincing or not. It's up to you to decide.

In the past 20 years, the butterfly effect has intrigued authors and filmmakers. For example, rent a copy of *Jurassic Park*. We've all seen that movie. Many, many millions of people have seen this blockbuster by Steven Spielberg based on Michael Crichton's novel from 1990. Do you remember that chaos was such a prominent part of the story? It was. I certainly remember it, because it showed what a chaos theorist looks like on screen. It was Jeff Goldblum, 6'4", dressed in black leather, looking like a rock star. Maybe not. Anyway, there was Jeff Goldblum looking like Mr. Chaos, playing a character named Ian Malcolm. In one scene, he decides that he is going to try to seduce Laura Dern, his love interest in the movie. I think she's also a scientist, maybe a paleontologist; I forget.

In any event, Jeff Goldblum is going to demonstrate the butterfly effect on her hand. Which reminds me, wasn't there a mention of the Venus butterfly in that old show *L.A. Law*? But that's something else. That was strictly sexual. This was just the honest-to-God butterfly effect. So he says to Laura Dern, "Let me have your hand." She puts out her hand, and he's holding it, and then he surprises her by dropping a drop of water on it and asking her to predict which way it will roll off.

Whether that flirtation succeeded, you can watch the rest of the movie to see, but the unpredictability of that droplet's trajectory is a metaphor, a harbinger for what is about to happen at Jurassic Park. Let me remind you that this is a story about a super-carefully controlled theme park, in which dinosaurs have been brought back to life by cloning them from ancient DNA trapped in amber. The engineers who have created the park and the paleontologists and the biotech people all say: There's absolutely no danger here. Everything's under control. We have the best scientists working on this, and it's failsafe. These dinosaurs cannot reproduce; they can't get out of their space; we're going to bring little kids from all over the world to see Jurassic Park.

Of course, chaos ensues, which is what the Jeff Goldblum character keeps warning everybody else: that in a system as complex as this, you can't control it. There are always unpredictable elements, and let's not assume that we know what's really going on at Jurassic Park or that we have the kind of control we hope.

The butterfly effect also plays a central role in other productions that you may have seen. For example, there's Tom Stoppard's play *Arcadia*, which I recommend to you. You may even want to read it if you can't see it anywhere because I think it has some of the best explanations of the basic ideas of chaos theory anywhere, better than scientific textbooks. Somehow that poet, that playwright, Stoppard, is such a genius that he manages to explain the butterfly effect, strange attractors, iterated maps—things that we will be learning about in the lectures to come—better than any scientist or mathematician has so far. So take a look at that if you're interested.

Another one that you might want to take a look at is the movie *Sliding Doors* from 1998. This is a pretty small film, and in my experience, not many people have seen it, but Gwyneth Paltrow is in that movie. She plays a character who's trying to catch a subway, and she goes to the train, and just before she gets through the doors, the doors slide closed, so she can't catch the train. That's in one version of her life. Then you see another version of her life where she gets to the train, and this time, as the doors are sliding closed, she gets in and catches the train. So what you see are two trajectories of a person's life, depending on whether she does or doesn't catch the train before the sliding doors close. In one version, she's on the train. She makes it home and sees her boyfriend in bed with another woman. That's one possible life. In the other version, she doesn't catch the train; the boyfriend is still in bed, but she doesn't know. So you can see what will happen with a seemingly insignificant thing, like whether you catch the train or not—how the whole outcome of your life can be radically different. This is an instance of the butterfly effect, a small change making a huge difference in a person's life. I thought it was a pretty respectable movie. You can see for yourself.

There's one movie I would caution you about, which is called *The Butterfly Effect* (2004), with the young actor Ashton Kutcher. I don't know if you're likely to see that, but probably the less said about that movie, the better. It does explore the concept of the butterfly effect, but I found it a bit grisly, and maybe that's enough. I don't want to put down the movie, and it may be to your taste, but I wouldn't recommend it. If you see that title and think, "This is something for me because I'm interested in chaos," maybe steer clear of that one.

These examples show how Hollywood and playwrights like Tom Stoppard have incorporated the idea of the butterfly effect into drama, but still you could ask, "Do I really need to be taking a course to learn about this? Everybody knows that little things can make a big difference. That's an ancient idea." There's nothing very surprising about that, and I have to agree with you. Yes, that's true. That is an ancient idea, and it certainly predates any work by Lorenz or anything that I'm saying here, even Poincaré. This is an ancient concept.

Lorenz himself, in his own writings, mentions precursors that are literary precursors to the idea of the butterfly effect that were important in his own life. He mentions that he, as a young man about to embark on his career as a meteorologist, was given a book as a Christmas present by his sister, called *Storm*. It's a 1941 novel by George Stewart. In *Storm*, there is a character who happens to be a meteorologist, which is why his sister thought of giving it to him. The meteorologist at one point remembers a remark from one of his professors that a man sneezing in China can set off a blizzard in New York. That sounds like the butterfly effect, and it's even a meteorologist who is saying it. So Lorenz certainly would have had that in his mind, but also, this was part of the culture of meteorology; that is, this remark was not due to the writer of *Storm*. Apparently, meteorologists had been saying things like this for a long time, that the slightest change could affect the weather. So that was out there in the culture. It wasn't really due to Lorenz.

A second interesting precursor that Lorenz mentions is a Ray Bradbury story, which I think a lot of people are aware of. It's a time-travel story called "A Sound of Thunder," from 1952. In that story, there's a critical scene where a traveler goes back in time and accidentally steps on a prehistoric butterfly and, by stepping on this prehistoric butterfly and killing it, changes the outcome of a present-day presidential election.

This idea that these small effects can cascade and cause enormous problems later on, that is, as I say, a truly ancient idea, and it goes back at least as far as a famous piece of verse that you might be familiar with. Some attribute it to Ben Franklin, but some others say that Ben took it from John Gower, who did publish, in 1390, the famous verse "For Want of a Nail." Let me read that to you in case you're not familiar with it or just to remind you how it goes. It's a story of the downfall of a kingdom.

> For want of a nail, the shoe was lost;
> For want of a shoe, the horse was lost;

> For want of a horse, the rider was lost;
> For want of a rider, the battle was lost;
> For want of a battle, the kingdom was lost!
> And all for the want of a horseshoe nail.

In other words, there was one nail that wasn't available to put into the horseshoe, ultimately causing the loss of a battle that led to the downfall of a kingdom. There is really nothing so surprising in the idea that small changes can make a big difference through this cascading of effects.

You might be wondering at this point, "What is new about the butterfly effect if it [has] been in all these novels, and ancient poems, and so on?" What's new is the recognition that this same kind of sensitivity to tiny changes can afflict even the simplest systems that you would think would be immune from them. That's the point—that we knew the butterfly effect was there in complicated real-world situations. We didn't know that it could be there in things like this double pendulum or other simple mechanical systems. In the case of our own lives, it was always clear [that] we couldn't predict the course of our fate, or wars, or things like that. There are just too many complexities, millions of unaccounted variables; we don't know what laws they obey or even if there are any laws at all. That, as I say, is not surprising, but with simple deterministic systems, like a double pendulum, or Poincaré's three-body problem, or Lorenz's artificial weather model, things like that, you would think, should not have this same kind of problem because we know all the variables, and we know the laws that determine their behavior. So where is there room for this kind of unpredictability? And yet it's there, and we still can't predict what systems like this will do in the long run. That's the news.

Poincaré certainly realized this; that is, what he found in his work, as I mentioned in Lecture Four about the three-body problem, he showed that horrible nightmarish tangle that I said gave the implication of chaos, that two states of a system that were indistinguishably close could lead to different fates for the three-body problem. He saw that in his imagination and his visualization of this high-dimensional state space, but he also wrote about it. At first, he wrote about it in this obscure way that I quoted in Lecture Four, but later, he wrote in an absolutely pristine, compelling, lucid way. I want to give you a quote from Poincaré. It's a little bit long, but it's so brilliant and so perfect that I want to read it all to you. I will point out as I'm reading it exactly what he's saying. He's trying to address the question of how chance can emerge in a deterministic world. We know that there is chance. Where does it come from, if we believe in Newton's laws?

So this is what chance is, he says: "A very small cause which escapes our notice determines a considerable effect that we cannot fail to see, and then we say the effect is due to chance." So think about that. Something so small, getting on the train or not, we don't even notice. It escapes our notice. It determines something very big that we cannot fail to see, like your boyfriend in bed with someone else or not. Then we say the effect is due to chance, whether you got on the train or not. Then he says, "If we knew exactly the laws of nature and the situation of the universe at the initial moment, we could predict exactly the situation of that same universe at a succeeding moment." If we knew *exactly* the laws of nature and the situation of the universe. You would need exact knowledge of two things to have perfect predictability and no chance. You would need to know exactly the laws, and you would need to know exactly the situation of the universe, the initial state. If you did have that, there would be no chance.

But then he says, "But even if it were the case that the natural laws had no longer any secret for us ..." I like that—that is, even if we knew the natural laws, if we had a God's-eye view of the laws of nature, even if the laws of nature had no longer any secret for us, "we could still only know the initial situation *approximately*." You can never know the state of the universe perfectly, only approximately. "If that enabled us to predict the succeeding situation with the same approximation, that is all we require, and we would say that the phenomenon had been predicted." In other words, if whatever approximation we had to make initially didn't really grow, we just had the same approximation in our answer, that would count as a genuine prediction. But he says, "But it is not always so; it may happen that small differences in the initial conditions produce very great ones in the final phenomena." That's the problematic case, when a little change blows up to make a big change. "A small error in the former will produce an enormous error in the latter. Prediction becomes impossible and we have the fortuitous phenomenon." That's how chance arises—as a fortuitous phenomenon from small uncertainties cascading into enormous ones. That's from Poincaré, a book called *Science and Method* from 1914.

That is a perfect description of the butterfly effect. He absolutely understood everything there was to understand about it back then. He wrote it as clearly as he could write it. I think you'll agree. If you parse that paragraph, you can't miss what he's saying. Still, everybody missed it. For nearly all other scientists, the recognition of this kind of unpredictably was very late in coming, not in 1914, much later, and philosophically hard to swallow. It was only around 1970 or '80 that chaos theory took hold. So this peculiar lapse of all these years, post-Lorenz it turns out—even Lorenz didn't convince them—led one of the most prominent scientists in the world, Sir James Lighthill, who held the same chair, the Lucasian Professorship of Mathematics, the same chair that Isaac Newton had held and that Steven Hawking holds today. Sir James was the holder right before Hawking.

Sir James Lighthill issued a remarkable collective apology on behalf of all scientists, saying we're sorry for misleading you about predictability for all these years. I want to read a quote from him. This is amazing, a scientist apologizing, on behalf of all of us, to the public. Here is his collective apology, taken from an article called (I like this title) "The Recently Recognized Failure of Predictability in Newtonian Dynamics." I want to comment on that later. It's not actually so recently recognized because, remember, Poincaré recognized it in 1914. But here [Sir Lighthill] is in 1986, apologizing for the recently recognized failure of predictability.

He says, "We're all deeply conscious today that the enthusiasm of our forebears for the marvelous achievements of Newtonian mechanics led them to make generalizations in this area of predictability which, indeed, we may have generally tended to believe before 1960, but which we now recognize were false." We thought everything was predictable from Newton's laws, but we now recognize that that was not really true. "We collectively wish to apologize for having misled the general educated public by spreading ideas about the determinism of systems satisfying Newton's laws of motion that, after 1960, were proved to be incorrect." We thought Newton's laws gave us total predictability, and by 1960 we knew that was wrong, and we were still lying to you for these intervening 26 years.

That was an important moment in the development of chaos theory because Lighthill was a great scientist but not a chaos theorist. He, as they say today, had not drunk the Kool-Aid. He was not a fringe guy who was going with this latest fad of chaos theory. He was totally rock solid, respectable, eminent, and his admission meant that chaos theory was for real and that everybody could join in now; this was not a passing fad. Though, as I say, he was late to the party. The greatest triumphs came in the first wave of chaos theory years earlier in the '70s and '80s, so he was on pretty safe ground saying that we could now admit this mistake. It had already been decisively shown to be a mistake by the early '80s. Still, his comments make chaos theory a respectable and even legitimate topic of study for the late adopters, and there were a lot of them.

That's an interesting question, then, in the sociology of science: Why were scientists so slow to recognize the butterfly effect? I've talked about why they were so slow to recognize chaos in the previous lecture. I talked about the frame of mind that you need to see things that are right in front of you and how so many scientists, electrical engineers, and mathematicians, and physicists had trouble seeing what their own data were showing them.

There's another reason why the butterfly effect, that one aspect of chaos, this unpredictability that flows from tiny errors magnifying into enormous ones, why was that so philosophically distasteful? I think because it was threatening to science itself. It really was. That is, science, as we conduct it, depends on being able to draw a conceptual box around a system that you're studying, that you're interested in. When I want to study the motion of this pendulum, I don't have to think about what's happening on Mars. I just analyze this pendulum. I put the box around it, and I can neglect tiny effects coming from outside the system.

People sometimes say everything affects everything else, but scientists hate that. No, it does not. Literally, it's true. There is a gravitational pull from every particle in the universe, but those are negligible. If we want to make progress and learn something, we can't believe everything affects everything else. It's a matter of degree. Some things have negligible effect, and we can ignore them.

The butterfly effect was worrisome because it suggested that the things you thought were negligible might not be, that they could come back and change your system. If that were true for every system, science itself would be impossible, but more than that, I would say, life would be impossible because everything would be up for grabs at every instant. So there seems to be something really repulsive about the butterfly effect. It has to be limited, and I think people sense that it was limited but were wondering exactly where is it limited? And so for the rest of the lecture, that's what we'll try to talk about, the limitations of the butterfly effect. Don't think it applies to everything at all times. That would be wrong. That's a misunderstanding, if you think that.

It certainly does not occur in systems of the type that most scientists had been studying since Newton. It's not there in systems that just gently relax to some equilibrium state, like a particle rolling down to the bottom of a curved surface and just sitting there. There's no butterfly effect there. If I start two particles close to each other, they will both just roll down, and they won't get farther apart. Boring, but there's no butterfly effect, at least.

[We see this] even in systems that have more life than that, that oscillate periodically, like two metronomes. If I start two identical metronomes, and they're both [going] tick-tock, tick-tock, and I get them in sync with each other so that you hear the clicks at the same time, and then I disturb one of them a little bit by flicking it with my finger, you will hear tick-tick, tick-tick, tick-tick. But that error will not grow. It will just stay like that for a very long time. Even if it does grow, it grows very slowly, only "linearly proportional to time" is how we would say it mathematically. So it's a minor thing that's not serious. The point is, the error does not snowball.

There are lots of systems like this. The tides, for example, are very predictable. Well into the future, months ahead, you can predict the tides. The return of Halley's Comet. We all know that's how Halley got his name attached to it. You could predict Halley's Comet, whatever it is, 80-something years in advance, and it will be there. The timing of eclipses—we can predict as far as we like into the future, or we can go back to the time of the Greeks or before. We are absolutely dead sure we know when eclipses occur. All of those things are very, very periodic and, hence, very predictable. That is, they repeat, which helps make them predictable. Tiny disturbances in those cases don't mushroom into major forecasting errors.

Lorenz made an interesting comment about this once, having to do with the tides, that you can't always tell which systems are going to have the butterfly effect or not, because think about what's involved in understanding the tides. The tides involve a huge, fluid mechanical system, the oceans of the Earth, enormous numbers of atoms and particles involved. In many ways, this problem is as complicated as the atmosphere. They're both giant, fluid mechanical systems. But the weather is unpredictable, and yet the tides are predictable. They're both big, fluid mechanical systems governed by partial differential equations. How come we can't predict one, but we can predict the other? Lorenz's point was: because one of them is periodic and the other one is not. That's the thing. The tides repeat; the weather does not. It's not determinism that's the important thing. It's determinism plus periodicity.

The title of Lorenz's famous 1963 paper that I'll be discussing in the next lecture is "Deterministic Nonperiodic Flow." It's the conjunction of those two things that is so essential—determinism plus non-periodicity. Remember, he wanted to make his weather non-periodic in the computer to test those models. It was by putting non-periodicity into determinism that he opened the door to chaos and the butterfly effect.

Another example of this would be the double-pendulum toy that we played with previously. Let me show you. What I've shown you already is that it has chaotic-looking behavior; that is, it whirls around crazily. That is, it's not periodic, but that also means, because it's deterministic and non-periodic, it should have the butterfly effect, according to Lorenz. So I can demonstrate that by trying to start these two identical double pendulums in a way that is [as] close to identical in their initial conditions as possible. That's why I've got this dowel here. I'm not going to cheat. Of course, there will be some tiny error. I can't do it perfectly. What I'm going to do is quickly pull this dowel away, and then you should see two things happen. First, the two double pendulums will stay close together. They really will. They will track each other. The unpredictability doesn't occur instantaneously. They will be predictable in the short term, and they will stay together because of their deterministic character. But they will very rapidly become unpredictable and different. They will diverge from each other after a certain amount of time. So let's try that. Maybe we'll be able to see this in slow motion; it will be dramatic if we can look at this later. Okay, ready? Look at that. They were together for a while. That came out. A live demonstration that worked; very good.

Now this could be dangerous. You recognize this is like Bruce Lee's nunchucks, so we don't want to start getting injured here with the double pendulum. They become unpredictable only after a certain amount of time, which we call the *horizon of predictability*. That's a great phrase that Lighthill introduced in his 1986 article. It's a beautiful article, very clear. I recommend it.

So it's the time that it takes for these tiny errors to double in size. That's a good definition of it. You're looking for the doubling of these errors. Once they start doubling, you're going to be in trouble. The question is, then, how long does it take? What is the length of the predictability horizon? That's the important quantitative issue here.

In our simulations of the Lorenz system, we had shown that the blue and the yellow curves stayed together for a while. You've seen this before, but I want to make a little different point about it. Remember, I start the Lorenz system. I see a yellow curve. This is going to be non-periodic. At this point, it [has] growing oscillations, then it starts to do what really look very non-periodic. There it is, doing that. Okay, so it's not repeating. And then, as you remember, what I do is start midway through that, only changing in the fourth decimal place. Now the blue curve is tracking the yellow one for a while, but now it has diverged. So that is an example of the predictability horizon in action. Blue and yellow were predictable and identical up to about here, where my mouse is right now, something like 1 time unit or maybe 2 time units. But certainly by 3 time units, yellow and blue couldn't be more different. For this case, the predictability horizon was something like 2 units of this abstract time. But in real systems, we can ask: What are we talking about? What time do we really mean? Like in a chaotic electrical circuit, the predictability horizon can be measured—how long it takes for two chaotic circuits to de-correlate and become totally different. It takes about a thousandth of a second, a millisecond—fast. For this double pendulum, we could, if we watch in slow motion, see how long they stay together, and just eyeballing it, it looked like it was on the order of a few tenths of a second, something like that.

For the weather, it's unknown, but it seems to be about a week or two that you can predict. For the entire solar system, which turns out to also be chaotic, as simulations have shown—we don't know this experimentally, but simulations, by many different groups, keep coming out with numbers on time scales of a few million years, like about 3, 4, 5 million years as determined by very careful, long-term integrations of the solar system in mathematical modeling. That's a long time—5 million years.

So it's because the horizon is so long for the solar system that the motions of the planets seem utterly predictable to us today. They are within the predictability horizon. There's no contradiction here. The solar system is chaotic yet predictable because we're within the predictability horizon. On the time scale of a human life or even the whole history of astronomy, they are predictable. When we calculate planetary motions hundreds of years into the past, no problem. But if we go thousands or millions or billions of years, like at the beginning of the solar system, say 4 billion years ago, at the dawn of life on Earth, that would be meaningless. You would have no way of knowing where the planets were at that point.

So the bottom line is you can never predict much longer than the predictability horizon, no matter how good your instruments become. Students are often confused on this point and ask me, "You keep talking about error. What if I don't make any error? What if my initial condition is exactly right?" You can't do that. You need an infinite number of digits. Real numbers involve an infinite number of decimal places, so there's always some error somewhere, even if it's out in the one-thousandth decimal place.

What that means is, because these errors grow exponentially fast, that even with 1000 digits of accuracy, the two trajectories will soon diverge; that is, the solution will end up being wrong. And the reason it happens so quickly is that the errors grow exponentially. If you try to make the initial condition more accurate, it's a losing proposition because, let me put it to you roughly like this: If I take 10 times more initial accuracy, I get one more digit. That only buys me one more unit of prediction time. If I try to make my prediction time twice as long, I need to go to 100 times more initial accuracy, or 3 times as long, 1000 times. This is going to kill you. You need exponential accuracy just to get a linear increase in prediction time. That's the problem: that the exponential growth of errors in chaotic systems overwhelms even the most meticulous observations or care.

It's time to end this lecture, and it's actually time to end a part of the course. So far in the course, we've emphasized the unpredictable and disorderly side of chaos. But there's something thrilling coming next, which is that chaos also has secret order. There's an orderly side to chaos, and the patterns in it will amaze you. See you next time.

Picturing Chaos as Order—Strange Attractors
Lecture 7

There's a cumulative order in a chaotic system, a kind of pattern ... but ... there's a kind of randomness, too. There is the butterfly effect and ... tiny differences get amplified as time goes on, exponentially fast, leading to the impossibility of predicting what this system will do far into the future. ... It's that confusing interplay of order and randomness that this strange attractor embodies.

In the last three lectures we've been focusing on the unpredictable side of chaos. But chaos is yang as well as yin, order as well as randomness. This lecture describes a way to visualize an amazing kind of order inherent in chaos. The resulting image is known as a *strange attractor*. Just as a circle is the shape of periodicity, a strange attractor is the shape of chaos. It's "strange" because its geometry is infinitely complex, and it's an "attractor" because the system is always drawn toward the behavior it represents. The existence of strange attractors is tremendously encouraging to scientists. By revealing the unexpected order within chaos, strange attractors offer hope that some forms of it might be partially predictable and controllable. So let's see what strange attractors are, how they were discovered, and why they matter.

We pick up with the hero of Lectures 5 and 6, the meteorologist Edward Lorenz of MIT. Having uncovered the butterfly effect in his artificial weather model, Lorenz next sought the mathematical essence of chaos. Could chaos arise in simpler systems? A colleague pointed him toward a model of convection, a rotating pattern of flow that sets in when a fluid, like air or water, is heated from below and cooled from above.

We can demonstrate all the phenomena of interest with a clever contraption—a tabletop waterwheel built for this purpose by Professor Willem Malkus of MIT. Water flows in steadily at the top and turns the wheel, much like the heat of the sun drives convection in the atmosphere. The waterwheel displays different types of behavior, depending on the amount of damping in the wheel. (We can vary the friction by adjusting a brake.) At high damping, the wheel settles into a uniform rotation. At low damping, the wheel rocks back and forth periodically, like a pendulum. In between, when the damping is moderate, the wheel turns chaotically. It rotates one way and then the other, reversing direction erratically and unpredictably.

Hoping to find order in the chaos, Lorenz sought to visualize the motion. The usual picture—a graph of a variable at different times—doesn't help. It just shows a complicated wiggle.

Instead, Lorenz made brilliant use of Poincaré's notion of state space (see Lecture 4). Recall that a "state" of a system is the collection of variables needed to determine the system's future. For Lorenz's convection model (or its waterwheel analogue), three variables determine the motion. Thus, the state space has three dimensions, with one dimension for each variable. That's wonderful, because the human mind can visualize three dimensions! (Lorenz's earlier weather model had 12 variables and so couldn't be visualized in this way.) Using computer simulation, we can graph the behavior of Lorenz's system as it moves from state to state, sailing like a comet on a trajectory through state space.

Let's start simply. Suppose the system settles down to rotating at constant speed. Then the picture in state space shows a trajectory spiraling down to a single point. The trajectory homes in on this point as if it were being attracted to it. Thus, we say that the system has settled onto an "attractor point."

Attractors are important. We'll be discussing them throughout this course. An attractor represents a system's natural, long-term mode of behavior. If you nudge a system off its attractor by disturbing it slightly, it will soon head back to the attractor. For example, if someone startles you by shouting "Boo!" your heart rate may speed up, but then it quickly relaxes back to normal. Your heart, like many other systems in nature, is carefully regulated to keeps its behavior within certain limits. So if an unchanging steady state corresponds to an attracting *point*, a periodic state (like the waterwheel rocking back and forth) corresponds to an attracting *cycle*. These simplest kinds of attractors—points and cycles—have very familiar geometry. Poincaré had mastered them 70 years earlier.

Now Lorenz was trying to visualize the attractor corresponding to a *chaotic* state. Poincaré hadn't done that. Lorenz's attractor has such weird geometry that it deserves the name *strange attractor*!

Let's try a few ways to explain what a strange attractor is. A computer simulation of the Lorenz system shows a trajectory spiraling around on a structure shaped like—prepare yourself for a twist straight from *The Twilight Zone*—a pair of butterfly wings. The trajectory makes a few loops around one wing before darting over to the other wing, jumping back and forth erratically. A 3-D plastic model of the attractor looks like a pair of surfaces that join at a hinge. Lorenz argued the surfaces could not actually join. Doing so would violate determinism. He concluded that what appear to be two surfaces must in fact be an "infinite complex of surfaces."

A 3-D plastic model of the attractor looks like a pair of surfaces that join at a hinge. Lorenz argued the surfaces could not actually join. Doing so would violate determinism.

This mind-boggling idea can be made more vivid by considering a metaphor involving a nightmarish parking garage, with two interconnected towers and infinitely many levels. This metaphor also illustrates four other key aspects of chaos on a strange attractor: the determinism of the motion, its non-periodicity, its confinement to the strange attractor, and its extreme sensitivity to initial conditions (the butterfly effect).

Strange attractors are important because they teach us that chaos is not the absence of order. It is a marvelously subtle state poised between order and randomness, with both aspects intermingled.

In this lecture we've focused on the abstract shape of chaos. Next we'll explore the structure of chaos in time and see that there is beautiful order in that as well. All of this will come back with great effect later in the course, when we'll see how the order in chaos is allowing it to be used for extraordinary applications in medicine, space travel, and communications technology. ■

Gleick, *Chaos*, 119–53.

Lorenz, *The Essence of Chaos*, 136–46.

Stewart, *Does God Play Dice?* chap. 7.

Strogatz, *Sync*, chap. 7.

Strogatz, *Nonlinear Dynamics and Chaos*, chaps. 9, 12.

Play with a Java applet of Lorenz's attractor here: http://www.aw-bc.com/ide/idefiles/media/JavaTools/lrnzr320.html.

For a simulation that ties the mathematical model to the physical problem of convection, try this one: http://www.aw-bc.com/ide/idefiles/media/JavaTools/lrnzphsp.html.

Questions to Consider

1. This lecture focused on strange attractors, which represent self-sustaining chaos. But attractors can also represent states of unchanging equilibrium, or periodically repeating cycles. The temperature in a house with a working furnace and thermostat would be an example of a system that has an equilibrium state as an attractor. (The temperature stays constant, and if disturbed, comes back to the desired set point.) Give other examples of systems with attracting equilibrium states.

2. What are some examples of systems with attracting cycles? (These are systems that relax back to some standard cycle after being disturbed.)

Picturing Chaos as Order—Strange Attractors
Lecture 7—Transcript

Welcome back to chaos. In the last three lectures, we've been focusing on the unpredictable side of chaos. But chaos is yang as well as yin, order as well as randomness. This lecture describes a way to visualize an amazing kind of order inherent in chaos. The resulting image is known as a *strange attractor*. Just as a circle is the shape that we would associate with periodicity—something repeating over and over, you think of as sort of coming back on itself, forming a circle—a strange attractor is the natural shape of chaos. It's "strange" because its geometry is strange; it's infinitely complex. And it's an "attractor" because the system that it describes is always drawn towards the behavior that it represents, as if attracted to it.

The existence of strange attractors is tremendously encouraging to scientists because by revealing the unexpected order within chaos, strange attractors offer hope that this sort of seeming chaos might be partially predictable and controllable. We'll see in subsequent lectures towards the end of the course some examples of how scientists have used the order in chaos, revealed through strange attractors, to control chaos in cardiac arrhythmias, and also in lasers that have gone into fitful pulsations, and so on.

Let's see what strange attractors are, how they were discovered, and why they matter. We'll pick up with the hero of Lectures Five and Six, the meteorologist Ed Lorenz of MIT. Remember our story: Having uncovered the butterfly effect in his artificial weather model, Lorenz next sought the mathematical essence of chaos. Could chaos arise in simpler systems? As he thought about it, he consulted with some colleagues, and a colleague told him, "Maybe, Ed, you ought to think about convection." Convection, of course, was familiar to Lorenz. It's a rotating pattern of flow that sets in, for instance, if you heat a fluid, like air or water, from above. Think of the Sun beating down on the Earth. The pattern of convection sets in when the hot air near the Earth's surface starts to rise because it's lighter. As it rises, eventually, it gets cooled off, and then it gets denser and starts to sink again. So you have these rotating patterns of air moving around. Those are convection cells. You can also see this heating water on the stove.

So convection was the idea. Let's look into that. I won't be showing you any convection demonstrations, but I will show you now all the phenomena of interest captured by an analogue model of a convection cell. It's a tabletop waterwheel built just for this purpose as an analogue of what Lorenz found. It was built by one of Lorenz's colleagues at MIT, a professor named Willem Malkus.

When I was a professor at MIT in the early 1990s, I loved this Malkus waterwheel, and I used to show it, every chance I got, to my chaos classes there. In fact, I loved it so much that after I moved to Cornell, I thought, "I've got to have a videotape of that wheel just to show again." So I made a pretty amateurish video of it, and I'll be showing you that in a minute.

First, let me try to orient you as to what you will be seeing. Here's a schematic of the waterwheel. The way it works is: Water is being pumped up through a kind of manifold, like in a car, and then sprayed out through many little nozzles in a sort of perforated hose, into the waterwheel structure, which is this whole structure sitting on the tabletop. I'm showing you a side view here. The confusing thing about the waterwheel is that—don't picture a big bucket. Picture a structure which has a thin rim around it, and the rim is partitioned into many little chambers, vertical chambers, thin slots, and the water is going to flow into those slots, effectively like a one-dimensional group of slots around the rim.

The water flows in steadily at the top of the waterwheel, as shown here. This is going to be the counterpart of the heat that's driving convection in the atmosphere. In other words, by pouring in the water at the top of the waterwheel, we'll make it top-heavy and tend to make it want to turn. This waterwheel is different from a regular waterwheel that you would see out somewhere on a farm or something. In a regular waterwheel, the water is added at the side to make the wheel turn in one direction. Here, the water is added at the top so that the wheel can spin equally well in either direction. There's a symmetry to the way this is set up.

The water enters these thin chambers around the rim and then drains out through the bottom of each of those chambers, which has a little hole in it. Then, it all collects in the bottom of the structure, where it gets pumped back up, eventually, through the manifold, to be poured back in. So it's a recirculating system, a self-contained device. It's very elegant. It doesn't spill any water on the table when you're doing the demonstration. Even I was able to use it successfully.

There is a way we can manipulate the system that is very important. We can adjust the amount of friction by turning a brake. The brake is shown here. It's just an ordinary brake. It puts strong viscosity on the wheel as it tries to turn. By adjusting the brake, we can make the wheel behave in very different ways. For instance, when the brake is set to be pretty tight, the wheel settles into a uniform rotation. I'm going to show you that now in the video.

I've given the wheel a little initial push. Notice that there's some water that has been colored with green food coloring. The water is coming in, and this wheel is just spinning steadily in one direction. You can see different amounts of water in the different chambers. But nothing will change over time, really. This is in a nice, steady rotation. That's the attractor for this system with the brake set the way it was, just a simple, steady rotation in one direction. I could have set it to rotating in the other direction. That would have been fine, too, by symmetry. It has two symmetrical rotations, in either direction.

On the other hand, suppose we loosen the brake. If the brake is loose, then the wheel has a totally different kind of behavior. It rocks back and forth periodically, sort of like a pendulum. Let me show you what that looks like. Change the setting on the brake, water is filling in the chambers, and now it starts turning in one direction. It goes twice in that direction; now, a third time. So it hasn't yet settled onto its attractor. In fact, it just seems to keep going in one direction. Now, it's reversed, and at this point, it's pretty close to its long-term attracting behavior, and I think it will just keep rocking back and forth.

Finally, if we set the brake in an intermediate amount of tightness, then we can get the wheel to show chaos. It will rotate one way and then the other, but it will reverse its direction erratically and unpredictably. Watch what that looks like. Now keep track of which way it's turning and then ask: How many times does it turn in that direction? All right, so it's reversed. It keeps going in that direction. It has made two turns that way, a third, now back, and so on. If you keep watching, you'll see that it makes a kind of unpredictable number of turns. It may be two, three, four in one direction and then two, three, four, five, or whatever in the other direction. That's what chaos looks like.

This waterwheel is a perfect mechanical analogue of Lorenz's convection model. They do all the same things, and they're chaotic in the same ways. That's what I mean by a mechanical analogue. Dynamically, they're the same. We can use this waterwheel to get an intuitive feel for what Lorenz saw in his convection simulations.

Now let's go back to Lorenz and his convection model. Hoping to find order in the chaos, somehow just sensing it might be there, Lorenz sought to visualize the motion of his system. If you were given the job of visualizing motion, you would probably think to make an ordinary graph. You would graph some variable versus time. You could do that. You could graph the spin rate versus time, and if you do, you'll find nothing very revealing, just some complicated wiggle corresponding to the chaotic motion.

Lorenz tried something else. He made brilliant use of Poincaré's notion of state space, which we discussed in Lecture Four. Let's remember what a *state* means. A *state* is the amount of information I need to give you to allow you to predict the future of a system using the laws of motion. To put it another way, it's the number of variables, the collection of variables, I need to give you to say what a system will do in the next instant, and then instant by instant, I can predict forward all the way into the future.

For Lorenz's convection model, it turns out that the state is described by just three numbers, three variables. The same is true for the waterwheel analogue. For the waterwheel, those three variables can be thought of roughly as being: how fast it's spinning (its spin rate), the gravitational torque on the wheel tending to turn it in one direction or the other, that torque being caused by the way the water is distributed around the rim of the wheel. So we've got spin rate, we've got torque, and finally, a measure of the top-heaviness of the wheel—how much of the water is near the top of the wheel, making it tippy like an inverted pendulum, and how much of the water is to the side or to the bottom of the wheel? It's some kind of ill-defined thing that I'll be calling top-heaviness.

It turns out these details are not important. You could do just as well by thinking of these as three variables, x, y, z, with no particular meaning, but if you want the physical picture, it's like what I just said. But what's really important here is that the state space has three dimensions, x, y, z, those three axes—one dimension for each variable. That's wonderful because our human minds can visualize three dimensions. Remember that Lorenz's earlier artificial weather model had 12 state variables, so there would be no hope of picturing it in this way.

Using computer simulation, we can graph the behavior of Lorenz's system as it moves through its state space, from state to state, sort of like a comet sailing through real space, except that this is an abstract point sailing through the abstract state space. The simulation I'm about to show you will possibly confuse you for a second until you get oriented. You're going to see three pictures at the same time.

The first is just an animation of the rotating convection cell or waterwheel. That will just be shown schematically as a dot moving around, either just rotating, or rocking back and forth, or alternating direction chaotically, so keep your eye on that dot. That will give you a physical picture of what is happening.

Then there's the usual representation of a time series, a graph of, say, the spin rate versus time. You will see a complicated wiggle.

The third is the good way of looking at things, the state-space way of looking at things. You will see a trajectory moving around in state space. Actually, not the whole state space, which is 3-dimensional, just two of the axes that we will pick out, sort of projecting the 3-dimensional state space onto a 2-dimensional plane so we can see it better. But, again, focus on the state-space version of what you're seeing so that you can try to understand what this abstract comet moving around in there, the trajectory, corresponds to in terms of the real motion of the waterwheel.

Let's start simply because this takes some practice. Let's suppose that we're in the simplest case, where the waterwheel is going to just settle into a uniform rotation in one direction at constant speed. I'll start the system somewhere in the plane of spin and top-heaviness, and you will see a transient behavior for a while until it settles down to its ultimate attractor. Let's see what that looks like. If you want, you can watch the dot moving around in one direction. You see, it keeps going the same way. You could look down at the yellow time series; there's something wiggling, sort of showing damped oscillations. But the best thing to look at is this blue trajectory in state space. Look at how it has started to spiral in onto a single point. Let me pause the system right there. What does that single point represent? That point is the attractor. In other words, this system has been attracted to that point in its state space, meaning it has settled into a behavior with a certain constant amount of top-heaviness and a certain amount of spin. That's why the spin time series has gone flat. The system has now found the spin that it wants to have.

This was the simplest possible kind of attractor, an attractor point. Attractors are very important, and we'll be discussing them throughout the rest of this course. An attractor represents a system's natural, long-term mode of behavior. If you nudge a system off of its attractor by disturbing it slightly in some way, it will soon head back, determined to get back to that attractor. For example, think about your own body. If somebody were to startle you right now—boo!—maybe your heart rate would go up for a second while you're scared. But soon it will come back to its natural rhythm, to its natural rate. It will quickly relax back to its attractor. Your heart, like many other systems in nature, is carefully regulated to keep its behavior within certain limits, and that's what attractors represent in their simplest form, when they're just an attracting point.

Now, if an unchanging steady state is represented by an attracting *point*, what would a periodic state look like? When we saw the waterwheel rocking back and forth, what would that look like in state space? Think about that. If it goes back and forth through the same sequence of states, always repeating its behavior, that means that in state space you will repeat to the same state; you will form a loop. The answer is, we should see an attracting *cycle*, a loop, corresponding to a periodic motion. Let's see about that.

In this next simulation, it's as if I set the brake to be in the regime where now we're expecting periodic reversals. Again, we start the system somewhere. It does something kind of wild in a transient way until it finds its way onto the attractor. I have a button on this program that allows me to clear the transients, to sort of get rid of this initial garbage until the system has found its way onto the attractor, so I'm going to do that. You can see that it has settled down already. At this point, in terms of the simulation, the animated point is just rocking back and forth. Down here in the time series, that has settled into a periodic oscillation. Up here in state space, the most unfamiliar, if I clear the transients, I can now see that the system has settled onto a loop. Beautiful. And it will just keep repeating, going around and around on that loop, the attracting cycle.

These simple kinds of attractors, a point or a loop, have very familiar geometry, just a point or, basically, a circle. Poincaré had mastered them already 70 years earlier. What Lorenz was now trying to do was to visualize the attractor corresponding to a *chaotic* state. Poincaré had not done that. Lorenz's attractor, we will see, has such weird geometry that it deserves the name *strange attractor*.

Let's spend the rest of our time trying a few different ways to get a grasp of what a strange attractor really is. Let me begin with a computer simulation, like we've just been doing. I'm going to set the equivalent of the brake on my waterwheel to a place where I expect to see chaotic reversals. You can check that by watching the little dot moving in the animation. What will the trajectory look like in state space? That's the question. So let's start. Prepare yourself for a twist, a twist almost out of *The Twilight Zone*. It's spooky, what you're going to see. Watch this.

We see the animation just doing its thing. There seems to be some sort of outward spiral here in the state space, a growing spiral. You can see growing oscillations down here in the time series. You may be wondering why I say *The Twilight Zone*. Soon, you'll see something eerie appearing in state space. Let me clear the transients to try to make what's happening a little more visible.

We see a growing spiral that will soon form a wing. It will look like a butterfly wing. That's the twist. That is, the strange attractor for the Lorenz system—remember, Lorenz, the discoverer of the butterfly effect—he is finding a strange attractor that looks like a pair of butterfly wings. Crazy.

What happens is the trajectory makes a few loops around each wing before it darts over to the other wing, and it jumps back and forth between them erratically each time the waterwheel reverses its direction.

Here's another way to think about what's happening. I have a 3-dimensional plastic model of the Lorenz attractor. This was made by my students as a gift, actually. I love this gift. It's made with a pretty fancy machine called a 3-dimensional printer. It takes plastic and can extrude it and make any 3-dimensional shape that you program into the computer, so this thing I'm showing you here is numerically accurate to the resolution of the machine. This is really what the Lorenz attractor looks like. Maybe now you can get a feeling for what it looks like in three dimensions.

I can show you, for instance, that it has what looks like a pair of surfaces joined together at a hinge. Lorenz noticed that in his computer simulations, and he noticed, by thinking about it abstractly, that there was something crazy about this picture, something really wrong about that hinge. He knew, on abstract grounds, these surfaces cannot join at a hinge. That's impossible. Why is that impossible? Think about what these surfaces represent. They're filled with trajectories. A trajectory is supposed to show you the evolving state of the system in its state space, so if the two surfaces ever joined, that would mean that there would be two different pasts that led to the same present. That's impossible in a deterministic system because in a deterministic system, like this one, you can imagine running the movie backwards. That would say that the past could go, running time backwards, into two different futures. That's impossible in a deterministic system; that is, it's deterministic both into the future and into the past. There's a unique past that brings us to today, and there's a unique future that follows from today.

So when [Lorenz] saw these trajectories apparently joining at the hinge, he said, "That's impossible. They must be coming close but barely missing." And so where it looks like 2 surfaces coming close together, they are not actually hitting; they're just coming close. It's almost like the sheets of mica—if you look at a rock, mica, made of many sheets. It's sort of like these sheets have come together, but they're separated, so where it looks like there are 2 of them, they don't actually join. Following them around on the eyes of the wings, those 2 surfaces, Lorenz concluded, have to come back and actually have a near encounter again, meaning that there are really 4 surfaces. The 4 have to make 8, and the 8 have to make 16. And from this kind of reasoning, he concluded that there actually have to be an infinite number of sheets, an infinite number of surfaces—he called it an "infinite complex of surfaces"—all joined together in some unbelievably intricate way that we can't visualize easily, to form the Lorenz strange attractor.

This mind-boggling idea may bother you. It bothered Lorenz. It bothers everybody when they first encounter it. Let me try to make it a little more vivid for you, and maybe more understandable. I'm going to try through a pretty strange metaphor in keeping with *The Twilight Zone* theme here. Think about a parking garage, the kind where there are many levels. Those many levels are going to represent the many sheets of the Lorenz attractor. To keep with the two wings, let's think about two towers in this parking garage. There are the two towers. It's an interesting garage in that you don't drive. You just get in your car and some kind of towing apparatus drags you out of the garage. The towing apparatus is to mimic the idea of determinism. You don't have any choice how you're going to drive. This apparatus just drags your car out. Meanwhile, maybe your friend is also there and has parked in the car right next to yours. Your friend is supposed to represent a nearby initial condition. I want to illustrate what the butterfly effect would look like on the Lorenz attractor.

So here you are, you get in your car, your friend gets in his or her car, and then the device starts pulling you through the Lorenz attractor, through the parking garage. At first, it's fun. You're both swirling around, going from level to level, and you're waving to each other; you're having a good time. But soon, the butterfly effect makes you diverge from each other. Your friend is off on another lobe of the attractor, and now you're just not together any more. You've exponentially separated. Meanwhile, the ride is starting to not seem as much fun any more because you can't get out. You don't get out of the garage. You're stuck on the attractor. So you go around; you shoot over to there; you're getting carsick; and you just keep going from level to level, never ending.

What I'm trying to show you with this analogy is that the strange attractor has four things that I want you to keep in mind. First [is] the determinism of the motion. You have no choice as to your future. The towing apparatus dictates everything. In the case of the real Lorenz system, the differential equation is dictating how it moves from state to state. It's non-periodic. You never come back to your initial parking space. You're confined to the strange attractor. That's the third point. You can't get out of this garage. And fourth, there's a kind of extreme sensitivity to initial conditions in that no matter how close your friend starts to you, you may stay together for a while, but soon you will have totally different fates. That's what chaos is like on a strange attractor.

Why are strange attractors so important? We've spent pretty much this whole lecture on them. Because they teach us something unexpected—that chaos, despite what the name sounds like, is not the absence of order. It's a marvelous, subtle state that's poised between order on the one hand and randomness on the other, with both aspects intermingled. That is, when I speak about the order here, I'm talking about the fact that in state space, we could have seen just the trajectories make a snarl like a ball of yarn. It could have been a thicket, something like that. But we didn't. We saw this butterfly shape emerge. That's a kind of order in the system. In fact, you can think of it as a cumulative kind of order. It's what the system does after a long time. We saw that the butterfly pattern didn't emerge immediately. We had to let the system run for a long time before the tracing revealed and exposed the butterfly wings.

So there's a cumulative order in a chaotic system, a kind of pattern in the chaos, but on the other hand, there's a kind of randomness, too. There is the butterfly effect and the fact that tiny differences get amplified as time goes on, exponentially fast, leading to the impossibility of predicting what this system will do far into the future. So it's that confusing interplay of order and randomness that this strange attractor embodies. It's all in there.

In this lecture, we've been focusing on the abstract shape of chaos. Next time, we'll be looking at the structure of chaos in time. As opposed to its shape in state space, we'll think about what chaos means from moment to moment, and we'll see that there's beautiful order in that, as well. When I say yin and yang, there is a lot of yin and a lot of yang. We're going to see plenty of order in strange attractors.

All of this will come back to great effect later in the course, when we see how the order in chaos is allowing it to be used. We no longer have to avoid chaos, as engineers have traditionally been taught to do, and scientists as well. We can actually use chaos and harness its remarkable properties for all kinds of extraordinary applications—in medicine, potentially in the control of arrhythmias, in dealing with epileptic seizures; in space travel, how to get to the Moon without using very much fuel (practically no fuel at all); in communications technology, in encryption. Suppose you want to send a private message. Can we use chaos to give us some secrecy? It turns out [that] we can because chaos sounds like noise. We can use it to cloak a message.

So all of these things come about as a result of this marvelous state, chaos, which we now understand at least something about as an ordered state, not a purely random one.

I will tell you more about that next time, and I look forward to it. Thanks.

Animating Chaos as Order—Iterated Maps
Lecture 8

Why ... is the Lorenz map so important? ... Because it's so simple. It's astonishing that there's a pattern like this lurking in chaos. ... [The iterated map] captures with great fidelity what is going on in the Lorenz system, and it ignores in-between parts. ... Iterated maps are a great boiling down of information in the differential equation.

This lecture continues our exploration of order in chaos. We're going to reveal a new kind of order by using the concept of an *iterated map*. To give you some intuition about this idea, let's contrast iterated maps with two other math ideas we've been developing—differential equations and strange attractors. In metaphorical terms, they're like three kinds of photography.

Suppose we want to film a dancer in a discotheque. (The chaotic system is like the dancer.) We could make a movie of the dancer. Here time flows continuously. Or we could make a time-lapse photograph, a blurry compendium of all his movements. Here all times are shown at once. Or we could turn off the lights and flash a strobe light on him, giving his dance a jerky, psychedelic feel. Here time is discrete.

A differential equation is like a movie (or more precisely, the logic behind a movie). It dictates how the system unfolds, instant by instant, as time flows continuously. But sometimes you don't want all this information. Attractors and iterated maps are two ways to boil the dynamics down to something simpler.

A strange attractor is like a time-lapse photograph. In Lecture 7, we used this technique to expose the long-term, cumulative shape of chaos (the strange attractor). It was as if we went out into the backyard and set our camera to take an overnight exposure of the stars, except that instead of seeing the stars moving in perfect circles—the simple kind of order that enthralled the ancient Greeks—we saw something much more modern and mysterious. Our time-lapse picture showed the orbits of the Lorenz system as it moved around in its abstract state space. Instead of forming a tangled mess, as we might have expected from a chaotic system, the orbits traced out a delicate, beautiful shape that resembled a pair of butterfly wings. This was the strange attractor.

Now, in this lecture, we are going to flash a strobe light on chaos, catching it at discrete moments at it winds around the attractor. The goal is to find a rule that animates the chaos from one snapshot to the next. The pattern we'll find (the "iterated map") is the engine of chaos. It drives the chaotic motion that we see in the Lorenz system. In later lectures, we'll see that iterated maps are much more than a quirk of the Lorenz system—they are pivotal to the whole science of chaos, underlying and orchestrating the chaos seen in everything from electronics to animal populations.

We are going to flash a strobe light on chaos, catching it at discrete moments at it winds around the attractor. ... The pattern we'll find (the "iterated map") is the engine of chaos.

The iterated map for Lorenz's system comes from looking at a computer simulation of his convection model. We watch the variable that Lorenz called z and track how it changes over time. It moves up and down but doesn't ever repeat exactly (as expected for a chaotic system). Whenever z tops out (reaches a maximum), we record its peak (how high it went). Thus, we observe the system only when it reaches a peak; that's the trigger for our mathematical strobe to flash. Does a peak in one snapshot predict anything about the peak that follows it? To find out, we graph one peak versus the next. As we repeat (or "iterate") this process for all the peaks in the record, the data points hop around haphazardly (because of the chaos they represent), but ultimately a pattern emerges. All the points fall on a curve that looks like a pointy hat, or an upside-down V. It's astonishing that such a simple pattern pops out of the chaos. (If the system were random instead of chaotic, the data would have scattered into a cloud, not a thin curve.) This curve is now called the *Lorenz map*. By iterating this map—using one peak to generate the next—we produce a chaotic string of numbers. This chaotic sequence is the heart and soul of the chaos produced by Lorenz's much more elaborate convection model. The math is no harder than pressing a button on a calculator, over and over again.

There are quite a few larger lessons here, so let's just focus on two for now and save the rest for subsequent lectures. The first lesson deepens an ongoing theme. Specifically, chaos is *not* the opposite of order. It's a mix of order and randomness, in the following senses: Lectures 5 and 6 showed that chaos is unpredictable and random-looking in the long run because of the butterfly effect. But it's predictable and orderly in the short run, because it obeys deterministic laws. Even in the long run, a kind of order persists, as embodied by the strange attractor (see Lecture 7). The system has to stay on the strange attractor, and hence the overall character of its long-term motion is predictable (even if the fine details are not). And now, in this lecture, we've seen that the middle-term dynamics are structured too. Knowing one peak, you can predict the next one by consulting Lorenz's map.

The second lesson is that the order in chaos involves nonlinear relationships, in which causes can produce disproportionate effects. The curve in the Lorenz map is not a straight line. It's an upside-down V. In the next lecture, we'll start to see why nonlinearity is so tremendously important (perhaps the most important idea in this course).

Unfortunately, as with Poincaré, Lorenz's discoveries about chaos fell like the proverbial tree in the forest, making a sound that no one heard. There were two reasons for the tepid reaction. First, the paper was published in a meteorology journal that few outsiders would read. Additionally, insiders didn't know what to make of it. Lorenz had butchered the hallowed equations for convection, hacking off important terms to reveal the essence of chaos but making his results seem artificial, or at best, hard to interpret.

In general, the scientific community was still not ready for chaos. Lorenz's seminal 1963 paper was cited only about once a year for its first decade. However, by around 1975, the paper started to take off, averaging about a hundred citations a year in the 1980s. In the next lecture we'll see what sparked the fire. ■

Stewart, *Does God Play Dice?* chap. 7.

Strogatz, *Nonlinear Dynamics and Chaos*, sec. 9.4.

Play with Lorenz's iterated map at this site: http://www.aw-bc.com/ide/idefiles/media/JavaTools/lrnzzmax.html.

1. Try to extract an iterated map from your bank account records. Make a graph of (x, y) pairs, showing your balance at the end of one month (y) as a function of the balance at the end of the previous month (x). Include as many pairs of months as you can, to see if a trend emerges on the graph. Do the data fall on a straight line, on a curve, or do they form an amorphous blob?

2. Suppose Lorenz's system had been cycling repetitively rather than fluctuating chaotically. What would he have found when he graphed the current maximum of z as a function of its previous maximum?

Animating Chaos as Order—Iterated Maps
Lecture 8—Transcript

Last time, we uncovered an amazing kind of order in chaos, which we call the strange attractor. The trajectories for the Lorenz system, chaotic as they are, nevertheless show a very organized pattern when plotted in state space. Remember how it worked: They traced out the moving-chaotically, delicate wings of a butterfly. Incredible. Well, not quite, as we saw. The butterfly would be a pretty strange one because it would have an infinite number of surfaces to its wings, but that's because the strange attractor, as we pointed out, is really an infinite complex of surfaces even though it looks like just two wings. But those surfaces are packed so tightly together, like the sheets of mica, that they just resemble two separate wings, seemingly just two surfaces right there. That is a great, unexpected, beautiful kind of order in chaos. Still, there's another kind of order in chaos, and that's what I want to focus on in this lecture.

We're going to be revealing a new kind of order by using the concept of an *iterated map*. Before we get into all the details of this, let me try to convey what this is all about by way of analogy. While we're at it, let me use that analogy to try to contrast iterated maps with two other math ideas that we've been developing so far in the course, which are the ideas of differential equations, which we've had ever since Newton and the clockwork universe, and strange attractors, introduced by Lorenz, as we discussed in the last lecture. By comparing these three, all with one analogy, I'm hoping that I can clarify the distinctions among these terms because, I admit, the jargon is flying thick and fast here—strange attractors, iterated maps—but you've got to understand these things. They are the bedrock of chaos theory, so let's try it with this analogy first.

Here's the analogy: These three big ideas, differential equations, iterated maps, and strange attractors, are all kinds of photography. They're sort of three different kinds of photography, and the distinction between them is in the way that they handle time. They take a very different approach to time. For instance, let's suppose we wanted to film a modern dancer. You've seen modern dance, right? For me, it's hard to make sense of these motions. They look a bit random, maybe even chaotic, so we're going to be watching this modern dancer, who represents, again, in my way of thinking, a chaotic system.

How would we capture the motion of this dancer? Let's try thinking about three ways we might capture it. First, simply, we could just take a movie camera and make a movie of the dancer. In that way of recording motion, time flows continuously—actually, frame by frame, so there is some discreteness—but let's pretend that we had a very tiny, tiny time interval, essentially zero, between the frames, so that we can think of time as just moving continuously. That's one kind of photography that would be natural.

Another kind that may be less obvious but that can be very interesting to look at is a time-lapse photograph. That is, imagine opening the shutter to capture all of [the dancer's] movements in one image, in a kind of blurry overlay. Here, time is treated differently in the sense that rather than being extended continuously, all times are shown at once, superimposed.

Or a third way is [that] we could turn off the lights and flash a strobe light on [the dancer] periodically—pop, pop, pop, pop. You've seen dancers under strobe lights because in discos that's a standard thing. When someone is dancing, illuminated by a strobe light, the dance has a jerky, psychedelic feeling to it. That way of treating time, thinking of time as existing only at the instants that the strobe light goes off, that kind of time would be discrete as opposed to continuous.

As we discussed in Lectures Two and Four, a differential equation is like a movie. More precisely, it's the logic behind the movie. It's the underlying principles that dictate what frame should logically follow from what in the movie. That is the differential equation description. It lets you see how a system is going to unfold instant by instant and frame by frame, which is a tremendous amount of information. That's the most information you could possibly have about a dynamical system. You record everything—it's all there—but sometimes you don't want all this information.

That's the point. Attractors and iterated maps are two ways of boiling down from differential equations to something simpler. A strange attractor is like the second thing that we mentioned, a time-lapse photograph. That's how you could think of a strange attractor. It's like a time-lapse photograph. It's abstract, of course. It's not really a time-lapse photograph of a system. It's an abstraction of that because it's a time-lapse photograph in state space. In Lecture Seven, we used this technique to expose the strange attractor in Lorenz's system, which represents the long-term, cumulative shape of chaos. Maybe you've done this experiment yourself as a kid. It was as if we went out on a beautiful starry night with our camera and opened up the shutter to take a picture of the stars overnight. If you've done that, you know what you see. The stars don't stand still. The stars appear to move in perfect circles around the North Star. Of course, the stars aren't really moving. It's the Earth that's turning, but pretend with me for a second that the stars were moving in these perfect circles. In this time-lapse photograph of the heavens, you would see concentric circles as all the different stars move around, and those circles represent the kind of simple, beautiful order that enthralled the ancient Greeks, who were in love with circles.

But we didn't see that kind of order when we took a time-lapse photograph of Lorenz's system in state space. Instead of seeing this ancient kind of order of circles, we saw something much more modern and mysterious. Our time-lapse picture showed the trajectories of the Lorenz system as it moved around in its abstract state space. But instead of forming perfect circles or a tangled mess, as you might have expected from a chaotic system, the orbits traced out another kind of beautiful shape, the pair of butterfly wings that I mentioned earlier. That was the strange attractor, the time-lapse photograph of the Lorenz system in state space.

In this lecture, we're going to flash a strobe light on chaos, the third way of capturing motion, catching it at discrete moments as it winds around the attractor. That's the new point of view, and that's going to give us iterated mappings. The goal is to find a rule that animates the chaos from one snapshot to the next. Just as a differential equation gave us frame-by-frame logic, an iterated map jumps forward through many frames all at once and lets us see how a frame far back in time is related to one quite discretely forward from that. So it's like leapfrogging over all the information contained in the continuous picture in the differential equation. You can sprint through time with an iterated map.

By the way, when I say *map*, I'm using the word here kind of as a verb more than as a noun. Don't think of a map as a map of the Earth or a map of the United States. I mean *map* as in to take something and map it on to something else. Think of it as a very active process. Maybe the word *mapping* would convey that to you more, which sounds a little more active. And that word is used, too. People speak of mappings, or iterated mappings. You could think of it as a kind of mathematical machine. The mapping takes the condition of the system at a given time—or, if you like, think of it as a frame—and then it maps it onto the next frame. Or if you want to stick with numbers, you feed a number into this machine, the machine does some mysterious manipulation to it, and then it spits out another number. That's what the map does. It's a machine that eats numbers and spits numbers out, and the reason it can be iterated—disgusting as this will sound—is that having spit out a number, wouldn't it be nice to feed that number back into the input hole in the machine, let it chew on it some more, and spit something else out? And we can keep going around and around in circles like that, iterating, eating our own—maybe I'd better stop with that—but you get the idea, that an iterated map is something that's going to take numbers and spit out more numbers.

We're going to see that there is a simple iterated map hiding in the chaos in the Lorenz system, and in effect, it's driving the chaotic motion that we see in this kind of "leapfroggy" way. Whereas a differential equation is inching forward instant by instant, the iterated map allows us to sprint.

In later lectures, we'll see that iterated maps are much more than a quirk of the Lorenz system. They're pivotal to the whole subject of chaos. They underlie and orchestrate the chaos seen in everything, from electronics to animal populations. Mastering iterated maps will be very, very helpful to us for the rest of the course.

Let's now take a closer look and see where this iterated map is in Lorenz's system and what it looks like. It comes from looking at a computer simulation of his convection model or, equivalently, thinking about the chaotic waterwheel that we discussed in the last lecture. We'll watch a variable, which Lorenz called z. The physical meaning of it really isn't important. Just think of it as some variable that changes over time. Here's what it looks like; z as a function of time produces a typical-looking chaotic time series. It [has] some wiggles that are growing and growing; and then, boom, it makes a spike; and then it goes down to smaller wiggles; and then they're growing again; and so on; and there's different numbers of smaller growing wiggles between each of these bigger ones. So it's chaotic in that way. It just moves up and down but doesn't repeat exactly, as you'd expect for a chaotic system.

Here is what we're going to start looking at: Whenever one of these oscillations peaks, whenever it tops out and reaches a maximum, I'm going to record only the peak value, which I've marked here for the nth peak. I'm calling it z_n, meaning the nth peak in this time series of z. And the question, then, is, for an iterated map, if we flash a strobe light right at that moment when z tops out, can I use that to predict the next peak, $z_n + 1$? Does a peak in one snapshot predict anything about the next peak that follows it? If it does, that's great because it means we can ignore everything that happened in between and leapfrog over that. We can just hop from peak to peak, sprinting through the Lorenz chaos. But we've sacrificed a lot of information there, it seems. We've sacrificed all the intervening time points. So can it really be that z_n predicts something about $z_n + 1$ just on its own? That's what we have to check.

It's not enough to make a graph for one instance of z_n and $z_n + 1$. That would be just one point. What I need to do is repeat the process, iterating the map for all the peaks in the record. I would go through my entire record, looking at each peak and seeing what it predicts about the next peak. I'm going to do that by making a graph with two axes, just like in high school algebra. One axis, the x-axis, will be z_n. Where did z_n happen to fall? The other axis, the y-axis, will be $z_n + 1$. Where does the successor point fall? We want to see if a pattern emerges in the graph of the $z_n + 1$ versus z_n. Let's try that. I already know the answer. Lorenz discovered this, that if you make such a graph— and Lorenz, in his original paper, explained why he was motivated to do it. Maybe I should just pause. Why would you think of doing this thing? What was the hint?

When you look at the Lorenz system—I'll try to just trace it with my fingers—remember that in state space, you saw growing spirals, and when they grew enough, the trajectory shot over to the other side and then grew some more and then shot back over. So this was the chaotic motion on the strange attractor, like this. Well, z measured the height of these growing spirals, so Lorenz realized that when the spiral gets too high, it gets wobbly and gets ready to jump. And so when z exceeds a certain height, it seems like it's ready to jump, and then it dives in near the middle of the wing and starts spiraling out from there. This led Lorenz to guess that maybe just knowing the maximum z, that is, the height as it's making a loop on the wing, when it reaches a max, maybe that's predictive of the next top of the spiral. And so he was just led—you know, he's a genius. He thought, "I'm going to look at this and see if there is a pattern." So he looked at $z_n + 1$ versus z_n, and when he did, he found the graph that I'm showing here, which is that all the dots fall elegantly on a very thin curve that looks sort of like the hat that a witch would wear in one of those old children's scary movies, or it's an upside-down V. Maybe I'll just say it that way. All right, there's an upside-down V. Now, so what? The dots fall on a V curve. Why is that important? Because it's astonishing! It's astonishing that there's a simple pattern like this. This was in a chaotic system. What is this simple curve doing there?

Maybe this is a good time to underscore how different chaos is from randomness. You've heard about randomness a lot; if you ever took a probability theory course, it was all about randomness. Is chaos just synonymous with randomness? No, no, no, no. It's not, because if this system were random, what would happen? *Random* means that given the current condition, many possible things could follow with some probability, and so what you would see is that the data points, rather than falling deterministically on this exquisitely thin curve, there would be a cloud of fuzz all over it. Another way to say it is that in a random system, whatever can happen, can happen next with some probability. And so you wouldn't just see one outcome; you would see a scatter of outcomes. You'd see a big blob of buckshot of points all over the place. Well, we don't see that because in a deterministic system, in contrast to a random system, only one thing can happen next. It's not the case that whatever can happen, can happen next.

Only one thing can happen next given current conditions. That's what we're seeing here. There's only one outcome, $z_n + 1$ for a given z_n. So this system is showing its underlying determinism through this single clean line of the V curve. This curve is now called the *Lorenz map*.

I should be a little more precise here. It's really the graph of the Lorenz map because remember, a map should be thought of as a machine, this thing that eats a number, z_n, and spits out another number, $z_n + 1$. The truth is, in practice, and even among professionals, there is a conflation of the two terms, so people speak of the *map* as being this picture, which is really the graph of the map or the output of the map, and they also use *map* to refer to the machine itself, the mathematical function that I'm showing you. We can deal with that.

The real question is, though—I'm telling you this is the Lorenz map, but let's see if we can watch it being created before our eyes. I want to show you a computer simulation of how the Lorenz map arises. And so, going to that simulation, as so often is the case in these simulations I've been showing you, I'm going to show you two graphs at the same time. One of them will be the creation of the Lorenz map. That will be the upper panel. The lower panel is going to be a garden-variety chaotic time series, z versus time. But remember what's going on—strobe light. We're creating an iterated map, so a strobe light is going to flash every time z reaches a peak. Zap. You're going to see a vertical line, and the simulation will stop. We then record the value of z, integrate the differential equation forward until the next peak, and then zap, the light goes off. That's what triggers the strobe light, whenever I reach the maximum. Then I'm going to start graphing $z_n + 1$ versus z_n. So let me do that now for you in the simulation. Here we are, upper panel: The current value of z is shown horizontally. The vertical axis is showing the next z, the successor. Down here, we're going to be seeing z as a function of time.

I have a little button here that says "step." What "step" means is, flash the strobe light; integrate forward for one amount of time—not a unit of time, but until the next peak is reached when our light goes off. Bam. You see the blue line indicating that the light has gone off. We started at a peak initially, and then we integrated until we reached the next peak, and then zap, the light goes off, and we have a flash. Now, notice a blue dot has appeared in the upper graph, and the reason is that the old value of z, which was about 31, is shown here above the 31, and the new value of z, at the next peak, is higher, something like 36 or 35, and so that is what's shown here on the vertical axis, 31 and 35. That's one dot in the Lorenz map.

Actually, we should be a little bit careful here because the Lorenz map only applies to the system once it's on the attractor. We've learned from the past that there are transient times for systems to relax onto their attractor. They have to be attracted to the attractor, after all. So this initial point in reality is not—there's no reason to expect it to be on the attractor. We're not there yet. We haven't settled down, so there's going to be some garbage at first, transient behavior that's not really natural for the system. It's just an artifact of how we started it. We're going to let that garbage sort itself out until the system has settled onto the attractor, and then we'll see the Lorenz map as it really is.

So let me step forward one more time. Bam, next maximum; notice another point has occurred. This one has a horizontal value of 36 or 35 because, remember, it is now considered the current z. This height of 35 is now the current z, and the next z is even higher, a little closer to 36. You get the idea. Now, I'm just going to step forward one step at a time. Let's keep stepping. The blue dots are being created. You get the idea. Of course, this is tedious to have to keep clicking the mouse, so conveniently, there is another button on the simulation called "iterate," and this is going to now start going by fast, as the computer will do the job of clicking for us. I'm going to iterate for a while, and then after the system is really on the attractor, I'm going to clear the transients, this initial stuff that we consider garbage, not really relevant. Here we go with some fast iteration. You see the points starting to fall on the curve. They're trying to trace out that V-shaped witch's hat. Now I've pretty well gotten onto the attractor. I was there long ago, but just to show you that it's going to settle down to this every time, I can clear the transients, and it's recreating it from scratch, and the same shape emerges every time. This is the attractor.

Notice that down here, the time series itself—this is what chaos looks like as a time series. You might say to yourself, "It doesn't look that chaotic to me. It looks very periodic. I see a lot of peaks and valleys." But notice that the peaks are not all the same height. Here is one that's very high, followed by one that's lower, higher, higher, higher, higher, and then lower, and then higher, and then lower, and so on. And there are troughs of different sizes, so it's the amplitude that is where the chaos expresses itself. Clearly, the Lorenz map is being formed as we keep going. So [let's] just let it fill itself in, that witch's hat that's just appearing out of the vacuum, so to speak. It's guaranteed.

Why, having said all this, is the Lorenz map so important, so significant? Because it's so simple. It's astonishing that there's a pattern like this lurking in chaos. Also, when I say it's simple, keep in mind that all I did was—I have this V-shaped picture which says, given the current z, I can tell you the next z significantly far ahead in time. That means that the operation of sort of leapfrogging forward through the time series is no harder than just repeatedly pressing the button on a calculator. If I had the Lorenz map programmed into my calculator, I could just say tell me a z, press the button, that's the next z, press the button, and I can fly through the Lorenz system. Specifically, by iterating the Lorenz map using one peak to generate the next, we generate a chaotic string of numbers that captures faithfully all the chaos in the Lorenz system. Maybe not all of it because it's missing the chaos in between the peaks. It's capturing the heart and soul of the chaos. That's the point. The iterated map is lightning fast. It captures with great fidelity what is going on in the Lorenz system, and it ignores in-between parts that maybe we don't care about so much. That's what I meant when I said earlier that iterated maps are a great boiling down of information in the differential equation. They are simpler, and they allow us to see farther and faster, or at least more easily—just punching a button on a calculator—and they are much less cumbersome than differential equations.

On the other hand, I guess I should say for honesty's sake that there are some scientists who don't have great respect for iterated maps because they know—and they're right—that the language of nature is differential equations. The true laws of physics and chemistry are written in differential equations, so by going to iterated maps, we are sacrificing some information. That's true. And the laws of nature are not expressed that way. That's also true. But iterated maps are so valuable in their speed and in their compression of information that we have to take them seriously and have to use them, limitations aside.

In the next few lectures, we'll be studying iterated maps much more deeply, and we're going to be leaving differential equations behind. They have served us well, but it's now time to let them rest. Also, this is historically what happened. The field of chaos theory developed through Newton, with his masterful differential equations; Poincaré; and Lorenz, but Lorenz brought in iterated maps, too; and since him, especially in the '70s and '80s, people leapt forward with iterated maps and gained great new insight that differential equations were just too unwieldy to provide. This is going to be a strategy that will produce some of the most amazing results. I would argue, possibly, *the* most amazing results in the whole subject of chaos theory [are from] the exploration of iterated maps.

What are the larger lessons here, besides this curious picture? There are quite a few of them, so I just want to focus on two for now. I'm going to save the rest for subsequent lectures. The first lesson deepens an ongoing theme in this course, which is that chaos is *not* the opposite of order. It's a mix of order and randomness in the following senses: Lectures Five and Six showed that chaos is unpredictable and random-looking in the long run because of the butterfly effect, but it's predictable and orderly in the short run because it obeys deterministic laws.

Even in the long run, a kind of order persists, as embodied by the strange attractor. A system that's chaotic, nevertheless, once attracted onto a strange attractor, has to stay there, and in that sense, its overall motion is predictable in a statistical sense. Its overall character is limited by the strange attractor, even though its fine details are not; that is, two trajectories on the attractor, representing two essentially indistinguishable ways we could start a system, will stay together for a while—they behave the same way; everything is predictable—but then they veer off, go off onto different parts of the attractor, never really to see each other again, maybe occasionally meeting but not for long. So if we watch chaos in that sense, we would say that there is short-term predictability but no long-term predictability and yet an overall predictability by virtue of being confined to the attractor.

In this lecture, we've seen that the medium-term dynamics are orderly, too, are structured. By *medium-term*, I mean not the very shortest term dynamics governed by the differential equation, but even leaping forward into the medium term, using iterated maps, there is great structure there, too. Knowing one peak in Lorenz's z time series, you can predict the next by consulting the Lorenz map. That's the first lesson: [There are] all these different kinds of order in chaos.

The second lesson is that order in chaos is always due to nonlinear relationships. There is a word I deliberately haven't used much, *nonlinear*. Big, big, big, big word. Nonlinear relationships mean that causes can produce disproportionate effects, not simply proportional, like in a linear system, but disproportionate. The word *linear*, as it suggests, has something to do with lines. We have seen that in the Lorenz system, the iterated map doesn't look like a line. It looks like an upside-down V. It looks like that. That's not what a straight line looks like. This thing is bent. It's a witch's hat. Why is that so important?

In the next lecture, we'll see why. We'll see why nonlinearity, innocuous as it seems, is tremendously important, perhaps the most important idea in the whole course. If you come away with nothing else, I hope you will come away with an appreciation of the mysteries of nonlinearity. That maybe doesn't make any sense yet. I understand. But we've got a long way to go, and by the end, I'm sure you'll take my point.

Unfortunately, as with Poincaré, Lorenz's great discoveries about chaos fell like the proverbial tree in the forest, making a sound that no one heard. You are probably shocked by that. This already happened to Poincaré. We understood why his work was ignored and why he himself didn't really carry it on. But now, 70 years have passed. Lorenz has discovered the strange attractor; he has found his iterated map; he's got computers. You would think people would get it. But they didn't. They still didn't get it. It's interesting in the history of science how these things work. Why the tepid reaction? We can speculate. I don't really know, and this is something for historians of science, but let me give you my own speculations.

One thing that is important to remember is that in the '60s, science was still quite specialized. People lived securely within their own disciplinary boundaries; there was not a lot of talking outside of disciplines, and Lorenz was a meteorologist. He published his work in a meteorology journal called *Journal of [the] Atmospheric Sciences*. It's not a general science journal. It's not a journal that anybody would read except for a meteorologist. There it was, a big breakthrough, and of course, with Lorenz's own personality, he wasn't blowing the trumpet. He just did his work and published his paper. Few outsiders would be aware that anything happened, that this thunderbolt had hit. Insiders, the meteorologists themselves and experts in fluid mechanics, didn't blow any trumpets either. In fact, they didn't even know what to make of it.

Lorenz had done something in his work that did not please the community. He took the hallowed equations for convection, which everybody studied and devoted a lot of energy to, and he butchered them, on purpose. He butchered these equations, hacking off important terms that weren't important to reveal the essence of chaos by making his results seem artificial, at least to the people who were wedded to this way of thinking about convection. Or at best, even the most generous interpreters didn't really understand. "What do these results mean, Ed? You're showing me chaos, but these equations don't describe anything that I know of." The waterwheel analogy was not known yet.

You can see why this would not have taken off. Finally, though, I think, here is the real reason. The main reason is that the scientific community was just not ready for chaos. To quantify that, let me show you the number of times that Lorenz's paper was cited, year by year, from the time he published it to the present, because citations are academic currency. That is money: How successful is your paper?

Here's what the citation record looks like: 1963, boom, the thunder. Nothing. Years go by, getting less than 10 citations a year, maybe 1. Some years it doesn't get cited at all. I'm showing citations as a function of time. For 10 years, nobody paid attention. Something started to happen around 1975. Now, Lorenz is being cited 10 times, 20 times. Now, he's up to 100 citations going into the '80s, and now, approaching the present, his paper gets cited close to 200 times a year. Incredible.

In the next lecture, we will see what finally started this fire. See you next time.

How Systems Turn Chaotic
Lecture 9

In common parlance, of course, *bifurcation* means a fork or a splitting. Mathematicians have borrowed the term to refer to situations where a dynamical system branches off into a new type of behavior, a new qualitative manner of behaving in the long run as one of its parameters is varied. That doesn't happen here, but it will happen in the logistic map.

Our detective story builds to its climax in the 1970s, with an unprecedented convergence of scientific disciplines. Researchers in mathematics, ecology, and fluid mechanics (who would normally have little to say to each other) found themselves asking the same question. Whether they were studying the dynamics of iterated maps, insect populations, or water flowing through a pipe, they all wondered how their systems made the transition from order to disorder, from stable equilibrium to regular oscillations—to wild, seemingly random fluctuations. In other words, how did a system turn chaotic? This is a new question for us in this course. We're no longer asking about the properties of chaos; we're asking about the *transition* to chaos. That turns out to be a marvelously fruitful question, because we'll see that the process of going chaotic has certain universal features, independent of whatever it is that's behaving chaotically. In other words, different things go chaotic in the same way.

Let's begin with the simplest possible iterated map. It describes the growth of money compounded annually at a constant interest rate. The "map" is the mathematical function that tells us how much money we have a year from now, given how much we have today. When graphed, the relationship is a straight line, so we say this map is "linear." By "iterating" the map, we can project into the future as far as we like.

To illustrate the transition from order to chaos, as well as the ideas of bifurcation and nonlinearity, we turn now to a famous iterated map known as the *logistic map*. Suppose you're a population biologist studying the agriculturally important problem of seasonal outbreaks of a certain insect pest. The logistic map tells you how the total insect population this year

can be related to the population of insects in the next generation a year f
rom now.

The graph of this relationship is hump-shaped, given by the parabola
$y = rx(1-x)$, where the parameter r is a constant that reflects the
population's intrinsic growth rate. The idea is that if last year's generation was
small, this year's will tend to be larger (because there were plenty of crops
for all the bugs to eat). But if last year's generation was big, overcrowding
and starvation may cause this year's generation to be small.

Iterating the map allows you to predict the generation size many years ahead.
What happens in the long run depends on the growth rate, or equivalently,
how steep the hump is. For low growth rate, the insects die out completely. As
we gradually turn up the steepness, nothing different happens until we cross
a threshold, or *bifurcation point*. Then the dynamics change qualitatively:
The population approaches an equilibrium level that stays constant from
year to year.

Turning up the growth rate even more yields a series of dramatic
bifurcations. First the equilibrium splits into a 2-year cycle of boom
and bust. Then that 2-year cycle splits into 4-year cycles; then 8, 16,
32, and so on. The cycle length becomes infinite at a critical value of
growth rate. Just beyond that point, the yearly population fluctuates wildly.
By increasing the growth rate, we drove the logistic map from order to chaos.
There's a lot more to this story, but let's defer the details until Lecture 10.

Although the logistic map originated in ecology, it holds lessons for all of
science. What the logistic map shares with so many other systems is its
nonlinear character. Simply put, the map isn't a straight line. It's bent and
hump-shaped. That's why we couldn't predict its dynamics with algebra and
had to use a computer. The same difficulty afflicted Poincaré in the three-
body problem; it's also governed by nonlinear equations. So are Lorenz's
weather model and convection model. Nonlinearity giveth, and nonlinearity
taketh away. It makes chaos possible but formulas impossible.

"Nonlinearity" sounds repulsive, I know, but you must try to fall in love
with the concept. It's the most important idea in the science of chaos.

Mathematician Stanislaw Ulam quipped that calling a system nonlinear was like describing zoology as the study of "non-elephant animals." Most animals are not elephants, and most systems are not linear! What's so special about linearity? It implies the whole is merely the sum of the parts. The banking problem gives a simple numerical example of this.

Nonlinearity arises whenever the whole is more (or less) than the sum of the parts. Whenever parts of a system compete or cooperate, and don't just add up, you have nonlinearity. The nonlinearity of the logistic map reflects the insects' competition for food and mates when they're overcrowded. Everyday examples of nonlinearity are the pleasures of music, the dangers of drug interactions, and the life-saving power of combination therapy for HIV.

For hundreds of years after Newton, physicists shunned nonlinearity. They didn't need it in electrical or civil engineering, or in quantum mechanics. Their mathematical methods couldn't handle it anyway. When nonlinearity was unavoidable, they tamed it by "linearizing" it (sweeping it under the rug).

Mathematician Stanislaw Ulam quipped that calling a system nonlinear was like describing zoology as the study of "non-elephant animals." Most animals are not elephants, and most systems are not linear!

But by the 1970s, a few mavericks in every field were ready to tackle nonlinearity, and with it, chaos. Which raises the question: Why *then*? Computers were part of it. Scientists could finally see chaos right before their eyes; it wasn't abstract anymore (as it was in Poincaré's era). Also, the term itself was catchy. "Chaos" was christened by the mathematicians T. Y. Li and James Yorke in 1975 in a famous paper about iterated maps.

In 1976, the theoretical biologist Robert May published an influential review article in which he made an impassioned plea that it was time to start studying nonlinear systems. He wrote, "Not only in research, but also in the everyday world of politics and economics, we would all be better off if more people realized that simple nonlinear systems do not necessarily possess simple dynamical properties." ∎

Essential Reading

Gleick, *Chaos*, 59–80.

Stewart, *Does God Play Dice?* 154–64.

Supplementary Reading

May, "Simple Mathematical Models."

Peak and Frame, *Chaos Under Control*, chap. 5.

Strogatz, *Nonlinear Dynamics and Chaos*, chap. 10.

Internet Resource

Explore the dynamics of the logistic map yourself. Go to http://www.student.math.uwaterloo.ca/~pmat370/JavaLinks.html to find some relevant Java applets.

One of the simplest to use is this one: http://www.geom.uiuc.edu/~math5337/ds/applets/iteration/Iteration.html.

Questions to Consider

1. Pick a number between 0 and 1000, and type it into a pocket calculator. Keep pressing the square root button repeatedly. What happens? Why?

2. What are some other familiar, real-world examples of linear and nonlinear systems?

How Systems Turn Chaotic
Lecture 9—Transcript

Welcome back to chaos. I can barely contain myself in this lecture because we're now approaching the climax of this part of our story. I've been describing it as kind of a detective story, but in fact, it's now going to take on the character of an international thriller, in that we have scientists in different disciplines, even on different continents, all scurrying around thinking about chaos, although the word doesn't yet exist. There's no name for the subject, but they're all fascinated by disorder. There is a kind of unprecedented scientific convergence going on, a convergence of disciplines. Researchers in math and ecology and fluid mechanics, who would normally have nothing, really, to say to each other, find themselves asking the same question, whether they're studying the dynamics of iterated mappings, or insect populations, or water flowing through a pipe in a way that's about to become turbulent. They're all wondering how their systems make the transition from order to disorder, specifically, from gentle, stable equilibrium to regular oscillations, and then to wild, seemingly random fluctuations.

In other words, how does a system turn chaotic? Notice, that's a new question for us because in the past few lectures, we've been talking about chaos but about the properties of chaos. Now, we're asking about the *transition* into chaos. That turns out to be a marvelously fruitful question because we'll see that the process of going chaotic has certain universal features, independent of whatever it is that's behaving chaotically. Different things can go chaotic in the same way, to put it roughly.

This lecture is jam-packed with new ideas. They'll be coming at you fast and furious, sort of consistent with what's happening in the subject itself. This is now the 1970s, and chaos is about to explode as a discipline.

While all these ideas are flying at you, I would like you to try to focus especially on three of them above all: iterated maps; bifurcations and the transition to chaos; and the third one, linearity versus nonlinearity. All three of those, I hope, will become clearer by the end of the lecture.

Let's begin with iterated maps. You can actually see an iterated map in a very concrete way if you have an old handheld calculator. You may have even played this game yourself. Take something like the square root button, if you have a calculator equipped with one. Most of them have it. Just type in whatever number you feel like—286—and then press the square root button. You'll get the square root of 286. If you're mischievous, you might keep pressing the square root button. You've probably done something like that. You're now iterating a map. You're iterating the square root function, which is a map. It's a rule that takes a number to another number, in this case, the number to the square root of that number. So that's really all an iterated map is. It's just this repetitive process of doing some function over and over, feeding its output back in as input.

We met a specific example of an iterated map—not this square root function, which is a sort of meaningless example—but we met a very meaningful one in the last lecture, in Lecture Eight, while we were flashing a kind of imaginary strobe light on the Lorenz system, the chaotic system that Ed Lorenz had investigated in connection with convection. While we were flashing the strobe, we were taking notice of the value of some variable that Lorenz called z as a function of time. Every time it reached its peak—flash!—the strobe would go off, and we would record the value of z and then try to use that to predict—flash!—the next peak value of z.

When we did that, following Lorenz, of course—it was his brilliant idea to do this—we found that the data of one maximum z versus the next didn't just fly all over the graph paper. In fact, they lay on a beautiful, simple V-shaped curve, like a witch's hat, an upside-down V, an amazing pattern in the midst of all the chaos. So that was the Lorenz map. That was our first iterated map, and we saw how it sort of naturally arose in that connection.

In this lecture, we're going to discuss two other iterated maps. One of them has a rounded top. Instead of a V, a pointy, cuspy top, like the Lorenz map, we're going to study one with a rounded top, and it will teach us even more about chaos than Lorenz's map did.

The other map is the simplest one you can possibly conceive of. It's just a function that's given by a straight line. It's intended as a pedagogical example to help you understand one of these other three big points I'm trying to make today, why linearity versus nonlinearity is such an important distinction.

Before we get into those two examples in any detail, let's remind ourselves of the bigger issue here, which is, why are we so interested in iterated maps? Because they are one of the two main types of dynamical systems, along with differential equations, which we spent many previous lectures discussing. Recall the distinction. Differential equations treat time as if it were continuous. They take a sort of cinematic view of the world, as if you were making a movie of reality, whereas iterated maps treat time as discrete, a kind of stroboscopic view of the world, flashing and illuminating the world only at discrete moments.

Until now, we've concentrated on differential equations for good reason. They are the language of science. They are the language in which the laws of nature are written, as Newton discovered, and because that's where mathematical chaos was first discovered by Poincaré and Lorenz. The trouble with differential equations is just how very complicated they are to understand. Although they do shed light on chaos and they are the fundamental description of nature, they can be mind-boggling, whereas iterated maps have the advantage of being much easier to understand. As I say, you can play with them just by punching a button on a calculator, so in that sense, they only require high school math. That's a big advantage. And iterated maps provide the simplest mathematical models for the transition from order into chaos, which remember, is what we are really interested in above all.

Let us turn to those two examples I hinted at a few minutes ago—the linear map and one with a rounded top. This first example arises when you put money in the bank and get a return on it at a constant interest rate. Just put it in the bank, and maybe they guarantee that it will grow at 8% a year, or 10%, or whatever. Let's look at an example like that.

This is a trivial problem, and it's meant to be. I'm just trying to show you the basic ideas of iterated maps in a context that I think you'll be familiar with. So we put money in the bank. Suppose we find some crazy bank that will give us 10% annual interest. That's nice. We like that bank. It's hard to find these days, but there have been times when it was 10%. The number is not too important. Let's suppose we put our paltry $100.00 in the bank, and after a year, how much will we have? [With] 10% interest, we will have $110.00. So there's $110.00 graphed vertically above the point showing $100.00—our balance today and then our balance next year as a function of our balance today. If we didn't have $100.00, if we only had $50.00, then we would have to put a point here, rise up from $50.00, and it would then become $55.00.

My point is that there is a straight-line relationship between the balance after a year and the balance today, given by this simple equation that y, the balance next year, is just 1.1 times the balance today, x. Okay, clear enough. Now, in effect, what we're doing here is strobing a light on our bank account. You might think of it that way. We've got the bank account, and we close our eyes and don't look at it for a year, and then the light goes off, and we see what we have a year from now.

Because next year's balance is proportional to today's, given by this straight-line relationship that I've just been showing you, we say that this is a *linear map*—*linear* for "line." You're probably not too impressed so far because you know already how to analyze something like this. You didn't need linear maps to know about your bank account. And that's right. They are easy to analyze if they're straight lines. In fact, you can project into the future. If I asked you how much money you would have 2 years from today, you could probably do that in your head. I have $100.00 today; I'll have $110.00 next year; I multiply it by 1.1 again. That will make $121.00 2 years into the future. If I ask you to calculate an arbitrary number of years into the future, even 100 years into the future, you could write down a simple formula for that. It would be 1.1^{100}, because you've multiplied 1.1 times itself 100 times, times the initial balance. So you can certainly project into the future with formulas, and you probably learned that at some point in your life.

There's another way to project into the future that will turn out to be more powerful because it generalizes better. It's the case of nonlinear maps. Here's how we can do it: I've got my initial balance here, $100.00. I go up to this line. I now have $110.00. How would I graphically figure out what I would have a year from now? What I could do is, think of the $110.00 as if it were occurring today. I mentally imagine myself one year into the future and call that today, which would mean that I have to put a little tick mark down here at $110.00, and then go up to the graph again, and that would then make the number $121.00, but I've just done that by moving vertically.

There's another way to see what's happening, which is graphically, like so: I draw a diagonal line, the line $y = x$. That makes it easier for me to do the calculation graphically that I just described. It shows me how far to move sideways because when $y = x$, that means that my balance a year from now, $110.00, can be shifted over to this line, and that now treats it as if it were today's balance, if you see what I mean. It allows me to project a year into the future. Then I bounce off that diagonal and go vertically to the curve. This red dot shows my balance a year from now. So this is an automatic way of going forward in time. If you do that over and over, you can predict your balance as far into the future as you want. We'll have the computer do that for us in a minute.

First, let me introduce our other example. It's the very celebrated nonlinear mapping known as the *logistic map*. This is a model that arises in connection with population biology. Suppose you are a population biologist and you're studying the very important problem in agriculture of understanding the dynamics of seasonal outbreaks of a certain bug that likes to eat crops. The logistic map tells you how the total insect population this year can be related to the population of insects in the next generation a year from now. By the way, we're thinking here of what biologists call *non-overlapping generations*, meaning that the insects might lay some eggs, the parents die, and a year later, the new insects hatch, but there's no generational overlap. The parents never get to live at the same time as the kids. It's just one generation and then the next, and there's no overlap. This kind of relationship, then, the logistic map is showing, given the insect population today, how many insects will I have next year?

First of all, notice that the relationship is a hump. In fact, it's a parabola given by this equation: y, the population next year, [equals] r, the growth rate of the insects, the per capita growth rate, meaning how many new insects come from a given insect, times x, the current population, times $1 - x$. Now that $1 - x$ is a little bit mysterious because if this were just the bank problem that we did a minute ago, it would just be y equals the growth rate of your money times the current [amount]. That would be a straight line, $y = rx$. But this $rx(1 - x)$, that's now a quadratic expression. There's an x^2 if I would multiply this out. That makes a parabola, and in particular—that algebra is not important—the thing that's important is this shape, this bent, hump-shaped nonlinear relationship. What is the biology behind that? Why is that sensible? Well, it's hump-shaped because if the population this year is small, then there is a lot of food to go around, a lot of crops. From the insect's point of view, it's a good year. There could be plenty of children next year with a lot to eat, so the number of insects in the next generation might even be larger. So there's an increasing function up to a certain point. But then you get to the situation where too much of a good thing is a bad thing. If the population is too big, then the parents are scrambling to find mates, scrambling to find food; there's overcrowding. So this curve starts to bend down, and things start getting bad for the insects (good for the farmer). The population will diminish if it's too large, so you're setting up a situation for booms and busts, large populations followed by small ones.

Now we want to take both of these examples—the linear case of the bank account and this nonlinear case of the bugs—and try to forecast far into the future. What will happen in the two separate cases? Let's do that by going to a different computer program.

Let me begin with the linear map, the money problem. Well, you know what will happen. This is a very nice deal you've got, 10% guaranteed interest every year. Your money is going to grow exponentially fast. Let me show that using that graphical iteration method that I described a few minutes ago.

I start with my $100.00. Let's just orient ourselves. There are two panels here. I'm going to be showing you two different ways to look at the results. First, the usual way, a time series—a graph of your money as a function of time, like any financial chart, just showing your money growing up as a function of time, exponentially fast. You'll see that on the left.

The more powerful way, and a novel way, is this graphic iteration that I'm trying to get you to cope with here, unfamiliar but very powerful. You'll see that on the right, showing your current account balance against next year's balance. So let me begin with my $100.00. I take the first step forward in time, and two things happen. From the $100.00, we now go up to $110.00, and these two graphs are linked sideways so that this dot corresponds to the same height as this dot, meaning that my $100.00, after one year, has grown to $110.00. Now I do the trick over here of bouncing off the diagonal and then vertically to the straight line. So I've done that; my money has grown accordingly. Let me step forward a few years at a time. You see on the left, the money is growing exponentially, and over here, it looks like we're climbing a staircase as we keep iterating this map.

Notice that there's nothing very mysterious happening. In fact, there's no surprise. Your money is simply growing exponentially. By the way, that is true no matter what the interest rate is. For example, suppose I clear the screen and find an even more attractive bank that gives me 20% interest. Do you see what just happened? As I slid the interest rate slider here—this is like turning a knob—notice that the line is rotating. I'm making the line steeper and steeper and steeper as I increase the interest rate. If I do the process over, my $100.00 grows faster. You can see that's a steeper exponential growth and the staircase is more dramatic. But nothing qualitatively different happens. For instance, we don't see oscillations in our bank account. We don't have good years followed by bad; we're just growing, and we certainly don't see chaos. Now, in your bank account, you might see chaos, but not under these conditions. You would just see exponential growth. We would say, to introduce a little bit of jargon that we're going to need, that this system shows no *bifurcations*. It shows no changes in its qualitative long-term behavior as we vary the interest rate.

In common parlance, of course, *bifurcation* means a fork or a splitting. Mathematicians have borrowed the term to refer to situations where a dynamical system branches off into a new type of behavior, a new qualitative manner of behaving in the long run as one of its parameters is varied. That doesn't happen here, but it will happen in the logistic map, as I'm going to show you now.

This becomes much more interesting when we go to the logistic map—in fact, mind-boggling. Remember, this is the model of population growth for the insects. Let's look at this simulation. As before, you are going to see two panels. On the left is a time series now representing population versus time, year by year, and on the right, again, graphical iteration. We see the diagonal that allows us to do the iteration, and there's the map. I have chosen a particular growth rate, that number r, the per capita growth rate for the insects. It happens to be 3.51. It doesn't really matter. We're going to investigate what happens at different growth rates.

Let's do this one for illustration. We started with a small population of insects. Next year, it's higher; bounce off the diagonal. Wow. Now we've got a lot of these bugs. In fact, we may have overdone it because look at what's going to happen next; anticipate it. Here we are at this height. That's a lot of bugs, and they're scrambling and crawling on top of each other, so that when we go forward a year over to the diagonal and then find the curve vertically, we're going to actually go down. Next year is going to be a smaller year, a smaller population. Do you see how this works? Let me keep iterating forward to help you understand how complicated this sort of picture can be. It's tracing out what looks like a spider making a web, and for that reason, this picture on the right is often called a *cobweb diagram*. What it corresponds to in terms of the time series, you'll see, is something that looks pretty complicated, and you might even say chaotic. We've had a boom and then a bust and then a bigger boom and a bigger bust, and this is oscillating in some crazy way, and it's not periodic. I can go forward in time even more. I've only gone a few generations by clearing the transients and letting it go and iterate faster. It's really not doing anything very repetitive here. It's complicated. Wait a second; now that I look at this what I just said was wrong, it actually *is* periodic eventually. Once the transients are cleared I can see it has settled into a periodic rhythm—it is repeating every 4 steps. That becomes obvious now that we look on the cobweb. This shows us that this cobweb construction can be helpful.

That's the logistic map. Now, is the behavior always this chaotic? No. That's the point, in fact. We want to use this example to study the transition from more orderly states into chaos. What will happen will depend on this growth rate, per capita growth rate—or equivalently, how steep the curve is. Notice, as I change this growth rate—I can do that by moving this slider—I'm making the parabola get taller or shorter. Here, it's quite squat. If I increase it, I can make it tall. Let me go down to a small value and watch what happens. When the growth rate is low, the insects simply die out, as you might expect. Watch. Over here on the right, we've gone to the curve, and then we're finding our way into a little channel into the origin, whereas the population as a function of time is just petering out.

On the other hand, if I were to increase the growth rate, I can make something qualitatively different happen. I can see a bifurcation. Watch. In this case, the population reaches a steady state. There is some steady level of insects over time. This growth rate is playing the role of what we call a *parameter*. It's kind of an adjustable knob, like when I did the experiments or showed you the video of the Lorenz waterwheel, I was tightening a brake. Or you could think of turning up the heat on a stove or any other sort of thing that's under the experimenter's control that can change the behavior of a system. Those are the parameters. Varying them can cause bifurcations, these dramatic changes in long-term behavior.

I want to show you the bifurcations in a much starker way, much more obvious, by graphing the time series of the insects under different conditions. So here we are again. Now you're seeing time plotted sideways, population of insects vertically. At the moment, the growth rate is small, so they're just dying out. Let me gradually start moving the growth rate up. Watch for qualitative changes in the yellow curve. It's changing quantitatively, but it's not changing qualitatively at the moment. The long-term is still that everyone dies out. I'll keep going. Watch for that curve to lift off the axis. There, it just happened. Did you see it? Something happened when I crossed a growth rate of one because now each insect is giving rise to more than one other new insect on average, so this can sustain itself. So there's an equilibrium level; it's just getting higher now. There are no bifurcations at this point; it's just getting higher, but nothing qualitatively changing.

Here's something interesting that starts to happen. We can overshoot and oscillate a little bit. Watch. Around here, you're starting to see initial oscillations. This is not considered a bifurcation because although something qualitatively different has happened here, the system is overshooting before settling to equilibrium. Bifurcation theory is only concerned with what happens in the long run, and in the long run, nothing qualitatively new has happened; we're still at equilibrium.

Now comes the next bifurcation. When I cross through this number 3, you will see something happen. Those up-and-down oscillations of high population and low are now starting to become more persistent, and at 3, now they have really set in. This system, in the long run, is going up and down every 2 years. It's having a cycle of boom and bust that's exactly periodic. It has what is called a *period-2 cycle*, meaning 2 years long.

This is where the story gets great because as I start turning up the growth rate, these bifurcations happen faster and faster. The 2-year cycle is going to split into a 4-year cycle. Watch. If my hands are good, I can make you see it. I'm just making the 2-year cycle more extreme. Nothing has happened yet, but soon, you'll see something happen. Keep your eyes on it. Did you see it? Look. That has now changed its qualitative behavior. It's now going up and down, repeating every 4 years. Watch. Very low, high, low, very high, very low. These very lows are all the same, but it takes 4 steps to repeat. So maybe you can guess the pattern. It repeated every year, then it repeats every 2 years, then it repeats every 4 years. You can guess, right? If I keep going, it's going to repeat every 8 years. Watch this. I think I did it. That's actually a period-8. Watch. Starting here at not very low, I go to high, very low. Can you see? This [rate] is the same as that. It took 8 years for it to repeat.

This is amazing. So now what we've done, by changing the growth rate, we've gone from one to 2 to 4 to 8, and I needed very fine motor control to get this to work because these bifurcations come faster and faster. This is what's so cool. The period-doubling is happening faster and faster and approaches a limit. It converges to a certain growth rate, and after that, who knows what's going to happen because we've gone to infinite period cycles. We're now at the onset of chaos, and just beyond that point, the yearly population starts to fluctuate wildly. By changing the growth rate parameter, we drove the logistic map into chaos from order. There is a lot more to this story, as you can imagine, but let me defer the details until the next lecture.

Although the logistic map originated in ecology, it holds lessons for all of science. What the logistic map shares with so many other systems is its nonlinear character. Simply put, the map is not a straight line, as we saw. It's bent and hump-shaped. Nonlinearity makes it impossible to find explicit formulas, like we did in the linear case for the map's long-term behavior. You have to use graphical methods or a computer. The same difficulty afflicted Poincaré when he was dealing with the three-body problem because the equations for that are nonlinear, too. They are nonlinear differential equations, but the key is that they were nonlinear. So was Lorenz's convection model, so was his weather model. Nonlinearity giveth, but nonlinearity taketh away. It gives the possibility of chaos, but it makes formulas impossible.

I know that normal people find "nonlinearity" a repulsive word. Mathematicians love it, but it may take some time to fall in love with it. But you have to try because it is the most important idea in the whole subject of chaos. The great Polish mathematician Stan Ulam once said that calling a system nonlinear was like going to the zoo and saying, "Look at all the very interesting non-elephant animals they have here." The point is that most animals are not elephants, and most equations are not linear. In fact, the linear ones are the strange ones. What's so special about the linear ones? Well, in a linear system, you have something incredibly simple that happens, abnormally simple, which is that the whole is exactly equal to the sum of the parts.

The banking problem that we did earlier gives a simple numerical illustration of this. If I deposit my $100.00 again at 10% interest, how much would that become after 2 years? We've done that in our heads, so it's easy, but what I want to show you here is that the whole is the same as the sum of the parts. You can do the problem two ways, and it gives the same answer. If you calculate the whole, you would say I had $100.00. I take that whole amount and multiply it by 1.1, it becomes $110.00; and I multiply that whole amount by 1.1, and it's $121.00 after 2 years. Fine. But if I go and take a different point of view, by thinking of the problem as having parts—that there are principal and interest—I could say, initially I had $100.00; that became $110.00, of which $100.00 was principal and $10.00 was interest. Now I calculate how much each of those separately will grow by multiplying them by 1.1 separately. The $100.00 becomes $110.00; the $10.00 interest becomes $11.00; add them together and I get $121.00. Simple, but what it means is, in a linear system, like this thing described by a linear map, you can just break problems into parts, calculate stuff separately and add it back together, and you get the right answer. And that is what doesn't work, in general. That does not work in nonlinear systems.

Whenever parts of a system compete or cooperate and don't just add up, you have nonlinearity, like in the case of the insect problem. They were competing with each other for food and for mates. That's why the curve was bent. Nonlinearity is everywhere in our world. It's in the pleasures of music. If you listen to your two favorite songs, you don't get double the pleasure because of nonlinear interactions in you head. You know about interactions. You're not supposed to take two drugs at the same time if they have dangerous interactions. On the other hand, those interactions can be synergistic, as in the case of combination therapy for HIV, where you can give three drugs in combination, a drug cocktail, and their total effect is much more powerful than it would be separately.

For hundreds of years after Newton, physicists shunned nonlinearity. They didn't need it in electrical or civil engineering, where the laws were basically linear, or even in quantum mechanics, where the governing laws were linear. Anyway, their mathematical methods couldn't handle it, and when nonlinearity was unavoidable, they tamed it by just "linearizing" it, pretending it was a straight line, in effect, sweeping it under the rugBy the 1970s, a few mavericks in every field were ready to tackle nonlinearity and, with it, the amazing possibilities of bifurcations and chaos, which raises the question: Why *then*? Computers were part of it. Scientists could finally see chaos right before their eyes. It wasn't abstract any more, like it was in Poincaré's era. The word helped when it was named a new field—chaos. It was a catchy name. People started to pay attention just for that reason. The name was christened in 1975 by two mathematicians, Li and Yorke, who were working on iterated maps.

There were people working in fluid mechanics who were thinking about the transition, not to chaos exactly, but to turbulence, like if you open up a water faucet. You see laminar, or a gentle flow, at first. If you open it more, gushing, turbulent flow. This was a classical, difficult, unsolved problem in fluid mechanics, and it was finally time to address it.

In my opinion, the real watershed moment came in 1976, when a great biologist, whom I'll tell you more about in the next lecture, Bob May, published an influential review article in which he made an impassioned plea that it was time to start studying nonlinear systems. Here's what he said: Students should be introduced to them early in their education because by only studying linear systems (which was the tradition), he said, "The mathematical intuition so developed ill equips the student to confront the bizarre behavior exhibited by the simplest nonlinear systems, such as [the logistic map]. Yet such nonlinear systems are surely the rule, not the exception, outside the physical sciences." Then he ended with this: "Not only in research, but also in the everyday world of politics and economics, we would all be better off if more people realized that simple nonlinear systems do not necessarily possess simple dynamical properties." There can be unintended consequences. Let me leave you with that thought. I will see you next time.

Displaying How Systems Turn Chaotic
Lecture 10

> In part, we want to study the orbit diagram because it's ... breathtakingly beautiful. ... [But] it's [also] a Rosetta stone for chaos in the real world. ... This is a way to decode the chaos. ... Mankind has been trying to do [this] for thousands of years, and we finally got it, at least for a certain kind of chaos.

To grasp the most surprising result in the entire subject of chaos (coming in Lecture 11), we first need to deepen our investigation of the logistic map. As we saw in the last lecture, the logistic map is a simple mathematical model with very complicated dynamics. Originally a model of the year-to-year fluctuations in an insect population, it's now a paradigm of chaos more generally. We found we could make this system behave in wildly different ways—from maintaining a stable equilibrium level, to oscillating in boom-and-bust cycles, and finally to bouncing around chaotically—just by tuning a single parameter.

Our goal in this lecture is to display all this fantastic richness (and more) in a single image known as the "orbit diagram." It's like a family portrait for the logistic map. But instead of a lot of cousins standing side by side, this family portrait shows all the different things the logistic map can do as we vary its growth rate. As mentioned, we want to study the orbit diagram not only because it's so beautiful, but also because it's a Rosetta stone for chaos in the real world, as we'll see in the next two lectures. By deciphering its secrets, we'll be able to make a series of powerful predictions about the route to chaos in everything from electronic circuits to dripping faucets.

So let's see how the orbit diagram is constructed. You can continue to think of it as a family portrait of mathematical cousins standing side by side, or it may be better to think of it as a series of vertical slices through some complicated object (like a CAT scan). By stacking the slices together, we reconstruct the whole picture. To create each slice, choose the associated value of the growth rate and then iterate the logistic map thousands of times with a computer. We're trying to see how the population is eventually going to behave. Since we only care about the long run, let's disregard what happens early on, while the system is still settling down to its ultimate behavior. After waiting long enough, we plot all the different values of the population that occur. This amounts to plotting the attractor for this slice of the diagram.

It turns out that the diagram divides naturally into two parts: an orderly part and a chaotic part. Let's begin with the simpler, orderly part. This part of the diagram looks like a tree. A trunk rises up from the ground and splits into two branches, each of which splits into two more, and so on, ad infinitum. The trunk signifies a population at equilibrium. As we increase the steepness parameter, the trunk splits (bifurcates) into a 2-year cycle of high and low populations—recall, we saw this kind of cycle in Lecture 9. Subsequent splittings represent cycles of length 4, 8, 16, etc. The period of the cycle doubles at each bifurcation.

There are two striking features of the tree. First, it looks self-similar; each portion of the tree resembles the whole. This self-similarity is one of the defining features of a *fractal* (to be discussed in Lectures 13 and 14). Additionally, the twigs of the tree shrink with each branching. As the structure becomes infinitely bifurcated, it seems to accumulate at an impenetrable wall, sometimes called the *accumulation point*. When theoretical biologist Robert May first encountered this amazing image, he was so bewildered that he left a note on a corridor blackboard for his graduate students, asking them, "What the Christ happens beyond the accumulation point?"

The whole diagram repeats in miniature, infinitely often. And … each of those mini-diagrams contains its *own* windows, with their own *mini*-mini-diagrams, on and on, like a hall of mirrors. We see all this from the simplest nonlinear system.

The answer is: chaos. But as you should expect by now, it's not formless—there's incredible order intermingled with it. Smooth, dark tracks run through the chaos. Vertical windows interrupt the chaotic scatter. Each harbors a cycle of some length. For example, there is a big window containing period-3 cycles and a smaller one containing period-5 cycles. Most astonishing of all, each window ends with its own little copies of the entire orbit diagram. The whole diagram repeats in miniature, infinitely often. And of course, each of those mini-diagrams contains its *own* windows, with their own *mini*-mini-diagrams, on and on, like a hall of mirrors. We see all this from the simplest nonlinear system. Just think what nature is capable of!

In the next lecture, we'll see there are further patterns in this picture—very precise numerical patterns. What's important about them is their universality.

The same patterns were found to occur in other maps and in differential equations, suggesting that they might also occur in real-world phenomena. In this way, the logistic map becomes more than a computer game, and the orbit diagram, more than a pretty picture. Together they provide a raft of scientific predictions, with great power for making sense of certain forms of chaos in the natural world. ■

Essential Reading

Gleick, *Chaos*, 59–80.

Stewart, *Does God Play Dice?* 154–64.

Supplementary Reading

Peak and Frame, *Chaos Under Control*, 169–76.

Strogatz, *Nonlinear Dynamics and Chaos*, sec. 10.2.

Internet Resource

Explore the orbit diagram yourself. A very nice Java applet is here: http://math.bu.edu/DYSYS/applets/bif-dgm/Logistic.html. Try to find the mini-orbit diagrams at the end of a big periodic window.

1. In what ways is the orbit diagram simple? In what ways is it complex?

2. Discuss more generally what it means for a mathematical or scientific object to be simple or complex.

Displaying How Systems Turn Chaotic
Lecture 10—Transcript

This is an exciting moment in our course because we're now on the verge of being able to understand the most surprising result in the entire subject of chaos, coming next in Lecture Eleven. But first, we need to deepen our understanding of the logistic map. Remember from the last lecture, the logistic map is an example of a simple mathematical model with very complicated dynamics. Originally a model of the year-to-year fluctuations in an insect population, we now view it as a paradigm of chaos more generally. Remember what we found: that we could make this system behavior in wildly different ways, from maintaining a stable equilibrium level, with the insects not changing their population from year to year, to oscillating in boom-and-bust cycles. They could be every 2 years. They could be every 4 years. In fact, the period was doubling: 2, 4, 8. In fact, we could make it double all the way up to infinity, and then just beyond that, it seemed to bounce around chaotically, so we could tune the system to go from order, from pure equilibrium, all the way up to chaos just by tuning a single knob, a single parameter, which was the intrinsic growth rate of the insects.

Let me remind you what the simulations looked like when we graphed the population as a function of time. Just to illustrate again what I've just said in words, here is a case where the population, shown vertically, is decaying as a function of time. These poor insects are dying out. They're not sufficiently fertile, or they're too overcrowded. As we start increasing their intrinsic growth rate, the curve changes its shape but not qualitatively. It still ends up dying out. When we cross this number right here, 1, something happens. The curve starts to lift off the axis, and now we have an equilibrium state. We can keep going; the equilibrium rises, rises, rises. No bifurcations yet. You'll see some preliminary oscillations starting to form, but it just means a damped oscillatory approach to equilibrium. It doesn't count as a bifurcation yet. A real dramatic change occurs when we cross this number. Now, the population alternates every 2 years. Now, the bifurcations start coming faster and faster. That period-2 cycle soon becomes a period-4. Keep your eye on it. Watch. There it is, period-4. It's now repeating every 4 years, very low, high, not so low, very high, very low. These four numbers are the same, so every 4 years we have a repetition, and if we go a little bit farther, we can make it repeat every 8, somewhere around here. You've got to have a very good hand because, as I say, the bifurcations come faster and faster, so as you're turning the knob, you might suddenly lapse into chaos. That's one of the things we're going to show very beautifully and graphically today, how we approach chaos and then fall into it.

Let me just tease you for one second with what chaos looks like—something like that. That is not repeating. It's all over the place, and I got that just by cranking up the knob a little bit more. The growth rate is now up to this number, 3.66.

Our goal in this lecture is a simple one: I want to show you a diagram that's going to blow your mind. That's it. I want to display all of this fantastic richness of dynamical behavior and more in a single image, known as the "orbit diagram." It's kind of like a family portrait for the logistic map. Think of each of the different choices of growth rate as being like a cousin. They are all logistic maps. They're related by being logistic maps, but they differ a little bit in their growth rate. So they're kind of like cousins, and the orbit diagram is like—we've had a cousins' club, and all the cousins are standing side by side for the family portrait. You can see what they all look like. They clearly resemble each other, but they're a little bit different. So we're going to do a mathematical family portrait of this cousins' club, showing all the information side by side, how the logistic map changes as we vary its parameters.

It's called the orbit diagram because it shows how the orbits of the logistic map behave in the long run as we vary the growth rate. An *orbit* is just another word for a trajectory, meaning how a dynamical system changes through its state space. I've been using the word *trajectory* throughout the course, but it's traditional to speak of this as an orbit diagram, so I will do that.

In part, we want to study the orbit diagram because it's just breathtakingly beautiful. But it's much more than that. It's a Rosetta stone for chaos in the real world. It is. This is a way to decode the chaos. This is what mankind has been trying to do for thousands of years, and we finally got it, at least for a certain kind of chaos, as we'll see in the next two lectures.

By deciphering its secrets, we will be able to make a series of powerful predictions about the route to chaos in everything from electronic circuits, with strange, fitful voltage oscillations; to heart cells that aren't beating in the normal way, that start to beat irregularly; to even something as mundane as a leaky faucet that instead of pouring out water in a continuous way has an annoying, chaotic drip, drip, drip, drip, drip, and so on. All of those things are going to be governed by the same law. It's unbelievable but true.

Let us see how this wondrous diagram, the orbit diagram, is constructed. I've given you this homey metaphor of a family portrait of mathematical cousins standing side by side. That may not be absolutely the best way to think about it to understand it. It may be better to think of it as a series of vertical slices through some sort of complicated object. It's kind of like, when you have a CAT scan done, your body is shown as a stack of vertical slices or other slices. So here, we're going to take a cut through the orbit diagram. It's almost kind of a living thing. We're going to chop it and look at it slice by slice and put the slices together to see what we've got. Each slice corresponds to a choice of growth rate. That's what I meant. Remember I said it was cousins distinguished by their growth rates? We're going to see all the cousins next to each other, all the growth rates, and what happens at each growth rate side by side. There's a lot to look at here, and as I say, your mind will be blown, so I want you to prepare yourself.

Let me try to orient you for what's about to come. First, in the simulations you're about to see, several things will happen at the same time. I'm going to go to that simulation now, and I'll try to be as deliberate as I can so you can follow this. I've got two panels here. On the left, nothing very mysterious. This is going to be like the sorts of graphs I've shown you where I show the population of the insects as a function of time, just a normal chart showing a variable bouncing up and down or going to equilibrium as a function of time. That will be the usual time series picture. As you can see, population [is] on this axis; time, on that axis. I haven't done anything yet, of course. I'm going to, but we're just getting ourselves prepared.

What's on the right? The right will be the orbit diagram, slice by slice. I'm not going to show you the whole orbit diagram just yet. That would be too blinding. That would be almost like Plato's poor people in the cave.

You don't take them out into the light right away. It would be too much to handle. So what you will see are separate slices to get you a feel for how this orbit diagram will be built up. Here's my slicer. I can move along here, and it creates a line. Let me say I choose this particular vertical line. That's a choice of a growth rate. See this axis, "rate of growth." Those of you who are trying to follow this with algebra, you may remember we called the growth rate r when I wrote down an algebraic description of the logistic map as a parabola, $rx(1 - x)$. So mathematically, I'm varying the number r, but for the rest of you non-algebraic folks, don't worry about that; it's just a growth rate.

So I choose a growth rate; fix it. Now what I'm going to do is click the mouse, and on the left side of the screen, you will see the prediction for what the population will do as a function of time when that growth rate is chosen. It will be fixed. So let's do that first experiment. I've chosen a small growth rate. We know what will happen; the population will die out. Let's watch it. What you see, now, are three things. Over here, the population as a function of time [is] not looking good for the insects. They're dead. After just 15 or 20 generations, they've gone extinct. The new thing to be paying attention to is—well, of course, there's this stream of numbers showing a lot of zeros. That's the long-term behavior of the insects—a whole lot of zeros as a function of time.

But the more important thing is what you're seeing in the second panel, which is that there is a yellow dot right there. That dot is encapsulating all the information of interest from the picture on the left. It's just showing the eventual behavior, which was zero. So that's the thing. We're going to boil down this whole time series. Rather than showing all of it, we're just going to show the attractor, the ultimate behavior. That's what we care about.

Let me show you some easy cases. If I have an even lower growth rate, it dies out faster—a whole lot of zeros, another yellow dot.

We saw in some earlier simulations that if we cross the magic number 1, the population can come to equilibrium. Let's click a number greater than 1. Population has come to equilibrium. At some level, that turns out to be .187. By the way, maybe I should say, what does .187 of an insect population mean? I haven't discussed this, and really, we don't have to think of this as population unless it helps you. We can. It's just an abstract math problem, really. But if you want to insist on the population interpretation where this originally came from, think of the field of crops that these insects are eating as being able to support a certain carrying capacity of insects, like maybe it's a million insects for that acre or something. We don't want to use the number a million because it's confusing, so we're going to be speaking in terms of—what percentage of the carrying capacity does the system attain? So .187 would mean 18% of the carrying capacity, whatever that is, 18% of a million. So the numbers will always be between 0 and 1, but that's what it means. It's as a fraction of the carrying capacity.

Okay, that was an aside. Let's get back to constructing our orbit diagram. What we saw here was that this population came to equilibrium when the growth rate was chosen in this slice, and accordingly, we plotted a yellow dot at the height .187 because that's the asymptotic level that was reached.

Let me go through and do more slices, and I'll start building up the orbit diagram by reconstructing my CAT scan. A bifurcation has occurred. Notice the zero has now come off the axis. You can see, I can do more and more, and there appears to be some kind of pattern emerging. The equilibrium level is rising as I increase the growth rate. Remember, we once said that as we approach the number 3—do you remember something happened at 3? At 3, the population alternated every 2 years. So let's see that. I'm just to the left of 3. There's a population that wants to alternate every 2 years but can't quite make it and ends up pooping out to equilibrium. But on the other side of 3, watch this. That population alternates every 2 years, as you can see, up, down, up, down, boom and bust, and I encapsulate all that information by just showing those two population levels, the two boom-and-bust levels, which are shown as these numbers here. A year when we had 7, 8, 7 was followed by a year with 5, 2, 9, which is followed by a 7, 8, 7, which is followed by a 5, 2, 9, and so on. That's what these two numbers are. Here is the 5, 2, 9, and here is the 7, 8, 7.

What we're going to do is the proverbial connecting the dots. If I click this button called "orbit diagram," I will show you the answer. Well, a piece of the answer, which is what happens when the dots are connected. There it goes. So the dots have been connected, and you see that there's an interesting treelike structure being formed here. It turns out that the orbit diagram has two paths with very different qualitative behavior in them. There's an orderly half before we go into chaos, and then there is a chaotic half. Let's begin with this simpler orderly half, the orderly part, which is commonly known as the *fig tree*. In part, it's called a fig tree because it looks like a tree, and in part, it's an attempt at mathematical humor because one of the great scientists who worked on this was named Feigenbaum, which in German means "fig tree."

Let us focus on the fig tree for now. What you see is a trunk rising up from the ground. It's a bit of a strange tree because it's going sideways, but don't worry about that. So it goes up from the ground. It bifurcates; that is, it splits into these two branches. And then notice, those two branches each split themselves. You've got to be good to see it. Can you see? I'm sorry about the vertical line there in your eye, but watch. As I go through with my CAT scanner, here I have two branches. Each of them splits right about here. Now, what would happen if I were in there? Well, of course, that's period-4. Remember, we saw that we go from period-2 to period-4. There are period doublings happening faster and faster. I needed very good hand control to see all of these. Now you're seeing numerically that period-4—that 8, 7, 9 is followed by 3, 7, 3, then 8, 2, 3, then 5, 1, 2, and then back to 8, 7, 9. This machine is spitting out repeating numbers every 4 years, and the population, just viewed in the commonsense way, is fluctuating every 4 years, repeating every 4.

As we increase this growth rate, what we're seeing is the trunk splitting into a 2-year cycle of high and low. We saw that in Lecture Nine. It's then showing subsequent splitting in cycles of length 4, then 8. I could even try to show you 8. Let me do that just to indulge myself. You might enjoy it. I have to go farther and zoom in a little bit on the diagram. Now I'm only showing growth rates between 3 and 3.57 or so. I think if I choose a number—well, I'm not exactly sure. Let me cheat by drawing the orbit diagram to start with. There it is. These are the period-2s. There are the period-4s. The claim is that if I'm in here, I should see period-8. Let's see if it's true. That's period-8. If I carefully look at this, you will see that there are 8 lines here, and simply, if you just look at the numbers, you can see they're repeating every 8 steps.

What happens on the orderly side of the diagram is what is known as a *period-doubling cascade*. We go through from repeating every 1 to every 2, every 4, every 8, 16, and so on, all the way up to infinity. And all of those numbers are converging to some limiting number in growth rate, somewhere around here at about 3.57.

There are a couple of things about this tree that I want to focus on. I just want to focus your attention on it. It looks self-similar. That is, each piece of the tree—you'll see two branches, which then split to make two more branches that look like the ones you just saw, except smaller. That self-similarity is one of the defining features of a class of objects that we're going to be discussing soon, called *fractals*. In Lectures Thirteen and Fourteen, we'll start focusing on fractals. But we're not ready to do that yet.

Let me, though, show you a whole different diagram, which is to illustrate this fig tree in a bit more detail. There it is. There's the whole fig tree. Well, again, just the orderly part of the fig tree, the fig tree referring to the orderly part of the orbit diagram. What I can do now is choose a piece of the tree and zoom in on it and blow it up to be full screen so that you can see the self-similarity. Imagine a microscope, and you're going to look [into it]. We see that this pair of branches is wider than this branch here, which is wider still than this one, kind of jammed up here. If I zoom in by drawing a little box, I say to myself, "What is in there? I am curious about that." Through the magic of computers, that box is now going to blown up to the whole screen. Keep in mind what you're currently looking at, and compare it to what you're about to see.

That sort of looks the same, right, qualitatively? So as the tree bifurcates, it creates itself again in replica but smaller. As the structure becomes infinitely bifurcated, right here at this edge, it seems to accumulate at a kind of impenetrable wall. I don't know if you can see this, but there are numbers down here. It says 3.57. There's a kind of wall there at a growth rate of 3.57, sometimes called the *accumulation point*. Naturally, you're wondering: What lies beyond the accumulation point?

When this problem was being studied in the 1970s, a great theoretical biologist named Robert May asked and wondered the same thing. When he encountered this amazing image in his research, he was so bewildered by this idea of an infinitely bifurcating tree that converged to an accumulation point that he left a note, in the corridor, on a blackboard, for his graduate students, begging for help, asking them to think about it. Pardon me for what I'm about to say. I'm quoting Robert May. He's from Australia, and maybe this is an expression used there. I can't say I've ever heard it, but the note he left on the board said, "What the Christ happens beyond the accumulation point?" I'm almost tempted to try my Australian accent on that one.

Anyway, Bob May was mystified, and if he was mystified, there's a real puzzle here. You want to know the answer, but I want to keep you in suspense for just a second because I want to tell you a little bit about Bob May, one of the great pioneers of chaos theory, along with people like Poincaré and Lorenz.

Bob is Australian. He was trained as a theoretical physicist, sort of an outsider coming into biology. He worked and has worked in ecology, in mathematics of infectious diseases, epidemiology, and he has become one of the world's great theoretical biologists. I actually had the privilege of knowing him a little bit when I was an undergraduate at Princeton. He was teaching there at the time. He later moved to Oxford and eventually arose to the position of science advisor to the queen of England, which I think may be a little unusual for an Australian; I'm not sure. In any case, he was science advisor to the queen. He's now the president of the Royal Society. He is now Lord May of Oxford, and I believe he may even be Baron May of Oxford. I'm not sure I know the difference, but it sounds good either way.

The man is small, wiry, a brilliantly witty speaker, an elegant writer, and feisty and competitive like you wouldn't believe. I have played tennis with him. I can tell you firsthand that he doesn't like to lose at anything, and he's a pretty scrappy tennis player.

More seriously, he wrote a very important review article in *Nature* magazine in 1976, which I quoted from in the last lecture, the one that ended with that impassioned plea that it's time to wake up and start studying nonlinear systems. That article brought nonlinearity and the logistic map to the widest possible audience of scientists. *Nature* magazine is read by everybody in every field, so when Bob May, one of the best writers on the planet, wrote this article saying, "Wake up," that was an alarm to the whole science community.

Back to our question: What happens beyond the accumulation point? The answer is chaos. But as you should expect by now, chaos is subtle. It's not formless, like the ancients thought. There is incredible order intermingled with it. Let me show you the entire orbit diagram, not just the orderly fig tree part, but the whole thing. Wow. That is the orbit diagram when we go all the way from 3 to 4 in growth rate. Let me get you oriented again. On the left, starting about here and moving back, that's the fig tree. We've already discussed that. You see the period-2 splitting into period-4, splitting into period-8, and then there is some kind of explosion of yellow dots occurring about here. That is the accumulation point. That's where order makes the transition into chaos.

This is the unbelievable thing. Look at that yellow mess. It's not a total mess. You see all kinds of structures lurking in it. First of all, there are smooth, bright tracks running through it. Here's one. Or maybe your eye can make out another one, like this. There is some kind of smooth curve running through the chaos. What is that doing there? Maybe I should just emphasize—why does a yellow mess means chaos? What this would mean is, at a particular choice of growth rate, the outcome could be any of these yellow dots in the long run. So you can be all over the place. That's what I mean by it being chaotic, and if we go back to our time series program, I can show you it's truly chaos. But let me not leave this picture because you're probably saying, "Why isn't he talking about the black stripes?" There are very, very conspicuous vertical black stripes in this picture, of different thicknesses, and they're interrupting the chaotic scatter.

Let me focus on those black stripes because what you're going to find inside them will expand your mind. Here's a very big one, a big, wide black stripe. To see what's in there, I can draw one of those stretchy boxes around it that will blow it up to the whole picture. Let me do that now. What is in this black box, this black stripe? Ready? What are we seeing here? Notice these three lines: 1, 2, and 3. I hope by now you understand how to interpret a picture like this. That means there is a period-3 orbit, something repeating every 3 steps. So this black stripe is called the *period-3 window*.

But even more astonishing, and maybe you can see that there is something sticking off of the period-3 orbit. Do you see what this is? It ends with its own little copy of the orbit diagram. The whole orbit diagram is right there at the end of the window. Let me convince you of that by blowing this up. Does that remind you of the original orbit diagram? It's right there, except it's not the original because it exists in this window of parameters between 3.841 and 3.857. It is a tiny diagram right there inside the bigger orbit diagram.

Are you ready for more? Do you see what's happening here, that it has its own period-3 window? If we zoom in on that, you know what's going to happen, don't you? That shows its own little period-3, which is now period-9, by the way, because we've only looked at a third of the original period-3, and that thing ends with its own orbit diagram. So this is a really breathtaking thing. This is fractal structure. The whole diagram repeats in miniature infinitely often. Of course, as I've said, each of these little mini-windows has its little *mini*-mini-diagram, like a hall of mirrors going down to infinity.

Let's take a breath and make sure we understand what this orbit diagram means. Let me reset to go back. I think you get it, but just in case. For instance, let me tune my growth rate to something like 3.69. Why have I chosen that? Because at a growth rate around here, I just see a yellow mess. So let's make sure that we really understand what happens at 3.69. You should see chaotic behavior if I go back to the time series program. Let me do that. I'm going to dial in 3.69. I'll do it fast because you can handle it, 3.69. That's pretty chaotic, as advertised. It works.

Let's move on, and now let me look inside one of the periodic windows to see if we really get what that is. There's one, for example, around 3.74. I'll show you that one. Let me go to my big orbit diagram at 3.74, which is right about here. Can you see there's a tiny black stripe? What is that tiny black stripe? If I zoom in on it to see what it is, it's a period-5: 1, 2, 3, 4, 5. I won't remark on these little orbit diagrams at the end of the window. They're there at the end of every window.

Let me go to 3.74 to confirm that it really is period-5. Zooming to 3.74. Do you see period-5? Look. It repeats exactly every 5. From here, I go up 1, 2, 3, 4, 5. Unbelievable. So what this means is that if I simulate the logistic map, as I just did, I will get something repeating every 5 steps. It works again.

Moving along, then, what I mean when I say that the orbit diagram is like a Rosetta stone is just exactly this: that by looking at the orbit diagram, I can predict all the periodicities I see at any growth rate. I can predict where chaos will occur. It shows everything a logistic map can do, all its cycles and chaos, in one amazingly beautiful image, all this from the simplest nonlinear system. Keep in mind, this is punching a button on a calculator for a simple parabolic, nonlinear map. So just think what nature itself is capable of.

I want to end this lecture with something dear to my heart, which is a kind of Torah of chaos, if you'll forgive me. It's a scroll showing the orbit diagram printed out on a printer with I don't know how many millions of dots. Can I ask for a volunteer to help with this? Thank you. This was printed overnight on a printer. These black dots show the chaos. Let me just point out a few things here. Here's the region of periodicity. This is where the period doubling has occurred. In here, we see the black just first beginning. This is the accumulation point. This is what Bob May wondered, what the Christ happens beyond it? Beyond it, we see blackness interrupted by all sorts of windows. For instance, here's a big, bright, open window. This is, it turns out, a period-6 window. The period-5 window that we looked at earlier is here, and then a big one, if I dare go, there's a period-3 window over here with its own little mini-orbit diagram in it.

In the next lecture, we'll see that there are further patterns in this picture, very precise numerical patterns. What's important about them is their universality. The same patterns were found to occur in other maps, not just the logistic map. As long as they were hump-shaped, they could have any algebraic form, and you saw the same qualitative picture. That led people to wonder, "Wait a second; if these can occur in any map with the right shape, maybe they can occur in differential equations, too. Maybe there are all these same phenomena, the period-doubling route to chaos." And that was a thrilling thought because if this would occur in differential equations, then it would occur in nature because we know that the differential equations way of looking at things is the correct physical description of nature.

So in that way, the logistic map will become more than a computer game, and the orbit diagram, more than a pretty picture. Together, they provide a raft of scientific predictions, with astonishing power for making sense of certain forms of chaos in the natural world. Next time, we'll see how all that led to chaos theory's greatest triumph.

Universal Features of the Route to Chaos

Lecture 11

[4.6692… and 2.5029]… are to chaos what pi (π) is to circles and periodicity. They are new fundamental constants of mathematics. They are the fundamental constants of chaos. … They're [also] … fundamental constants of nature because chaos is in nature, and this description we're giving, as we will see in the next lecture, applies to nature.

W e're now ready to behold the most incredible and disconcerting result in the whole subject of chaos. We'll see that what happens in the logistic map (from Lectures 9 and 10) happens *universally* in a very wide class of systems. In fact, the logistic map displays such universal features en route to chaos that we have good reason to expect they must also occur in nature. Our goal is to understand exactly what we mean by *universality*, why it matters, and why its discovery was the most stunning breakthrough in the new science of chaos.

Let's begin by explaining the feature known as the U-sequence ("U" for "universal"). Take a look at the orbit diagram (from Lecture 10). Focus on the prominent vertical stripes. These are the "periodic windows" for the logistic map. They represent parameter values where the system behaves periodically, not chaotically. In other words, the map generates numbers that repeat after every 3, 4, 5, or some other whole number of iterations. This kind of attractor is called a *stable cycle*. The biggest window is the one for period-3 cycles (meaning those cycles that repeat every 3 steps). To the left of it is the window for period-5 cycles, and even farther left, period-6.

The sequence seems strange: 6, 5, 3. What's the pattern? Let's step back a bit and look at *all* the stable cycles, including those in the tree portion of the diagram. Since there are infinitely many stable cycles, let's only look at the ones with short periods, say period 6 or smaller. Then, starting from the left of the diagram and moving to the right, the periods of the stable cycles appear in the following sequence: 1, 2, 4, 6, 5, 3, 6, 5, 6, 4, 6, 5, 6. (Most of the windows after period 3 are invisible at this scale, but they are there.)

Now for the shocker: This same weird sequence of numbers occurs for *any* iterated map, as long as it is hump-shaped like the logistic map. What's so amazing about this is that only the rough shape of the map matters. Its precise algebraic formula is totally irrelevant—even though changing the formula would certainly alter all the numbers that come out of it. Somehow, the periodic windows don't care about those details; geometry trumps algebra here.

Now we'll look at an even more incredible kind of universality, one that *does* involve specific numbers. (That's a selling point, because scientists regard numbers as strong predictors.) Focus on the tree part of the orbit diagram. Look at successive wishbones in the picture and measure their heights and widths. The wishbones shrink in both directions as we approach the onset of chaos. What's the rule governing how *quickly* they shrink, in both directions? Each wishbone is about 4.7 times smaller (measured sideways) and 2.5 times smaller (measured vertically) than the one to the left of it. Roughly the same shrinkage factor applies all along the tree. As you approach the onset of chaos, the shrinkage factors converge to very specific, very peculiar numbers: 4.6692… and 2.5029… .

American theoretical physicist, Mitchell Feigenbaum (b. 1944) used a pocket calculator to uncover a stunning kind of universality in chaos--that diverse systems go chaotic in the same way.

Here comes the universality. Suppose you redo the measurements for another hump-shaped map whose algebraic form is completely different from the logistic map's (for instance, the hump of a sine wave instead of a parabola). Then the numerical details of the tree will change. The widths of the wishbones will change. But the shrinkage factors again turn out to be 4.7 horizontally and 2.5 vertically. And near the onset of chaos, they converge to 4.6692… and 2.5029…, exactly the numbers seen before!

This quantitative form of universality was discovered by physicist Mitchell Feigenbaum (b. 1944), a colorful character. Currently a Professor of Mathematics and Physics at Rockefeller University, he has intense eyes, a mischievous grin, and the swept-back hair of a Romantic composer. His seminal discoveries came in 1975 while he was working at Los Alamos in the Theoretical Division. His working habits were odd, even for T-Division: He roamed the streets in the middle of the night; lived on cigarettes and soda; and kept to a 26-hour day, his sleep-wake schedule rolling in and out of phase with the rest of society every few weeks. His discoveries about universality were published in 1978, but only after several journals rejected them.

> **The implications of Feigenbaum's ideas are shocking. … The logistic map has no scientific content, no laws of nature built into it. It seems like pure numerology.**

Feigenbaum predicted that the universal features he'd discovered in iterated maps must also occur in nature. After all, he'd found the same patterns in differential equations. And since the laws of nature are expressed by differential equations, the universal numbers 4.669… and 2.502… should appear in real-world chaos too. The only caveat was that the system had to arrive at chaos via a cascade of period-doubling bifurcations, just as in the logistic map. Otherwise all bets were off.

The implications of Feigenbaum's ideas are shocking. What makes them so unbelievable is that the logistic map has no scientific content, no laws of nature built into it. It seems like pure numerology. This breaks the rules of how to do science. Even quantum mechanics and relativity had not dispensed with the idea of basing predictions on fundamental laws of nature. And yet, miraculously, subsequent experiments on a variety of real systems conformed to Feigenbaum's predictions (which we will see in Lecture 12). It was as if the old Pythagorean dream had come true. The universe is not made of earth, air, fire, or water—it is made of number.

Another startling quality of Feigenbaum's work was its cross-disciplinarity. He'd applied a Nobel Prize-winning method in condensed-matter physics to a math problem a child could play with on a calculator—and from this came predictions about the route to chaos in heart cells, electronic circuits, and convecting fluids! ∎

Essential Reading

Gleick, *Chaos*, 157–87.

Stewart, *Does God Play Dice?* chap. 10.

Supplementary Reading

Cvitanovic, *Universality in Chaos*.

Strogatz, *Nonlinear Dynamics and Chaos*, sec. 10.6.

Questions to Consider

1. Does the universality of Feigenbaum's predictions make them powerful (because they apply so widely) or weak (because they are so generic and unspecific)?

2. After Feigenbaum achieved his breakthroughs in the late 1970s, he was rarely heard from again and has published relatively little. Why might this be? Enlarge the discussion to include other creative individuals who did the same thing, from scientists to writers, artists, musicians, filmmakers, and inventors.

Universal Features of the Route to Chaos
Lecture 11—Transcript

We're now ready to behold the most incredible and disconcerting result in the whole subject of chaos theory. Let's summarize where we are in our detective story. In the last two lectures, we've been discussing the logistic map. Originally, it was a simple model of population dynamics for insects, but now the logistic map is viewed more broadly as a paradigm for all the amazing things that can happen when we allow ourselves to think about nonlinear systems—nonlinearity being the most important and yet least understood part of the entire landscape of science. It's a place where scientists long feared to tread because, as the old cartographers used to say, "Here be dragons." This is the unexplored part of the scientific world.

We have seen, in the case of the logistic mapping, that its nonlinear shape— remember, it had a hump shape, a parabolic shape—gave it marvelously rich dynamics. Depending on the intrinsic growth rate of the insects, the population could go extinct if the growth rate was too low; or it could build up and then reach a stable equilibrium level; or show boom-and-bust cycles, alternating every 2 years, or 4, or 8, or 16—this incredible period-doubling cascade as we increase the growth rate. We saw this system going to 16, 32, and so on, up to periods of infinite length, essentially, all accumulating at a place that we called the "accumulation point" at the onset of chaos. Just beyond that, the dynamics could be chaotic, and as we increase the growth rate still further, the chaos was suddenly interrupted by new periodicity, periodic windows that appeared as vertical stripes in the orbit diagram, for every possible period we could think of. It's an incredibly rich story, and you're probably thinking, as you should be, "That logistic map is pretty special." But that is the surprise. It's not so special—just the opposite. In many ways, what happens in the logistic map happens *universally* in a very wide class of systems. The logistic map displays such universal features en route to chaos that we have good reason to expect they must also occur in nature, in reality.

Our goal in this lecture is to understand exactly what do we mean by *universality*, why would it matter, and why its discovery was *the* most stunning breakthrough in the new science of chaos, in my opinion anyway.

Let us begin by explaining the feature known as the U-sequence, "U" for "universal." This comes from work that was done at Los Alamos by three scientists named Metropolis, Stein, and Stein. The U-sequence appears in the orbit diagram that we've looked at already in Lecture Ten. Let me remind you what that orbit diagram was. Remember that fantastic icon of chaos? We had our fig tree. These correspond to period-2 cycles; I increased the growth rate; I see a splitting, a bifurcation, into period-4, period-8, then this burst of yellow chaos interrupted by these black "periodic windows," the thickest one being here, representing a period-3 behavior. Remember, those windows represent places where the parameter values, where the system behaves periodically, repeating, not chaotically. In other words, the logistic map generates numbers that repeat. It could be 3, or 4, or 5, or some other whole number of iterations. This kind of attractor, then, of course, is called a *stable cycle* because no matter how we start, the system will be attracted to that kind of behavior if we're in the appropriate window.

The biggest window is the one that I'm indicating right here, very wide, for period-3, meaning cycles that repeat every 3 years, if we're thinking of the insects, or more abstractly, every 3 steps. The U-sequence has to do with the order in which these periodic windows occur. That's what I want to focus on now.

So let's look. Here is period-3. It's big. See if you can see any other big stripes. Your eye is probably noticing; here's one, and there's one there. The question is, what are the periods corresponding to those? Remember, we can see this more clearly by zooming in, so let's first look at this stripe, this periodic window. What is going on in there? When I blow it up, I see 5 spindly little lines here, which I hope now you remember mean periodicity. These are 5 population levels that will repeat. So this is a period-5. There are 5 of them: 1, 2, 3, 4, 5. In fact, we studied this example because if you can read at the bottom of the screen, you'll see that the growth rate is 3.74. See down here? 3.74. We actually explored that in the previous lecture, and we confirmed that it really does give period-5 behavior.

So let's go back to the original orbit diagram. What we just showed was next to the big period-3 window, there was a period-5 window. Next, there's this conspicuous one. What is that? Let's zoom in on that. I'm going to be doing a little numerology here. Just bear with me. I'm trying to lay down the numbers that occur—3, 5. Maybe you're thinking of a pattern. Are you guessing that this is 7? A reasonable guess. It's not; it's 6. Look at those numbers—1, 2, 3, 4, 5, 6, and you can look hard, [but] there's no seventh point. It's not a period-7; it's a period-6. That is a strange thing. The sequence is 3, 5, 6. What's the pattern there?

Let's step back a bit and look at all the stable cycles to try to see the larger scheme of things, including those in the tree portion of the diagram. I'm going to pan back again. Here I am. Remember, this is the so-called fig tree. I see in here period-2. Period-2 is next to period-4, which is next to period-8, and so on. That part seems understandable. That's just period doubling. We understand that. Yet that 2, 4, and 8, those are to the left of the 6, 5, 3 that we already look at, so we're trying to make sense of this. That's the point here.

We want to understand, what is the order in which these periods occur? Two is before 4, is before 8, is before 6, is before 5, is before 3. That's mysterious. In fact, if I were listing all the numbers of periods going up to 6, let me make an exhaustive list for the whole picture, and I won't count anything with a period longer than 6. Let's just focus on the short cycles of length 6 or less. Here's what I claim: that they occur in the order—get ready—1, 2, 4, 6, 5, 3, 6, 5, 6. I feel like I'm Rain Man here. Now I've lost my place. I think I said 6, 5, 6, 4, 6, 5, 6. It's hard to figure out what that pattern is, but that is the pattern. Many of those you can't really see. That is, some of them we have mentioned. One is off the picture. That would just be something repeating every year. That was equilibrium, and we're not showing it in this graph, but that was the branch that's not shown that bifurcated to the period-2. So there was 1, 2, 4, 8; this one we said was 6; here's 5; here's 3. So far, I've mentioned the beginning of the U-sequence, but then I said that strange group of numbers, 6, 5, 6, 4, 6, 5, 6—where the heck are those? Looking at this, you probably can't see any windows after period-3. That's reasonable because they're very small, but they're there.

To convince you of that, let me just blow up what looks like a featureless mess. Let's see if we see any windows. I guarantee you that we will. Look at that. They're in there. So what's this one? That's one of the period-5s that I advertised. We're going back again like prospectors looking for gold. We start panning around in here. Have I looked at this one yet? I don't think so. What's that? That's the period-4 I mentioned. They're all in there; they're very thin; you can't see them. Look at the numbers involved. This is for a growth rate between 3.958 and 3.966. That's fine. They're very close together, but there is a period-4 sitting in there. So we've got all these mysterious numbers. Let me give it to you again. The sequence was 1, 2, 4, 6, 5, 3, 6, 5, 6, 4, 6, 5, 6. That could be a good number to use on your cell phone or a good password. I don't recommend it, actually, because it's a well-known sequence. It's called the *universal sequence*, the U-sequence.

That was the U-sequence up to period-6. In fact, I could have included period-7 and higher, but I'm not going to. You've probably had enough.

Here comes the shocker. What is the point of all these weird windows in this weird order? This same weird sequence of numbers occurs for any iterated map, as long as it has the general shape that the logistic map has, having a hump with a rounded top. That's the thing that I'm trying to club you over the head with. It's amazing. For instance, let me show you a head-to-head comparison of the orbit diagrams for the logistic map that we've been studying and a different map, which also has a rounded top. The way I form that other map is that I take a sine wave, which you will remember from high school trigonometry. Remember, a sine wave looks like this. Let me just take one hump of it, just one arc of that sine wave, and that mathematically is a totally different function, that trigonometric function, totally different from the parabola we've been studying. The parabola uses only x and x^2; it's quadratic. The sine wave is not expressible in terms of x and x^2. If you know about infinite series, you could write it that way. But anyway, the sine function has no relation to the parabola except that it has the same general shape if I only look at one hump of it.

Now let me show you the orbit diagram for both of them. What I'm showing you, then, are orbit diagrams compared head to head for what I'm calling the sine map and the logistic map. Look at them carefully. Are they absolutely identical? You can see that they are not absolutely identical. Look at some differences. For instance, everything is scaled so that this is the fairest possible comparison. I'm not doing some kind of distortion to try to trick you. This branching where the period-1 splits into period-2 occurs numerically earlier than it does down here. That is, this one happens later. This period-2 going to period-4 happens much earlier above than it does here. You would have to go this far in the logistic. The sine map seems to be doing everything sooner. Also, look at its period-3 window, the big white stripe. The period-3 window for the sine map seems to be earlier and much wider than it is for the logistic map. All of these things are quantitative differences, numerical differences—the various windows and the branchings are shifted left to right; things are widened. But those are quantitative differences, not qualitative ones. Those are issues of numerical detail.

What is astounding, and maybe what was your first impression in looking at these pictures, is how similar the two diagrams look in terms of their overall architecture. Their layout is totally identical. That's blindingly obvious, I hope. Where this one has a period-3, you can find a period-3 for that one. They have the same period-doubling route to chaos. They have the same pattern of bands. That's my point: that qualitatively, they are identical. All that matters is the hump shape, in particular, the sequence in which the periodic windows occur, like 6 before 5 before 3; that's happening down here. This is 6 before 5 before 3. That is what I meant by the U-sequence. The sequence of periodic windows is absolutely the same for the sine map, for the logistic map, for any map with that same general shape.

Let's pause for a second to scratch our heads and think about that. That's not the math that you ever learned in school. In school, you learned that numbers matter; quantitative differences matter. Yet here, they don't. What's amazing is that only the general shape of the map matters. That seems to determine everything. Any numerical information beyond that is irrelevant. As long as the map has just one rounded hump, the U-sequence can be proven rigorously, mathematically, to follow as a logical necessity. Anything else numerically sort of falls out as chaff, not of interest. Somehow, the periodic windows don't care about the algebraic details of the mapping. It's as if geometry trumps algebra in this case.

That is the first kind of universality that I wanted to emphasize today, and it's often known as *qualitative universality* for reasons that I've suggested, that it's a qualitative feature of orbit diagrams that's the same—quality. But now comes an even more incredible kind of universality. This is what I consider the apex of this early part of chaos theory. It's a kind of universality that is quantitative. Scientists like that because they like numbers; they help them make strong predictions.

Let's focus on the fig tree part of the orbit diagram. Here we go again to the fig tree, and notice what you see as a kind of wishbone-on-top-of-wishbone structure. We're going to look at successive wishbones. Here's one big wishbone, and then each of its branches splits into a tinier wishbone. You can see that.

What we want to do now is compare the widths of the wishbones, one to the next. Notice that the wishbones are getting smaller in both directions as they approach this accumulation point. For instance, this wishbone is sort of roughly this wide compared to the width of the next one, which is that wide. Meanwhile, they are also shrinking in the vertical direction. We had the total width appear like that, compared to a skinnier width on these other wishbones.

We want to see if there is some rule governing how *fast* they shrink, both in the sideways direction and also in the vertical direction. Well, there is a rule, and it is that each wishbone is about 4.7 times smaller, measured sideways, than the wishbone to the left of it. It's about 2.5 times smaller measured in the vertical direction. What is amazing about this is that the same shrinkage factor applies all the way down the tree. Remember, we said that this tree is infinitely bifurcated and approaches an accumulation point. If we try to measure these shrinkage factors in successive wishbones, we'll find that it's always the case that this one is about 4.7 times skinnier in its sideways direction, all the way down.

This might remind you of a game that kids play, where they challenge each other: What happens if you walk halfway to the wall and then halfway to the wall? You'll never get there. That's technically called *Zeno's paradox*. It's true that if you do go halfway to the wall and halfway to the wall, you would approach the wall as the limit of an infinite series. What's happening here is very similar, except that the number involved is not 2. We're not moving halfway to the wall. We would be moving—well, since the number is 4.7 that we keep talking about, it's not quite right to say we're moving $\frac{1}{4.7}$ of the way to the wall. In fact, it's the opposite of that. It would be $1 - \frac{1}{4.7}$. In fact, it amounts to something like $1 - \frac{1}{5}$, okay, rounding off the 4.7 to 5; $1 - \frac{1}{5}$; that's $\frac{4}{5}$. We're moving about $\frac{4}{5}$ of the way to the wall. That's why you see in this orbit diagram that these bifurcations come faster and faster and faster. Remember when I was trying to tune the knob, I had trouble because they come at you so fast.

At the very onset of chaos, right at the accumulation point, those rates, those factors, those geometric ratios in this geometric series, have been measured precisely, and they converge to very specific numbers, very peculiar numbers, which turn out to be not really 4.7, but 4.6692... . In the vertical direction, the number that I had been calling 2.5 converges to some number 2.5029... . There are more digits after that. Those two numbers are to chaos what pi (π) is to circles and periodicity. They are new fundamental constants of mathematics. They are the fundamental constants of chaos. Not just fundamental constants of math; they're going to turn out to be fundamental constants of nature because chaos is in nature, and this description we're giving, as we will see in the next lecture, applies to nature.

Why are there two of them instead of just one, like pi? Of course, because the orbit diagram has two directions, two separate things. It has a parameter direction, the growth rate, and it has a vertical direction showing the population or the "state variable," we could call it. So there are those two things with scaling, different ratios and different directions.

Now, I haven't said what's universal yet. I've only said that there are these peculiar numbers. Those peculiar numbers were known even in the 1960s, but no one made any sense of them. They were just numbers; so what? Who cares about 4.66? It doesn't sound like pi; it's not $\sqrt{2}$; it's no familiar number, so no one knew what to make of it.

Suppose we redo the measurements for some other orbit diagram. Like we've seen, if I go back to my diagram here; remember, I showed you that the sine map has a qualitatively similar orbit diagram to the logistic map—qualitatively, same general layout but with numerical differences. Remember, this branch occurs before that and so on.

Here is the unbelievable thing. What if I measure the ratio of the wishbones for the sine map? As I do the same game I just played, asking how much longer—in fact, I wasn't indicating it as a sine map in what I've been drawing up here, but let me now be pointing at the logistic map. Here are its wishbones. How much longer is that than this? Keep in mind that the two endpoints have changed, right? This endpoint here does not compare well with the one in the sine map, nor does this endpoint compare well with this down here. So the numbers have changed for the endpoints, and yet I'm now going to ask about, not the endpoints, but the length in between the endpoints. I'm going to look at the ratio of these successive lengths.

Despite all those endpoints changing, the ratio of those widths does not change. The shrinkage factors still turn out to be the same numbers—4.7 horizontally and 2.5 vertically. Near the onset of chaos, they converge to 4.6692… and 2.5029…, exactly the numbers seen before. That's *quantitative universality*, discovered by Mitchell Feigenbaum, who at the time was working at Los Alamos.

Let me tell you a little bit about Mitch Feigenbaum. He is a colorful character. You can read about him in James Gleick's wonderful book *Chaos*. Mitchell Feigenbaum is currently a Professor of Math and Physics at Rockefeller. He has intense eyes, mischievous grin, and the swept-back hair of a Romantic composer. You think you're looking at the reincarnation of Beethoven. His seminal discoveries came in 1975 when he was working at Los Alamos in T-Division, Theoretical Division. His working habits were odd, even for T-Division. He roamed the streets in the middle of the night, lived on cigarettes and Coca-Cola, and kept to a 26-hour day. His sleep/wake cycle would roll in and out of phase with the rest of society every few weeks. That's the way he liked to work.

Here's the story of how he discovered universality. He was studying a logistic map. He was interested in the transition to turbulence, and he thought the logistic map might be a kind of simpler model of that phenomenon. As he was roaming the streets one night, he got the idea, "Maybe I should calculate the lengths of these wishbones." He did; he tried to develop a theory, and he noticed this geometrical convergence of the 4.7 that I mentioned, and he tried to explain that number. In fact, he tried to think, "Maybe that number is related to pi or $\sqrt{2}$," and he tried to combine all the mathematical constants he could think of to see by just monkeying around if there was a way he could get that number, 4.669... . He said he spent about a month on it and he couldn't figure out any way to do it.

It's worth saying, at this point in his career, Feigenbaum was a bit of an outsider. Everyone knew that he was the smartest person in the room, but he hadn't accomplished very much. He was at a point in his career where he was looking for something big, and other people would often come to him when they had reached an intractable point in their problem and ask, "Mitchell, could you help me?" And he could always solve their problem. Do you know the kind of person I'm talking about? But he himself was having a creative logjam, and he had trouble doing anything of his own that was new. Yet he was the smartest guy at Los Alamos.

Here he was, feeling unproductive, and walking around, and he wants to think about something deserving of his brain, so he's thinking about chaos, which was a totally fringe subject at that point. He's doing this crazy game with a calculator. This is the era when handheld calculators became widely available, in the early '70s. He had a Texas Instruments calculator; he could program it; he put in the logistic map, and he noticed this funny number, 4.7. He spent a month trying to explain the number, and it didn't seem like he was going to help his career working on this calculator problem. He got discouraged, and then he happened to be eating lunch one day, and he talked to a man named Paul Stein. Maybe you remember the name Stein because I said earlier in the lecture that the U-sequence was discovered by Stein, Stein, and Metropolis. That's the same Stein—Paul Stein—and Mitch asked Paul about, "I'm studying the logistic map. I'm seeing this number. I'm confused about it." Stein says, "Remember, Mitch, that the kinds of things you're seeing don't just occur in the logistic map. They occur in the sine map. Remember the paper I published about the U-sequence? You should look at other maps and see what the number is for them."

Feigenbaum, then, starts calculating the shrinkage factor for the wishbones for the sine map, and he gets 4.669... . He said that his hairs stood up because that was the same number that he had seen. He remembered it because he had spent a month trying to calculate it.

So that's the discovery of quantitative universality, and the story, as [with] so many things in chaos, doesn't have a simple ending. Feigenbaum tried to publish it, and his papers kept getting rejected. No one knew what the heck he was talking about. He couldn't get it published. It seemed like a meaningless calculator game, and years went by before he could finally publish it. When he did, if you try to read Feigenbaum's original papers, I have to wish you good luck. They are almost impenetrable. This is not Bob May writing. This is Mitchell Feigenbaum, and you're in the presence of someone brilliant. May is brilliant, too, but Feigenbaum is obscure, hard to read.

In 1978, Feigenbaum published his works, and he also explained those mysterious numbers using a technique from physics, a very high-powered, Nobel Prize–winning technique called the *renormalization group*, which had been used in particle physics and then also adapted to condensed-matter physics problems of magnets, and superconductors, and liquids, and so on. It had won a Nobel Prize for Ken Wilson in the 1970s, who had been applying renormalization group to the physics of phase transitions in magnets and superconductors.

Renormalization means that we're going to take a picture and renormalize it—blow it up or scale it in some way to make it look like some other picture. You see that's what we've been doing. We saw the self-similarity of the fig tree. Feigenbaum recognized the connection and applied this method to his problem in chaos, and with that, he could explain the mysterious numbers—4.66. It came right out of his renormalization analysis. This was a huge conceptual breakthrough because it meant that the Nobel Prize–winning work on phase transitions was now being linked to the problem of the onset of chaos, long an unsolved problem. The common thread was that both types of systems were looking self-similar right at the transition point, either in the phase transition or right at the onset of chaos.

Feigenbaum further predicted that the universal features he had discovered in his calculator experiments, because they were universal, they must occur in nature. He checked that they occurred in differential equations. The same numbers came out. Differential equations describe nature, so it should be there. Since these universal numbers were expressed as precise numbers—that is, it wasn't just a qualitative thing—it was the sort of thing experimentalists could go after. They could do an experiment, see if they saw a period-doubling cascade, and then check if the wishbones were shrinking at the predicted rate.

The only caveat in all this was, when I say universal, I don't mean everything period-doubles, and not everything obeys Mitchell Feigenbaum's numbers. The system has to have some underlying iterated map in it that leads to a period doubling to chaos just as in the logistic map. Otherwise, all bets would be off. So this only applies when a system period-doubles en route to chaos, has an accumulation point followed by chaos.

Let me close by emphasizing why I think this is the most stunning breakthrough in the whole subject. The implications of Feigenbaum's ideas are absolutely shocking. What makes them so unbelievable is that the logistic map has no scientific content. It has no laws of nature built into it. It's just pure numerology. It's like I told you I'm going to do something in quantum mechanics by pressing the square root button on my calculator. You wouldn't believe it. But that's about what Mitchell Feigenbaum did—not for quantum mechanics but for chaos, and not with the square root button but with the logistic map button. This breaks the rules of how you do science. Even quantum mechanics and relativity hadn't dispensed with the ideas of basing predictions on fundamental laws of nature. Yet miraculously, subsequent experiments on a variety of real systems conformed to Feigenbaum's predictions, as we will discuss in the next lecture. It was as if the old Pythagorean dream had come true. The universe is not made of earth, air, fire, and water—it is made of number.

Another startling quality of [Feigenbaum's] work was its cross-disciplinarity. He had applied a Nobel Prize–winning method in condensed-matter physics to a math problem a child could play with on a calculator that came from population biology, and from that came predictions about the route to chaos in systems as diverse as heart cells, electronic circuits, fluids undergoing convection, and chemical reactions showing chaos.

Next time, we'll discuss some of those experiments in greater detail.

Experimental Tests of the New Theory
Lecture 12

When [a team lead by Harry Swinney at The University of Texas] pushed the system past the accumulation point ... to the onset of chaos, they could see chaos ... but more than that, the chaos was interrupted by periodic windows. It's incredible. ... They saw that in reality. ... And the order of the periodic windows was the magic universal sequence, 6 came before 5 came before 3. All in all, a great confirmation of the theory.

The moment of truth for the science of chaos comes in the early 1980s, with the first experimental tests of the new theory. In case after case, the theory's predictions were confirmed. It was a heady time. Researchers were starting to make sense of the borderland between order and disorder. But as with any theory, the new ideas apply only in certain restricted settings—here, for systems with only a few variables, not millions. The lecture concludes with a balanced assessment of where the theory works and where it breaks down.

One elegant experiment involved an oscillating chemical reaction. Such reactions are a favorite demonstration in chemistry classes, because they change color with clocklike regularity. Under some conditions, the reaction can switch from clockwork to chaos.

Harry Swinney (b. 1939), a physicist at The University of Texas, led a team that tracked the chaos by recording the reaction's bromide ion potential as a function of time. The peaks and valleys in the time series vary erratically. But is this genuine deterministic chaos or merely randomness caused by sloppy experimental technique?

To find out, they graphed their data in state space, and out popped a strange attractor. Remarkably, it looks like one wing of Lorenz's butterfly attractor (discussed in Lecture 7).

Within the chemical strange attractor, the dynamics were animated by an iterated map, just as Lorenz's had been. The map has a single hump, like

Lorenz's map and the logistic map (from Lectures 8 and 9). When the experimenters varied the flow rate of chemicals feeding through the reaction, they observed the period-doubling route to chaos, just as in the logistic map. Within the chaotic region, the system also has periodic windows, and they occur in the order predicted by the universal sequence (from Lecture 11). All in all, a great confirmation of the theory!

Other researchers did similarly careful experiments on convecting fluids and electronic circuits. They confirmed the universal number 4.669... that Feigenbaum had predicted for the period-doubling route to chaos.

But there are important caveats. First, the experiments are difficult. You can rarely measure more than four or five period doublings, because noise in the experiments obliterates the fine structure of the high-period cycles, which makes it hard to tell precisely when a bifurcation has occurred. Also, it's hard to vary the control parameter precisely enough. (Recall that the spacing between bifurcations shrinks roughly 5-fold each time, so each new data point requires a 5-fold improvement in the experimenter's ability to tune the parameter.)

Most importantly, tremendous care must be taken to suppress all other possible instabilities that could occur. Otherwise, when you drive up the flow rate or turn up the temperature, the system may go unstable in ways you don't want. It often starts to vary in *space* as well as time. For example, in the chemical experiment, the system is continuously stirred so that it can't set up any spatial patterns. In the convection experiments, scientists used liquid mercury instead of water; the advantage was they could keep the convection rolls perfectly straight by aligning them to a constant magnetic field. Otherwise the rolls would start to wiggle and eventually degenerate into seething turbulence. Feigenbaum's theory can't handle such complications. It only applies to systems following a nice, stately period-doubling route to chaos.

Still, back in 1983, all this was a remarkable triumph for the emerging science of chaos. None of these phenomena could have been understood just a decade earlier. Moreover, no one would have even thought to ask the questions. And that is perhaps the greatest triumph of all.

To assess the strengths and limitations of chaos theory more clearly, consider where it resides on a metaphorical "thermometer of complexity." Just as on a real thermometer, the coldest temperatures are at the bottom. This is where order lives, the realm of regular, well-behaved systems that Newtonian science mastered and that many great scientists are continuing to study today. Turn up the heat a bit by stressing the system or forcing it to feed back on itself. Where order first begins to break down, we enter the realm of chaos—but not total chaos. Instead, we see a chaos that lives much closer to regularity than mayhem.

Next we come to systems that vary wildly in space as well as time. Think of turbulence in a roiling pot of water, or fibrillation in a quivering heart. Still, we have hope of understanding them—the laws of fluid mechanics are known, and the laws of cardiac electrophysiology are becoming clearer every day.

Finally, we reach the boiling point—the most complex, unpredictable systems we can imagine, with billions of interconnected variables. Think of the global economy, human society, the brain. Here we don't know the laws. Plus they might be changing or nonexistent.

> **In their enthusiasm, some people in the '80s tried to apply [chaos theory] to much hotter systems—the stock market, or turbulence, or warfare between nations. Such overreaching provoked a backlash that also went too far.**

The point is that chaos theory lives in a very cool part of the thermometer. In their enthusiasm, some people in the '80s tried to apply it to much hotter systems—the stock market, or turbulence, or warfare between nations. Such overreaching provoked a backlash that also went too far. Finally, notice that as the mercury rises, we also retrace the historical development of chaos theory and point toward the problems at the cutting edge.

In the next half of the course, we'll look at the most exciting advances in chaos theory that have occurred since the 1980s, including some recent ones that take us much higher on the thermometer of complexity. But first, to prepare ourselves for the wonders that lie ahead, we need to become better acquainted with fractals, the infinitely intricate shapes that we've already encountered on several occasions. ■

Essential Reading

Gleick, *Chaos*, 191–211.

Stewart, *Does God Play Dice?* chaps. 9–10.

Supplementary Reading

Cvitanovic, *Universality in Chaos*.

Roux, Simoyi, and Swinney, "Observation of a Strange Attractor."

Strogatz, *Nonlinear Dynamics and Chaos*, sec. 10.6.

Questions to Consider

1. What observations could you make to distinguish between a deterministic but chaotic system and a system that is random?

2. Do you believe that massively complex systems like the brain, the global economy, and human society obey natural laws in the same sense that pendulums and planets do?

Experimental Tests of the New Theory
Lecture 12—Transcript

The moment of truth for the science of chaos comes in the early 1980s with the first experimental tests of the new theory. In case after case, the theory's predictions were confirmed. It was a heady time. Actually, I remember it. It was a very thrilling time. Researchers were starting to make sense of the borderland between order and disorder. In this lecture, we're going to discuss some of those thrilling triumphs. But, as with any theory, the new ideas apply only in certain restricted settings—here, for systems where only a few variables are active, not millions of them. We'll discuss those limitations in the last part of the lecture. The goal is to give a balanced, honest assessment of where the theory works and where it breaks down. This cautionary note is important because some early advocates of chaos theory, understandably, in their enthusiasm, overreached and provoked a backlash of skepticism that persists, unfairly, to this day, at least in some quarters.

One elegant experiment involved an oscillating chemical reaction. Oscillating reactions are a favorite demonstration in chemistry classes because they change color with clocklike regularity. In a minute, I will show you a video of a chemical oscillator, a kind of chemical clock, and I would like to thank Irv Epstein from Brandeis University, who helped me. He's a great chemist, working in this area of oscillating chemical reactions, and this was done in his lab.

Here's what you will be seeing. Before I show it to you, let me get you oriented. First, you will see a beaker, as usual in a chemistry experiment. Here is a beaker with a clear chemical solution in it. At the bottom, you will see something that looks like a little lozenge of some kind that's spinning around. That's a magnetic stirrer. That's not part of the experiment. All it's doing is stirring the system in a kind of inconspicuous way to keep it well mixed. Underneath it, there is a device that's providing a rotating magnet that you won't see, but you'll just see the effect of this little lozenge-looking thing spinning around at the bottom of the beaker. So, pay no attention to that. That is not the important point, except insofar as the stirrer keeps the whole system very thoroughly mixed so it can't show any spatial structure of any kind, no spatial patterns. That will mean that the whole beaker will do whatever it does all at once. Otherwise, if we did have spatial patterns, as would occur if we turned off the stirrer, then the behavior of the system will be more complicated than we want and actually more complicated than the theory we've developed so far can deal with. So to that extent, the stirrer is important. It keeps things simple.

The experiment will start when chemicals are poured into a beaker. So I'll do that, and watch carefully. The chemicals are poured in. It reminds you of chemistry experiments you've done or seen. There's a little color change. We see some yellow, and maybe that's it, you might think. The show's over. That's what often happens in chemistry lab. The best you could hope for is maybe a bang or a stinky smell. Now something has happened. It suddenly turned blue. That's better than usual.

It's just gone back to yellow. You've got an oscillating reaction. It's a chemical clock. It will go back and forth between yellow and blue periodically. I'll let that run a little bit more. This is not trick photography. This is real. This is really what happens, and it's dramatic and stunning, especially if you've never seen it. Under some conditions, reactions like this can switch from clockwork, as we're seeing here, to chaos. In other words, instead of switching between blue and yellow or any other two colors with this periodic behavior that we saw, we could see something more complicated. We could see chaotic oscillations that don't repeat.

These were investigated by a team lead by Harry Swinney at The University of Texas. Harry is a physicist, not a chemist, and this was a bit of a stretch for him. I give him credit for taking on this work. He led a team that did a careful study of a chaotic chemical reaction and brought his tremendous experimental ability and aptitude to this question in chemistry that had been a thorny one. Just a word about Harry: He's curious about spatial patterns of all kinds or, I should say, patterns in general. They could be patterns in time, as they will be here, because remember, we're going to keep the system well mixed, so it won't show spatial patterns. Earlier in his career, Harry did work on spatial patterns—for instance, the Great Red Spot on Jupiter. That's a spatial pattern, and he did a lab model of that using a rotating fluid and looking at the vortices and how they formed together to make a single big equivalent of a rotating spot. He also did pioneering work with Jerry Gollub, another physicist, on the question of what happens in rotating fluids when they start to make the transition to turbulence. And that work was one of the early works that led to the idea that strange attractors, which we discussed in connection with Ed Lorenz, might have something to do with the transition to turbulence.

As I say, Harry was not particularly trained in chemistry, but he and his team thought, "Well, we have experience with other experimental systems; why not? Let's take a stab at these oscillating reactions and see if we can make them go chaotic." He devised a way to do that by flowing fresh chemicals steadily through the reaction. It wasn't isolated in a beaker, like the reaction we just saw, where that sort of thing doesn't have any fresh material coming into it. It's just sitting in the beaker, isolated from the rest of the world, and the problem with that is when you are a closed system, the second law of thermodynamics dictates your fate; you are on your way to equilibrium. Entropy is going to just keep increasing. You can't sustain patterns for long. In fact, oscillating reactions at first, when they were discovered in the 1950s, were thought to violate the second law of thermodynamics and thought to be impossible. How could these things oscillate? They have to go to equilibrium. But that was a misunderstanding at the time. People had not realized that reactions could oscillate en route to equilibrium. There was no need that they have to go monotonically to equilibrium; they could oscillate. So there was no violation of the second law, just as there isn't in the video that I showed you a second ago.

Swinney and company tried this more elaborate setup where they flowed in fresh chemicals all the time, keeping the system far from equilibrium at all times, sort of like a living thing, like we ourselves are kept far from equilibrium by the food we eat, by the energy of the Sun. By varying the flow rate at which the chemicals were pushed into the system and taken out, they had a controlled parameter at their disposal, analogous to the growth rate in the logistic map or to the way we tightened the brake on the Lorenz waterwheel. By varying this flow rate, they hoped they could drive the system into chaos, and they found that they could.

To track the chaos that they were observing, they recorded something that—I have to say I'm not a chemist and I can't really tell you much about what this means—but they recorded something called the *bromide ion potential*. That sounds like chemistry. And they measured it as a function of time. Here's what it looked like.

You see something that clearly looks—at first, your eye might think that it's periodic. You see the bromide potential going up and down, up and down, but notice that the peaks and the troughs change in terms of their height. This trough is lower than the next one; the next one then goes down again. But then they're followed by an even deeper one, and so the question of the heights of the troughs and peaks—those vary chaotically. This is actually something chaotic, although your eye might at first think it's periodic. But hopefully by now, at this point in the course, you have enough experience to recognize—this is what chaos looks like. It doesn't look like a blur, or a big mess of scribble, or spaghetti all over the page. It looks like something periodic but with a variable amplitude, kind of.

Still, the question arose at the time: Is this genuine deterministic chaos that is controlled by deterministic laws with no randomness of any kind in them— no uncertainty, no noise, just pure determinism, like Newton's laws, except for chemistry—is it that leading to chaos? Or is it just good old-fashioned randomness caused, perhaps, by sloppy experimental technique or maybe vibrations coming from the environment, from cars driving by, or who knows what?

To find out, Swinney and company would have liked to draw a state-space diagram for this system. Remember, we talked about state space. This is Poincaré's great idea of looking at a system abstractly by looking at all the information needed to predict what it will do in the next instant. If we could see in that state space, as in the case of the Lorenz system, which we did when we were thinking about the waterwheel a few lectures ago, there we could trace out a strange attractor that looked liked the butterfly wings in that state space. That was the dream. Could this be done with real experiments?

Unfortunately, that was impossible to do for this chemical experiment for a reason that may not surprise you that much. There is a lot of complicated chemistry involved here, with dozens of chemical reactions involved, many dozens of intermediate chemicals, some of which are not known and certainly many of which are not measurable. That's a problem in trying to analyze a system which we would say has many degrees of freedom, many chemicals involved. Or if we were trying to represent its state space, it would be very high dimensional. It would have 50 axes or something like that, 100 axes. It's going to be trouble to draw that on graph paper. So what can you do? In fact, in this particular experiment, all that could really be done was to measure this bromide ion potential. What can that tell you? Can that give you enough information to see that the predictions of chaos theory are really right? It turns out that having access to just one time series, like this one, can be sufficient.

That was shown by a mathematician named Floris Takens. He found a way to get by with just that amount of information. Takens proved that if you plotted a variable like this against itself but delayed by a certain amount of time or, possibly, against two copies of itself, delayed by some time and maybe delayed by a longer time, if you did something like that—a delayed plot, just of this one variable looked at at different times in its history—that that might substitute for actually having genuinely different state variables. It might seem like cheating. How can that work? It's the same information over and over, but it isn't, and Takens was able to prove that the strange attractor or other state-space drawing that you might make from such a procedure of time delay would give you a faithful representation of the actual attractor in the system, maybe not in its numerical details but in its overall architecture. This is a technique that is now called *attractor reconstruction* or sometimes called *attractor from a time series*. Because all you are using is the time series to get the information you're looking for.

Swinney and collaborators knew about this and implemented the technique of Takens. They graphed the bromide ion potential against itself with just one time delay. Watch what they found. Bromide at a certain time on the horizontal axis against itself at a later time. Maybe you can read that it says, "$B(t_1 + T)$"; that's the time delay, T. When you graph this variable against itself at the later time, you see something that looks a lot like one wing of Lorenz's butterfly attractor. That's very reassuring because that's what chaos looks like, and that is not randomness. That is deterministic chaos. Very good news. Pushing farther, they wondered, is this chaos orchestrated by an iterated map the way that Lorenz's was? Yes, it is.

They were able to construct an iterated map by making measurements off of the strange attractor. The dots are actual data points. The curve is what we call to guide the eye or to guide the mind. I'm trying to get you to believe that there is an underlying iterated map here. And notice that its shape is a curve with a single hump. It's not like the Lorenz map, which had, if you recall, a sort of witch's hat, a pointy, cuspy top. This doesn't look like [that], to the extent that—the data aren't super clean, but they're pretty clean—and it looks more consistent with a rounded top, just like the logistic map which we discussed in Lectures Eight and Nine.

Once you see something that looks like the logistic map, you know what's coming next. Now we're opening the door to period doubling. So the experimenters wondered, "What if we vary our available control parameter? What if we change the flow rate of the chemicals feeding through the reaction? Can, by some miracle, we see a period-doubling cascade? If so, will the separations of the wishbones obey Feigenbaum's magic numbers? Will we see a universal sequence?" All these things we discussed in the past lecture, the one previous to this. Here is what they tried.

First they saw, yes, we could see period doubling. Here is what period doubling looks like in the chemical experiment. You see on the top left something that is repeating over and over, not changing its amplitude at all. That's period-1, meaning the equivalent of a steady behavior for this chemical system. Just a pure oscillation, like a chemical clock. But then, look at the top right picture. It's now repeating every 2. It has a high peak and then a lower peak, and then back to the same high value and then lower, and so on. This is period-2, like we saw in the logistic map. Pushing even farther, they could get something that was period-4, and so on. They could see the period-doubling cascade.

Furthermore, when they pushed the system past the accumulation point of the period-doubling cascade all the way to the onset of chaos, they could see chaos, just as expected, but more than that, the chaos was interrupted by periodic windows. It's incredible. Remember when we unrolled the scroll in the previous lecture on the orbit diagram and we saw all the periodic windows interrupting the black mess? They saw that in reality; they saw that in chemistry. And the order of the periodic windows was the magic universal sequence, 6 came before 5 came before 3. All in all, a great confirmation of the theory.

Other researchers did similar careful experiments on all kinds of different systems—fluids undergoing convection, electronic circuits showing oscillations that became chaotic. They confirmed Feigenbaum's universal number, 4.669..., that had been predicted for the shrinkage factor as we approach the onset of chaos. Here's a table of such numbers, as measured in real experiments. What I'm showing is a series of physically totally different systems—water, mercury undergoing convection, electronic circuits of different types, diodes, transistors, a superconducting device called the Josephson junction.

What was measured was a certain number of these wishbones in the orbit diagram, and typically, it's very hard to measure more than, as you see, four or five of them. That shouldn't surprise you because remember that each wishbone is 4.7 times longer than the wishbone that follows it. When I was trying to do it manually here with a totally clean environment of a computer simulation, I had trouble controlling the mouse enough to be able to see more than out to like period-8. I didn't even try to go to period-16. It takes a very steady hand.

Experiments are even more difficult than that, of course, because they're not as clean as a computer. You can rarely measure more than four or five period doublings. That's just the way it is. Noise in the experiments, inevitable in any real system, obliterates the fine structure of higher-period cycles, and it makes it hard to tell precisely when a bifurcation has occurred, where the wishbone begins, because they become so tiny. It's also hard to vary the control parameter precisely enough. I myself had motor control difficulties doing that, that control parameter being like the growth rate in the logistic map, the flow rate through the chemical system, or it could be a flow in a fluid, whatever. There is always some available control parameter, and it can be very hard to measure it precisely enough to detect and control all these bifurcation points.

Most importantly, tremendous care must be taken to suppress any other instabilities that might occur. This is an important point. Otherwise, when you drive the flow rate of the system up higher or you turn up the temperature, the system may go unstable in ways that you don't want. It often starts to vary in *space* as well as time. It might set up some kind of spatial pattern. For example, in the chemical experiment, Swinney and his team kept the whole thing continuously stirred, so it can't set up any spatial patterns; it's mixed.

In the convection experiments, which I haven't really described yet, a similar amount of care was taken. Those experiments were done by a man named Albert Libchaber, another brilliant experimentalist, in the same league with Swinney, the sort of hands and technique of a watchmaker. Libchaber and his team did the study of convection in liquid mercury. It's convection in a tiny box. They took liquid mercury instead of water, heated it from below, and watched it start swirling into what are called *convection rolls*. Here's what a convection roll looks like. It sort of looks like fluid swirling around, and it makes a long, straight roll.

The problem is, if you start driving this system harder and harder by turning up the heat, you can get instabilities you don't want. The rolls might start to wiggle. They don't stay straight. They could eventually, if uncontrolled, degenerate into seething turbulence. That's not what chaos theory is about. We're trying to keep the rolls straight. We don't want spatial structures, so that's why liquid mercury was used, because liquid mercury is responsive to a magnetic field. That was the brilliant idea of Libchaber, that by using liquid mercury he could align a strong magnetic field with the rolls and keep them straight. So he suppressed those instabilities and allowed himself to focus on the period-doubling instabilities that chaos theory can describe.

Feigenbaum's theory could not have handled the seething turbulence or anything of this other sort, the wiggling in space. It only applies to systems following a nice, stately period-doubling route to chaos.

Still, in the years between 1980 and 1983 when all of this was happening, this was a remarkable triumph for the emerging, nascent science of chaos. None of these phenomena could have been predicted or really understood just a decade earlier. No one would have even thought to ask the questions. That is perhaps the greatest triumph of all—the chaos revolution of the 1970s and

'80s launched science on a journey it needed to take. We have seen that this was a pressing issue for centuries and that it had been suppressed, put under the rug. Now it was time to take that journey, a journey into the richness of the everyday world around us, a world of nonlinearity and complexity.

To assess the strengths and limitations of chaos theory more clearly, let me give you a metaphor. Consider where chaos theory, the type that we have been developing so far in the course, resides on a metaphorical "thermometer of complexity." On a real thermometer, the coldest temperatures would be at the bottom, and that's what I'm trying to indicate here. I realize this is a bit of a silly sketch, but I hope it will help you think about where chaos theory sits in relation to the problems ahead. The bottom of the thermometer is where order lives. This is the realm of regular, well-behaved systems that Newtonian science mastered and that many great scientists are continuing to study today—the world of order, the Newtonian clockwork world and all that followed from it.

For a long time, that was the only part of science that existed in physics and math. But now you see there are more places to go on the thermometer. In the old days, we didn't even think about points higher up. They were forbidden. They were no-man's-land. But the chaos theorists were the first to turn up the heat. They turned up the heat a bit by stressing the system in some way, forcing it out of equilibrium, forcing it to feed back on itself, something to make it take that transition into chaos.

Now look at the three columns next to the thermometer of complexity on the left. I try to categorize things by their complexity in various ways. Are they complicated in time? Are they complicated in space? Are underlying laws known that govern these systems? Then, on the right, I try to list words that we associate with the different kinds of problems that occur when we have various ingredients in this menu. Notice, at the very bottom, we would say the behavior in time is simple, the behavior in space is simple, and the laws are known. That's the world of order.

Moving up one step, we get to the world we've been studying—the world of chaos. But it's not total chaos. It's not the colloquial concept of a riotous confusion. It's a chaos that lives right next door to regularity. It's just past the brink. This is where the breakthroughs that we have been talking about occur, the breakthroughs of the 1980s, the culmination of the story we have been telling so far. On the one hand, these breakthroughs inspired a giddy sense of triumph—deservedly so, for all that had been discovered about chaos and especially about the transitions from order into chaos. Humanity had never understood much about chaos at all until this point. This had been totally virgin territory, unexplored, a scientific no-man's-land. But you have to appreciate, this was just the first step. This is a long journey that we will be taking for the next century and maybe beyond.

Look at what defines chaos. The behavior in time is complex, yes. That was the news. We could now handle temporal complexity of a certain type, but we have made sure that the systems were still simple in space. That's what I was talking about—stirring that chemical reaction, lining up the rolls with the magnetic field in the convection experiment. You have to keep it simple in space, or the theory, as developed so far, doesn't work and doesn't apply. And the laws were known. These are either systems governed by chemical reactions, where we know the kinetic equations, or by Newton's laws for mechanical systems. So the laws were known; spatial behavior, simple, but time, still complex.

Next up on the thermometer are systems that we allow to start varying wildly in space. We're turning up the heat more. Now, they're varying wildly in space, as well as time. Here, think about turbulence in a roiling pot of water or fibrillation in a quivering heart. We don't understand those systems very well yet, but we do have hope of understanding them because the laws of fluid mechanics are known. They've been known for 150 years. In fact, fluid mechanics is a good way—studying turbulence is a well-known quagmire. That is a monster-hard problem. Heisenberg, the great Heisenberg of the Heisenberg uncertainty principle, started working on turbulence at one point in his career and realized that wasn't a good idea.

Turbulence will ruin you scientifically. That is probably the hardest problem in classical physics. It is still unsolved, and Heisenberg got out of there for quantum mechanics, which by comparison, was easy. I'm being a little bit silly, but it was. Turbulence is still not very well understood, whereas we have nailed quantum mechanics.

We do have hope of understanding turbulence because the laws governing it are known. They've been known for 150 years. It's sort of like what Poincaré faced with the three-body problem. We know the laws, but we're just not able to see our way into their consequences. That's what remains.

Also, when I mentioned fibrillation, that's a little bit higher on the complexity thermometer in the sense that although the laws of cardiac electrophysiology are not completely known yet, we're learning more about them every day. Biology and mathematics and physics are combining to help us understand the arrhythmias of the heart, so we'll get there.

Then above that, we see what I'm just calling "complex systems." This is at the boiling point. These are the most complex, unpredictable systems we can imagine, with billions of interconnected variables. Think about everything that we don't understand at all: the global economy, human society, the workings of the human brain. We know little bits. We know a little bit about how neurons work. We know what certain regions of the brain are doing, but we don't really understand the functioning of the whole brain. Of course not. That's going to take us a long time.

In problems like this, we don't even know the laws. We don't even know if the laws exist or if they might be changing. The point is, the parts of chaos theory that we've been developing so far live in a very cool part of the thermometer.

In their enthusiasm, understandably, some people in the 1980s tried to apply this mild form of chaos theory to much hotter systems. Papers were written about chaos in the stock market, chaos in turbulence, chaos in the heart for real fibrillation, even models of warfare between nations using chaos theory. That's wrong. There's no chance that [the theory] can apply to such complicated systems. Not yet.

So that kind of overreaching provoked a backlash that also went too far because people started to think, "This chaos theory is a lot of nonsense and hype. I always thought it smelled bad. Now I really think it smells bad." That's what mainstream scientists were apt to say if they weren't too impressed or willing to enjoy other people's success with all the hype that was coming about chaos theory. So there was a backlash, and those of us who are chaos theorists have not appreciated that backlash because it's really not fair. There were some tremendous triumphs that are genuine. I hope I've convinced you of that. But, fair enough, there were also some excesses, some overreaching, and we need to understand that, too.

Finally, I think it's an interesting point to notice that if you look at the complexity thermometer from the point of view of history, you can see the path that we must take in science. You can see the historical development of the subject itself right there on the thermometer. Go up the thermometer, and you're pointing your way to the problems at the cutting edge. We have to move from order to chaos, through turbulence and fibrillation, up to mastering complex systems. I may not live to see it, but that's where science is heading. I'm sure of it.

In the next half of the course, we'll look at the most exciting advances in chaos theory that have occurred since these glory days of the 1980s. A lot has happened. You won't be able to read about it in James Gleick's wonderful book because that book was written in 1987. This is what has happened now in the past 20 years. I can't wait to tell you about it, and I'll be telling you about some recent discoveries that take us much higher, to the dizzying heights near the top of the thermometer of complexity.

But first, we need to prepare ourselves for the wonders that lie ahead. We need to become better acquainted with something that we've seen slipping into the story, fractals. We've seen them, these infinitely intricate shapes that we have already encountered on several occasions. We'll take that step in the next lecture.

Fractals—The Geometry of Chaos
Lecture 13

> This baker may ... [put] a little pat of butter in [the pastry]. That pat ... represents nearby initial conditions. That is, there is a tiny blob consisting of states that are all close together, particles close together. ... What would happen if I were to do my process of rolling, folding, rolling? ... It will cause mixing. But mixing is chaos.

W e've now reached the midpoint of the course, and it's time to change gears slightly. The next six lectures are devoted to fractals—beautiful, intricate shapes with endless self-similar structure. We'll see that fractals *are* beautiful, yes, but they're also much more than that: They're crucial to the science of chaos. One could even say that they're the geometry of chaos. And they're changing the way we think about complexity in all its forms.

By that, I mean that fractals are useful all the way up and down the metaphorical "thermometer of complexity" (from Lecture 12). For example, we'll see how fractals shed light on such poorly understood phenomena as earthquakes and stock market fluctuations. Both are illuminated by fractal statistics—a very different kind of statistics from the orthodox kind you might have learned about elsewhere. Thus, fractals are going to play an essential part in the science of the 21st century as we strive to unravel systems of ever-increasing complexity.

Although fractal geometry might seem like a new topic for us, it really isn't. We already encountered fractals in Lecture 10, when we noticed the self-similarity of the fig tree, with its infinitely bifurcated structure. And later in that same lecture, we spotted the incredible miniature copies of the entire orbit diagram at the end of each periodic window. Then, in Lecture 7, we puzzled over the infinitely many layers of Lorenz's strange attractor, all packed together like sheets of mica. But in all those cases we mentioned fractals en passant, whereas in the next few lectures we're going to put them at the center of our attention.

First, we consider the cause of the infinitely layered structure of strange attractors, which is the same as the cause of chaos: a repeated process of stretching and folding in state space. Consider an analogy with the way that pastry is made, the very flaky kind with thousands of paper-thin layers, as in croissant or strudel. Think of state space as being like dough. The particles of flour represent all the possible states, all possible initial conditions.

The differential equation (or iterated map) does roughly the same thing to state space that the baker does to the dough. Rolling the dough flattens it. In the analogy, this compresses state space into a sheet that later forms the paper-thin layers of the attractor. Rolling also stretches the dough lengthwise. This has the effect of spreading nearby bits of dough apart—which is analogous to chaos (two nearby initial states diverge rapidly toward different fates). Finally, repeated folding brings distant parts of the sheet close together, creating many flaky layers.

This intuition about stretching and folding occurred independently to three scientists in different fields (math, chemistry, and astronomy), all of whom played important roles in chaos theory. The first person to have the idea was Stephen Smale (b. 1930), one of the world's greatest living mathematicians. If we were doing a pure math course on chaos theory, he would be one of the heroes. In the 1960s, he invented something called "the horseshoe mapping" (which stretches and folds a rectangle into a horseshoe shape, just like the baker does to the dough), and he used it to prove very important theorems about chaos.

Next came Otto Rössler (b. 1940), a chemist. Working in the mid-1970s, he knew about Lorenz's work and wanted to concoct an even simpler example of a chaotic differential equation. He constructed an example that has just one nonlinear term to Lorenz's two. Rössler's inspiration came from watching a taffy-pulling machine as it was stretching and folding taffy. The strange attractor produced by Rössler's system has been found in many real

This intuition about stretching and folding occurred independently to three scientists in different fields (math, chemistry, and astronomy), all of whom played important roles in chaos theory.

systems. (In fact, it was the strange attractor for the chaotic chemical system we discussed in Lecture 12—it looks like one wing of Lorenz's butterfly.)

Finally, astronomer Michel Hénon (b. 1931) demonstrated that stretching and folding generate a hierarchical structure that is "self-similar." Recall, self-similarity means that small pieces of a structure resemble the whole. It's one of the defining features of a fractal. Hénon devised his mathematical version of the pastry-making process because he wanted to visualize the layered structure of Lorenz's strange attractor. Hénon's mapping takes a solid rectangle and then flattens it, stretches it, and folds it into a horseshoe shape. If we iterate long enough, we should see the ghost coming out of the mist. The object that appears is the Hénon attractor. From a distance it resembles a boomerang. Zooming in reveals three sets of parallel lines, arranged in groups of 1, 2, and 3 thinner lines per set. Further magnifications show that this structure repeats endlessly, at smaller and smaller scales. It is the basic structure of strange attractors—what Lorenz meant by an "infinite complex of surfaces."

Self-similarity is actually a very old and universal idea (now being mathematized by fractals). Artists with a keen eye for nature have long been fascinated by it. Think of Leonardo da Vinci's sketches of turbulent whirlpools, or Hokusai's woodblock print of *The Great Wave off Kanagawa*, or Escher's visual explorations of infinity in a confined space. Once you start to look for fractals, you'll notice them everywhere in the world around you, and even inside you, from mountains and clouds to the branching patterns of your own blood vessels.

In the next five lectures, we'll define fractals more precisely. We'll also see how they illuminate a diverse set of complex phenomena in nature, technology, business, and art. ■

Essential Reading

Gleick, *Chaos*, 83–118 and 144–53.

Mandelbrot, *The Fractal Geometry of Nature*.

Schroeder, *Fractals, Chaos, Power Laws*, chaps. 3, 11.

Stewart, *Does God Play Dice?* chap. 11.

Supplementary Reading

Liebovitch, *Fractals and Chaos Simplified for the Life Sciences*, pt. I.

Strogatz, *Nonlinear Dynamics and Chaos*, chap. 11.

Internet Resource

Play with Hénon's mapping and strange attractor here: http://www.cmp.caltech.edu/~mcc/Chaos_Course/Lesson5/Demo1.html.

Questions to Consider

1. What other artists have touched on fractals in their work?

2. Does mixing always rely on chaos? Discuss the process of shuffling cards or stirring milk into coffee in these terms.

Fractals—The Geometry of Chaos
Lecture 13—Transcript

Welcome back to chaos, or should I say, "May I introduce you to fractals?" We have now reached the midpoint of the course, and it's time to change gears slightly. For the next six lectures, we'll be devoting ourselves to the fascinating world of fractals. You've heard of fractals, of course. Everyone has. [Fractals are] those endlessly intricate shapes with self-similar structure all the way down as far as the eye can see. Dazzling, multicolored images of fractals seemed to be everywhere in the late 1980s—on T-shirts, on coffee mugs, postcards, screen savers. There was even a cartoon where one woman says to her best friend, "We decided to redo the living room in fractals."

One of my goals in the next part of the course is to show you that fractals *are* beautiful, yes, but they are so much more than that. They're crucial to the science of chaos. Indeed, we could say that fractals are the geometry of chaos. They're also changing the way we think about complexity in all its forms, wherever it occurs in the natural world. By that, I mean that fractals are useful all the way up and down the "thermometer of complexity" that I described at the end of the last lecture in a metaphor. Certainly, they apply to chaos. That's in a relatively cool part of the thermometer, but they also apply higher up, into the hotter realms of turbulence and even into the most complex systems we know.

For example, we will see the way fractals are shedding light on phenomena as wild and poorly understood as earthquakes, stock market fluctuations, and congestion of traffic on the Internet. All these are illuminated by fractal statistics, a very different kind of statistics from the normal, orthodox statistics that you may have learned about in a college or high school course. Thus, fractals are going to play an essential role in the science of the 21st century as we strive to understand and unravel systems of ever-increasing complexity.

Although fractal geometry might seem like a brand new topic for us, it really isn't. We've encountered fractals a few times already in earlier lectures, such as in Lecture Ten. Remember we saw a structure we called the fig tree, having to do with the period-doubling route to chaos? When we looked at the fig tree, it had an infinitely bifurcated, self-similar structure. There was a branch that split into a wishbone, and each of those wishbones split into two more wishbones, and those split into two. This self-similar, bifurcating wishbone structure, what we called the fig tree, was an instance of a fractal, something where each small part of the structure resembles the whole.

Later in that lecture, we saw an even more dramatic example of a fractal, in which we were looking in the regions we call periodic windows, which were the open stripes in the orbit diagram. We saw that at the end of each of those open stripes, there was a magnificent little homunculus, a little copy of the whole orbit diagram sitting right there, with its own homunculi inside of it, and so on, all the way down to infinity. Unbelievable.

In Lecture Seven, we saw a similar, infinitely intricate structure, except that we couldn't see it. That's something I want to deal with in today's lecture. The thing we talked about at that point was Lorenz's strange attractor. Do you remember that? That was a butterfly-shaped object, and when I was talking about it, I said Lorenz made an abstract argument that although it looks like it has just two wings, apparently just two surfaces coming around and meeting each other at a hinge, he knew that was impossible. They couldn't meet at a hinge because that would mean that two trajectories were touching and intersecting, which would violate determinism. From a given state, there can be only one past and one future, so you can't possibly have these surfaces touching. That would be a contradiction, a mathematical impossibility. Lorenz concluded that where the two surfaces appeared to meet, they just have a close encounter. Then, as they swirl around on the wings again, they have to have another close encounter, which means that the two must really be four, and the four must really be eight, and so on. From this kind of argument, he concluded that these apparent two surfaces were really infinitely many surfaces packed tightly together in sheets like mica.

What we would like to do in today's lecture is try to understand that invisible [substructure] because you certainly couldn't see it. If I hold this up and you looked as closely as you can, there just seems to be two surfaces there. You won't see any mica substructure. I want to try to make the argument today why we know it has to be there and actually help you visualize it.

In all the cases that I mentioned so far in previous lectures, we were just sort of throwing out the idea of fractals en passant, whereas in the next few lectures, we're going to be putting them right at the center of our attention.

What I would like to spend most of this lecture doing is trying to give you some intuition about this enigma having to do with the Lorenz attractor and the more general question of why chaos goes hand in hand with the infinitely layered structure that we see in strange attractors. That question will occupy us for quite a while, but it will lead us naturally into another topic, which is a broader discussion of this idea of self-similarity that I mentioned a few minutes ago, self-similarity being crucial. That is one of the defining features of fractals.

My hope is that with this lecture—although I'm suggesting that fractals are somehow a new topic for us, this should provide a very natural connection, a segue from chaos into fractals because they are intimately related, as we will see through the rest of the course.

The first issue, then: What is it that causes the infinitely layered structure of strange attractors? The style of argument I'm going to give you is that there is a certain process operating that causes them both at the same time. The thing that causes chaos causes fractals. They go hand in hand because they are two sides of the same coin. What is this process I'm talking about? You may get the feeling that I'm hungry because I'm going to give you an analogy that is about something delicious—pastry. That is, by thinking about the process by which a chef makes pastry—the very flaky, layered kind, like you might use in a croissant or strudel or something like that—understanding what the baker is doing is going to help us understand, believe it or not, where chaos comes from and also why fractals come out simultaneously. It's something that the baker does without thinking about it.

Here is the analogy: The dough that the baker is dealing with is going to be analogous to state space. Actually, I'm wondering if I should tell you about the baker first and then do state space. I'll just keep going the way I am. So we've got the dough, and I'm claiming [that] dough is analogous to state space in the sense that dough is made of billions of flour particles and state space is going to be made of billions of states. Now, what are states?

We talked about this, going back to Lecture Four, but possibly it's worth reminding you.

State space: It's a great abstract conception that Poincaré came up with that we discussed in Lecture Four. When you're trying to understand what is happening in a differential equation, you discuss things in terms of the states of the system. A state is all the information you need to predict what will happen in the next instant. Using the laws of motion, the differential equations, if I tell you the state—like in the case of a pendulum, we saw that it was two numbers, the initial position and the velocity—if I give you those two numbers, I can predict the position and velocity of the pendulum forever afterward. In that case, state space was a 2-dimensional plane, two axes, one for position and one for velocity. So I want you to try to think of state space as a palpable, doughy thing. This plane that I'm talking about is no longer an abstract plane. It's a nice, flat piece of dough. That's the image.

Now, what does a differential equation do to the dough? It's easy to understand what a baker does to the dough, but let's try to understand. How can a differential equation do anything? It's an abstract mathematical concept. Can it do anything? Yes, it can. Use your imagination. The differential equation takes states—think particles of dough—and as time moves forward, each state moves to its successor state. That is, the dynamical system we're describing evolves—the pendulum moves—so you go from one state to the next state. That would mean that a particle of flour has to move to some new location. If you do that to all the particles, it's kind of like taking the dough and moving every particle somewhere, which is like deforming the dough. You're doing something to the dough. You're kneading it, stretching it, and folding it.

Let me show you that in terms of a picture to make this a bit more vivid. Let's go back to pastry, which is much simpler. Take a look. I'm claiming that the process used in making pastry, from a mathematician's point of view, is these three operations: You start with the dough, you do something to flatten it and stretch it. I have never cooked, but I'm guessing the way it's done is you take a rolling pin, start rolling out the dough, and that has these two effects. It flattens the dough, and it stretches it. Then, my hypothetical chef folds the dough. Even if this isn't true, pretend, okay? Bear with me. He would fold the dough into something like a horseshoe shape. Then, what I call here *reinject* just means take the thing that you just made, this horseshoe, put it back on your table, and start going at it again with the rolling pin, flattening it out and stretching it some more. Keep doing that.

That's an iterated process. By rolling the dough, you've flattened it, and in the analogy I'm about to try to elaborate, that would squash state space flat. What that will do is create all the sheets, the flat sheets like the mica sheets that are later going to make up the paper-thin butterfly wings of the attractor. What does rolling do? Rolling, in addition to flattening, stretches the dough. That produces chaos. This is causing chaos in the sense that two particles of dough that were near each other are now getting stretched apart. Stretching apart means that two initial conditions that were indistinguishable start to get spread apart. That's like saying two systems [that] started the same way start to have different fates. That's what chaos is about.

Here is a more vivid way to see that that's really chaotic as opposed to just some gentler kind of stretching. Think about, this baker may want to make this pastry taste better by putting some butter in there, so he puts a little pat of butter in. That pat of butter represents nearby initial conditions. That is, there is a tiny blob consisting of states that are all close together, particles close together—they're now butter particles. What would happen if I were to do my process of rolling, folding, rolling? Can you see that eventually the butter would be uniformly spread through the entire structure? It's obvious, right? It will cause mixing. But mixing is chaos. That means that this blob that started as a concentrated thing has eventually gotten mixed and could be anywhere. The particles have gotten sheared and spread all through the whole structure, which is like saying that on the Lorenz attractor or any other strange attractor, start as close together as you want in the tiniest bit of butter, so to speak, and you will end up anywhere on the attractor. Your fate will be unpredictable. There is a kind of ungainly pun waiting to happen here about butter and butterfly, and I think I'll just avoid that for my own good. If it occurred to you, congratulations.

What do we conclude from this? That by this repeated stretching and folding, we did two things at the same time. We mixed the initial conditions, or the states causing chaos. We also created a very heavily layered structure. I haven't emphasized the layers. There are infinitely many layers if we do this infinitely often. You can see that in this next picture. I've only shown you, at first, one horseshoe, one fold, but if I go farther by taking the horseshoe, that is, the bent, flattened slab of dough, imagine subjecting that to one more round of rolling and folding. You would get something that looks like this second structure. Whereas before it had two fairly thick layers, it now has four thinner layers. If I repeat that process again, it now has eight much thinner layers. This is sort of what Lorenz was talking about, this doubling of layers getting thinner and thinner. If you carry this on ad infinitum, you will get an infinitely layered structure with everything all mixed up in it. You get chaos in infinite layers, for free, simultaneously.

This intuition about what's causing the fractal structure of strange attractors, stretching and folding, independently occurred to at least three different scientists in three different fields: math, astronomy, and chemistry. All three played important roles in the development of chaos theory and were, as far as I know, working completely independently of each other, no contact whatever. It's another nice example of the interdisciplinary convergence that so often occurs in chaos theory.

The first person to have this idea was Stephen Smale, one of the world's great mathematicians. He's still living. He is sort of the modern-day heir of Poincaré, just an unbelievably good mathematician. He won the highest prize we give for math, the Fields Medal, many years ago. He's intensely geometric, like Poincaré, and has worked in many different fields, also like Poincaré. If I were doing a pure math course on chaos theory, Stephen Smale would be one of the two heroes right there alongside Poincaré.

In the 1960s, Smale invented something called "the horseshoe mapping." That's why I kept using the word *horseshoe* in describing the shape. That comes from Smale. It looks like a horseshoe. Smale invented it to try to understand and prove rigorous theorems about chaos. Up until now, we have proven nothing. Everything has been computer simulations, intuition, scientific arguments, and so on—which is fine, but not for pure mathematicians. They like rigorous proof. I should say "we." There's an aspect of me that is pure mathematician, too. So, yes, when I have that hat on, I would insist on rigorous proof, and then I would turn to the work of

Stephen Smale. In choosing to give him a somewhat minor role in this, I didn't feel that he was necessarily the easiest way for you to understand the subject, and in fact his work had relatively little impact on most scientists for that reason. Mathematicians, unfortunately, mainly talk to themselves.

A second person, then, was Otto Rössler, a German chemist who was working in the mid-1970s. At that point, Lorenz's work was already well known, and everyone was trying to understand it. Rössler wanted to concoct an even simpler example of a chaotic differential equation than what Lorenz had written down. Specifically, Lorenz's equations, even though I haven't written them down for you algebraically, there are three separate equations with what are called *nonlinear terms* in them. It turns out he had just two nonlinear terms. That's about as mild as you can get as far as ingredients to produce chaos; that is, it was known you would need three equations, three differential equations—you couldn't get it with fewer—and you need at least one nonlinearity. Lorenz had two. So that was what Rössler wanted. Could [he] produce an even smaller system, three equations with just one nonlinearity? That would be the absolute minimum. Could it be done? He pondered this for quite a while, and the breakthrough came when he was at a carnival looking at a taffy-pulling machine. When he watched this strange machine looping and folding and twisting, stretching and folding the taffy, he somehow figured out how to think of the equations that would do the analogue of that process, wrote them down, and miraculously, they did produce a strange attractor simpler than Lorenz's. Nowadays, we know it as the Rössler attractor.

The Rössler attractor, as you see, was made in a sort of fit of inspiration from looking at—again, notice the food theme. Noticing the taffy, he did not have any particular scientific application in mind. Remember, Lorenz was thinking about convection, and Rössler was just thinking about taffy. He was not thinking of anything scientific. So it's interesting that nowadays we have seen that the Rössler attractor is out there in nature. In fact, you've seen it without realizing it. When we were discussing the chaotic chemical reaction in Lecture Twelve, I pointed out a structure that I said looks like one wing of Lorenz's attractor. Maybe you remember it. That one wing is the Rössler attractor. So it was always there, waiting to be discovered, and it does exist in this real chemical experiment and in many other examples of chaotic systems in nature.

Finally, the third person was an astronomer, Michel Hénon, a Frenchman. His is the work I want to focus on now because it's beautiful and vividly revealing of this microstructure of strange attractors, this infinite layering that I have tried to convey through the analogy. Hénon did computer experiments that will let us see this infinite structure directly. What's more, there is a nice feature of his work, which is that it brings us to self-similarity. His calculations showed that there weren't just infinitely many layers. They were arranged in a hierarchical way, a way that repeats and was "self-similar," so that each small part of the hierarchy was reminiscent of the original whole.

I want to refer you to this previous slide that I showed, having to do with the horseshoe mapping. I pointed out that when we started with the horseshoe and then created four layers and then eight, there is a kind of hierarchical structure that should be visible here. Notice that the original structure has one big gap. A *gap* will refer to the white part between the layers. We started with one big gap. Then we had one big gap and two smaller gaps. Then, after one more round, that same structure is there but at two different scales. Notice there is a huge gap surrounded by two smaller gaps, ignoring the very smallest gap. Suppose your eyes weren't good enough to see it or your resolution wasn't sufficient. Then, if you now focus only on this scale, one-half of the structure, and look there, your eye sees the same type of pattern we saw to the left: a big gap. Now assuming you're at this small scale, your eyes are better. You see this medium-sized gap surrounded by two even smaller gaps. So there's a kind of repetitiveness to the structure built into the pastry-making process. There is self-similarity there to begin with. Maybe this doesn't [strike] you as math, and so I think when you see the Hénon map, you will be completely convinced about what I'm saying about self-similarity.

Self-similarity is a kind of symmetry. That's a cool way to think about it. It's a symmetry under magnification, in which small pieces can be magnified to look just like the whole. It's one of the defining features of a fractal.

You might be wondering, "What's an astronomer doing thinking about this? Why would Hénon be interested in this question to begin with?" It's sort of an amusing story here. He devised his mathematical version of the pastry-making process because he wanted to visualize the alleged infinitely many sheets in the strange attractor of Lorenz. Remember, I keep saying that there's somehow this mica-like substructure with all these sheets. Hénon heard about that when he was at a lecture, just like you are all sitting in a lecture, being given by one of his colleagues, Yves Pomeau, and Pomeau was carrying the news of the Lorenz system to the French scientific community. Hénon had not heard about this before and thought it was very interesting about all these infinite layers. "How come I can't see them? I want to see them." So he tried to concoct an example simpler than the Lorenz system that would allow him to directly visualize what he became very curious about after hearing Pomeau's lecture.

Here is how Hénon's mapping process works. He takes a solid rectangle, just like the dough, and then stretches it. He has the same idea. [He] takes a solid rectangle, stretches it, folds it into a horseshoe shape, much like Smale did—but Hénon, being a scientist, is not afraid or embarrassed to use a computer, and he is just going to show the results, plot them. He's not necessarily trying to prove anything. He just wants to understand. It was very advantageous. Everyone else could understand what he was doing, too, whereas Smale stuck to the realm of abstraction, and so only mathematicians could understand it.

Hénon, then, with his process, now writes down equations for this process so he can implement it in the computer. Let me now show you a simulation of what Hénon's mapping does. Here is what you are going to see: You will see a plane with two axes, an x, y plane, and there will be a yellow point at one place on the plane. You can think of that as a speck of butter—yellow, butter—starting somewhere in the baker's dough. Remember, the plane is the dough. Then we're going to implement these equations, the Hénon mapping, and start stretching and folding the dough by iterating the map. You won't see any stretching and folding happening. All you'll see is the action that this process would have on the speck. It will be [as] if the dough were invisible but there was an illuminated, glowing speck. Just watch.

Where does that thing go as the process unfolds? When we repeat the process many times, you'll see that speck popping all over the place, and we'll make a time-lapse photograph, the way we always do when we want to show an attractor. We will keep all the specks on the page. The first striking thing you will see is that the speck is hopping around crazily with no obvious pattern. That's chaos. So it has this aperiodic, crazy-looking hopping, and then the miracle is that out of that crazy hopping a structure will start to appear before your eyes; as James Gleick put it in his great book on chaos, "It will appear like a ghost out of the mist." I love that. You will see the ghost appearing out of the mist. Let's see if we can see all of that.

There's the plane, and it doesn't matter what the axes are. Just think of them as abstract, x and y. For those of you who are enamored of algebra, and I know that some of you are, you may want to play with this on your own computer. Here are equations for the Hénon map with parameters a and b in them. You can type in the a and b. I'm not going to get into that, but it's certainly possible if you're curious. We can talk more about that.

Let me begin by showing the first yellow speck. There it is. You see a yellow dot. Now I do the Hénon map to it, and watch where the dot goes. Keep your eye on it. There's another one. It's hopping around. It seems to be jumping back and forth across. Now it's getting farther away. That's the stretching process having its way. Now it's down there. You see what I mean? It's hard to figure out what's happening. It's hopping around pretty crazily. If we iterate long enough, we should see the ghost coming out of the mist, so let me click the "iterate" button and have the computer do this much faster than I can with my finger.

You see something coming out, I imagine, which is something roughly resembling a boomerang. It looks like it has sheets to it, which is encouraging, because, remember, we're using this example to try to demonstrate the multi-sheeted structure of a strange attractor. This object that's appearing before your eyes and is now pretty well formed is the Hénon attractor. By the way, it's always the same. If I dare do this, I'm going to clear the screen and try starting from a different point, and we should see the same structure appear even though the individual points at the beginning are quite different. That's not the same pattern we saw originally, but the overall gestalt will be the same when I let it go. That's the point. It's an attractor. It will attract everything no matter how you start, as long as you're in the black region.

There's the Hénon attractor. What I want to show you is that up close, the boomerang has substructure. I'll drag the little box over part of the structure, and that will allow us to zoom in. We'll blow up the part that's inside what I'll call the operating theater. That will be blown up to the full screen, and you'll see some interesting structure. Let's see what it looks like. Suppose I choose this region. It doesn't really matter where I look. That's going to be blown up. You see something that looks like it has two layers, which is amusing. There is some other little bended piece sticking up here, but fine. Here is what looked like two layers, and we're not sure if we're seeing that right because it does look a little bit like this line is thinner than that line, doesn't it? So let's zoom in on this line because you know it's not a line. It has got to have substructure. This is going to be fractal, right? Do you see all that substructure? It has one little, thin line next to 2, next to 3, if you can resolve that. I won't belabor this, but if we continued to zoom in, you would see that the 1, 2, 3 pattern repeats all the way down. But notice that here on the outer edge of what looks like the rings of Saturn, there appears to be a thicker line. If we were to zoom in on that, we would see it's not a line. It's a 1, 2, 3 but smaller scale. Do I dare? I'll dare. The computer may get slow because we are now at a very, very small scale. You can see there is substructure there. Do you see the 1, 2, 3 reemerging? Good. I better stop that before it starts doing something I don't like. So that's the basic structure of strange attractors. This is what Lorenz meant by an "infinite complex of surfaces."

Self-similarity, this infinite repetition of architectural theme, is now associated with fractals, but in fact, as you have probably realized, it's a very old, ancient idea that is just only recently being mathematized but has been with us forever. Artists with a keen eye for nature have long been fascinated by it. Think, for example, of Leonardo da Vinci with his pictures of flowing water. These are his sketches from the 1500s—flowing water. What I would like you to see in that is that in thinking about the nature of turbulence— da Vinci is looking here at water pouring out of some sort of channel and then it's spilling down here. Notice the eddies. There's a big whirlpool of this scale, but then there are tinier eddies of a smaller scale, and tinier and tinier eddies. He realized that turbulence had a sort of infinite, self-similar structure of eddies.

Or think about this famous picture, a woodblock print from Hokusai, *The Great Wave*. It's actually often described as a view of Mount Fuji. There's Mount Fuji in the background. Of course, Mount Fuji is such an awesome structure that it was an interesting creative choice to make it look so tiny here, so dominated by this wave. These poor fishermen are being tossed around by this wave. But what I want you to really look at is the fractal structure of turbulence. Look at the way he depicts the wave, with its ominous, almost like claws coming out, the water breaking into little branches. Even the spray appears to be fractal, with big dots and little dots.

Finally, if you're thinking about infinity in a confined space, the natural person to think of would be Escher. Here is his diagram known as *Smaller and Smaller*, with these interlocking lizards, geckos or whatever they are, a black one next to a white one next to an orange one, moving in the same motif, smaller and smaller, descending into infinity in the center of the picture. Here, combining a kind of symmetry under magnification also with a symmetry with rotation, a gorgeous exploration of symmetry, both fractal and a more conventional rotational symmetry.

Once you start to look for fractals, you'll see them everywhere in the world around you, even inside yourself—from mountains and clouds to the branching patterns of your own blood vessels.

In the next five lectures, we'll define fractals more precisely, and we'll see how they illuminate a diverse set of complex phenomena in nature, technology, business, and even art. See you next time.

The Properties of Fractals
Lecture 14

The key point is that fractals behave oddly when you magnify them.

L ast time we discussed fractals as the embodiment of chaos. Now we look at fractals more generally, beginning with their two most distinctive features (their inexhaustibility of structure and their self-similarity) and then touching on some larger issues:

- Where else do fractals arise, outside of chaos theory?

- How did the science of fractals come into being?

- Answering these questions will help us appreciate the unique relationship between fractals and chaos, as well as the place of fractals in the broader scientific landscape.

To contrast fractals with familiar (non-fractal) shapes, imagine looking at a fractal under a microscope. As you turn up the power, inexhaustible fine structure is revealed. In contrast, Euclidean shapes (like circles or spheres) start to look featureless as you zoom in. Furthermore, the small features of a fractal resemble the larger ones. In this sense the structure is "self-similar." For an edible example, consider a head of Romanesco broccoli, whose florets look like mini-Romanescos.

Strictly speaking, however, only mathematical fractals are truly inexhaustible and self-similar. Real objects (like the Romanesco) are only approximately fractal. The repetition of structure isn't perfect, and it stops below some smallest scale. Thus, the issue is the range of scales over which an object can be usefully regarded as fractal. The wider the range and the more faithful the self-similarity, the more fractal it is.

Self-similarity can be viewed as a kind of symmetry. An object has a symmetry if its stays the same despite a change. For example, a sphere has rotational symmetry—it stays the same if you rotate it. Our bodies have mirror symmetry—the left side looks like the right. Likewise, a fractal has scale symmetry—it stays the same if you magnify it. This type of symmetry hasn't been studied much until recently.

The concept of *scales* is important for distinguishing fractals from non-fractals. Ordinary objects tend to have just a single characteristic scale, given by the size of their smallest features. Below that scale they look bland and unstructured. In contrast, fractals vary over such a wide range of scales that they are described as being *scale-free*. For example, geologists are taught to place a hammer or a coin next to any craggy rock formation before photographing it. Otherwise, without these clues to set the scale, the viewer will be unsure whether the image shows a mountain, a cracked pebble, or a magnified speck. Likewise, the jagged charts of stock prices look remarkably similar, whether shown minute-by-minute over the span of a day, or day-by-day over the span of a year. There's something liberating about being scale-free. Maybe that's why we find such beauty in the appearance of clouds, cracked rocks, trees, and frost on a window pane.

But the scale-free structure of fractals also poses confusing challenges of measurement. Suppose you were asked to the measure the surface area of a football field. It's 100 yards long and 50 yards wide, so you answer, "5000 square yards." But that neglects the surface area of each blade of grass. In this problem, there are (at least) two scales—that of a human and that of a blade of grass (which, if an ant were making the measurement, would be the more obvious scale). The point is that in problems with multiple scales, it's meaningless to report a single answer. The answer you get depends on the resolution you use. The right approach is to quantify exactly *how* the answer depends on the resolution. We'll elaborate on this in the next lecture, using Norway's coastline for illustration.

As the examples of broccoli, coastlines, stock prices, and mountains should suggest, fractals have real-world applications far beyond the strange attractors and other abstractions we encountered when visualizing chaotic dynamics. The ideas of fractals have now been incorporated into virtually every branch of science. But the subject was largely unknown before the mid-1970s, when it was synthesized, popularized, and christened by Benoit Mandelbrot (b. 1924), a polymath then working at IBM and now associated with Yale. Mandelbrot's interests ranged from commodity price fluctuations to computer graphics, and from the structure of galaxies to the firing patterns of nerve cells. Besides his own pioneering contributions, he drew on older, scattered work from an amazing range of disciplines—everything from pure mathematics to architecture—and thereby created the new field of "fractal geometry," which he named for all that is fragmented, fractured, broken, and irregular. There's an irony here. Earlier mathematicians had created fractals as pathological counterexamples, monsters to chasten their intuition of what a curve or surface could be like. These examples were regarded as weird and unnatural. Yet nature has been using fractals all along!

Is fractal geometry part of chaos theory, or vice versa? They are parallel developments. Both deal with irregularity. But fractals are mainly about irregularity in *space*; chaos is about irregularity in *time*. Each developed independently, but also with important intersections (such as the fractal shapes that arise in the visualization of chaotic systems, as we've discussed).

Fractals are mainly about irregularity in *space*; chaos is about irregularity in *time*.

Still, that's not why the public associates chaos with fractals. Their association in popular consciousness is something of a historical accident. Both subjects became media darlings at the same time. Chaos theory was confirmed experimentally in the early 1980s, just before the first spectacular images of fractals began to appear. And perhaps their simultaneous appearance was a reflection of a zeitgeist in the world of science, a readiness to finally confront the disorderly, jagged side of nature and to search for the hidden patterns within it. ∎

Essential Reading

Gleick, *Chaos*, 83–118.

Mandelbrot, *The Fractal Geometry of Nature*.

Schroeder, *Fractals, Chaos, Power Laws*, chaps. 1, 10.

Stewart, *Does God Play Dice?* chap. 11.

Liebovitch, *Fractals and Chaos Simplified for the Life Sciences*, part I.

Strogatz, *Nonlinear Dynamics and Chaos*, chap. 11.

1. Some objects found in nature are Euclidean (like crystals) and others are fractal (like ferns). Give several examples of each type.

2. Real objects are fractal only over a limited range of scales. What other scientific concepts hold only over a limited range? Is this true of all scientific concepts?

The Properties of Fractals
Lecture 14—Transcript

Welcome back to fractals. I'm really looking forward to this lecture because, for once, if I suddenly get hungry, I just might munch on one of the examples. You'll see what I mean soon enough, but before I subject you to that spectacle, let's recall where things stand.

Lecture Thirteen: We tried to gain some intuition about why fractals and chaos were so intimately related, why strange attractors have this infinitely layered self-similar structure at the same time as they arise as the accumulative result of chaos in a state space. Now we want to step back a bit from this point of view. We're not going to be so much living in state space any more but now in real space, the real world. We'll try to look at fractals more generally.

First, we want to clarify their two most distinctive features. You might even think of them as their defining features: (1) their inexhaustible richness of structure and (2) their self-similarity. Then, we'll also touch on some larger issues, such as where else do fractals arise besides state space? Outside of chaos theory, that is, where do they arise? Secondly, a historical point, how did the science of fractals come into being? Answering these questions will help us appreciate the unique relationship between the subjects of fractals and chaos, as well as the place of fractals in the larger scientific landscape as a whole.

Let's begin by clarifying what's distinctive about fractals by comparing them with familiar, non-fractal shapes and processes. The key point is that fractals behave oddly when you magnify them. Here is a schematic of a fractal shape. It's some kind of wiggly thing. I've drawn a little box around a piece of it to suggest that this is a region that I want to magnify and blow up into the next figure. What appeared here to be sort of a rounded, curvy feature, under magnification, you start to see it has wiggles on it that weren't visible at the earlier resolution. If we blow up one of those rounded features at that resolution, then we see that that rounded feature itself has wiggles on it. So that would be typical of a fractal, to see some kind of theme—in this case, wiggles—repeating at smaller and smaller scales. The structure is inexhaustible. Finer structure is revealed no matter how far down you look.

In contrast, Euclidean shapes, by which I mean things that you would have studied in high school geometry: circles, straight lines, squares, spheres, things like that. Imagine them in their most perfect form. I'm not talking about a real circle, say, with a loop of string. I mean an abstract, Platonic circle, a perfectly thin line formed into a circle. So think of the perfect circle or the perfect sphere. A shape like that, if you look at that closely enough, if you imagine zooming in with a microscope on a shape like that, it starts to look smoother and smoother, and more and more featureless. That's what I'm trying to indicate here in this lower panel, a typical non-fractal shape—I'm using the circle as illustration. Again, if I blow up the curved region, it now starts to look straighter but also smoother. Then, if I blow up that region, pretty soon it really starts to look featureless. You will just eventually see a line with nothing more to see. The show's over. That's the first property: that fractals have structure that goes all the way down.

You might be wondering, at this point, and maybe we should clarify it before we go any further—anything real, like I spoke of the piece of string as a kind of imperfect realization of the Platonic perfect circle. If I had a loop of string, you might say that has structure all the way down, too. It's true; it does. I could look at the string and then if I magnify in, pretty soon I would start to see little fibers of the string and there would be structure there. Then I could look at the fibers under a microscope and see things there. Conceptually, if you had an electron microscope, pretty soon, you'd be seeing things at the molecular scale, so you could argue that this whole distinction is bogus. Everything has structure all the way down. I agree with that, but it doesn't make the distinction bogus. We're still in the realm of mathematics where I'm trying to make the distinction between idealized mathematical fractals and idealized non-fractals, these perfect Platonic shapes. Just bear with me and try not to think too much about the fact that real things do have atomic structure and so on. That's true but beside the point here.

The second defining feature of fractals, then, is that the small features resemble the larger ones. In that sense, the structure can be said to be "self-similar." For an edible example—this is what I was warning you about—let me first illustrate this—did you notice, by the way, this theme of food? In the previous lecture, I was concerning myself with taffy-pulling machines and pastry, which I know, I didn't necessarily give the best culinary discussion of, but still, food. What's with food? So now it's a head of Romanesco broccoli, considered by some to be a kind of cauliflower, so if you go to your supermarket, you can order it under either variety. But here's what it looks like. If you haven't seen it, you'll be amazed. By the way, everyone calls it "Romanesco," so that's what we'll call it. Ask your grocer for Romanesco, or sometimes Romanesque, or the Italian word, Romanesca. Anyway, it's this unbelievable fractal vegetable. Look at this thing. From a distance, you just say, "Wait a second; it just looks like a cone." But of course, you can see it has all kinds of florets coming off of it. It has subcones, smaller cones, and those cones have cones. In fact, they're not really cones. The way to think of it is that this whole shape is kind of a fractal version of a cone. It's a very bumpy cone, and each of the florets is a mini-Romanesco. If we were to zoom in on this mini-structure, you would see it's pretty much a faithful copy of the original larger structure. What is even more stunning is that that mini-Romanesco has little florets on it. If you were to zoom in on one of those, again, [you would see] the same structure. The repetition goes on for at least three levels.

Seeing this in person is even more dramatic, and it just so happens that I have one. Here is a Romanesco. Take a look at that. It's incredible, and as I say, it illustrates the scheme of self-similarity. There are tiny, tiny florets near the top, bigger ones at the bottom. The structure is architecturally identical all the way around and also at different levels.

Strictly speaking, any real fractal, like this one, as opposed to this idealized Platonic—I shouldn't say Platonic because Plato never thought about fractals, but you know what I mean—idealized, perfect mathematical fractals. Those are truly inexhaustible and self-similar, whereas real things like this are not. They couldn't possibly be for all kinds of reasons. Real objects, like the Romanesco, are only approximately fractal in various ways. For one, the repetition of structure is not perfect. It can't repeat all the way down to the molecular scale. It can't repeat at the subatomic or quark scale, of course not. There is some lower limit where this particular architectural motif, a bumpy cone, has to stop. Fair enough; that's true.

I should probably say there's one other issue as far as the imperfection goes. There is the limitation of scales; that's one thing. There's another limit, which is that the replication may not be perfect. The similarity itself may not be perfect. That is, if I took one of these small florets and could somehow magically enlarge it to be the size of the whole object, it wouldn't necessarily be a perfect copy either. So the self-similarity may not be perfect, and the inexhaustibility may not be perfect. Both things can break down.

So the issue is: Over what range of scales can it usefully be regarded as fractal? Does it hold over 3 levels of florets, or can we go down 10 levels? It becomes a quantitative issue. It's a matter of degree, to what extent this fractal approximation is a useful thing for thinking about a real object. Likewise, when the self-similarity is not perfect, we can say [that] the features resembled the original whole. How much do they resemble? We could try to quantify that. In some cases, the best that we could say is that there is a kind of statistical resemblance; that is, the shape is roughly reminiscent in some average sense but not more than that.

Keeping all that in mind, then, we're going to use the term *fractals* to refer to this whole class of objects, from the most perfect ones to the ones that are only very roughly, crudely statistical. The more scales we have over which the self-similarity holds, the better, and the better the replication, also the more it's a fractal.

In most real examples, what you find is that typically you would have about 2 orders of magnitude in size over which you would see self-similarity. Now that maybe doesn't sound like a lot. Two doesn't sound like a big number, but 2 orders of magnitude means 2 powers of 10. It's a factor of 100 in size, which I can't indicate with my arms very easily. I suppose if I go like that. That's a small distance, and then maybe 100 times bigger is that. That's about what you would see in most real examples. I'm not sure if that's true of this broccoli. It might be. That is, the scale of the whole object. Here it is. If I were to go 100 times smaller, can I find any floret that resembles that broccoli? I think I can. Of course, I want to say "yes." There are tiny, tiny ones at the very tips, which look like they're trying to be little mini-Romanescos. It's a ballpark number. Two orders of magnitude is what you see in practice.

For many people, that's not a very convincing thing. They say, "That's not a very good range of scales. I would like to see more." In some examples, we will see more. When I show you some data relating to fractal frequency of earthquakes, you will see 8 orders of magnitude. In a later lecture, I'll be discussing fractals inside our own bodies and a certain scaling law that holds across different organisms, from microbes to the hugest animals—to whales. There, we're going to have something holding over 27 orders of magnitude. So there are laws that are very convincingly wide-ranging, some which are less so.

That's a serious issue that plagues the whole subject of fractal analysis. It comes up a lot. I don't want to bicker about it a lot, but I want to remind you that it's there and it will come up again in Lecture Eighteen when we talk about Jackson Pollock's drip paintings. Fractals have been used to analyze the structure of his paintings, and there has been a controversy recently in regard to a certain set of paintings that were discovered in 2005 alleged to be by Jackson Pollock. Are they real, or are they fake? Fractals give us a possible answer to that question, at least one insight into the authenticity of these paintings. So we'll discuss that analysis in Lecture Eighteen, and this issue of "Is it really fractal or not, and over what range of scales?" will come back strongly at that time.

An interesting way to think about self-similarity, as I mentioned briefly in the last lecture, is as a kind of symmetry. I didn't really say what a *symmetry* means. I think you probably have an instinct for what a symmetry means, but the way a mathematician would put it is that an object has a symmetry if it stays the same despite a change. It's sort of a cool way to think of it. Something that stays the same despite a change has a symmetry. For example, a sphere clearly has some kind of symmetry. It has rotational symmetry because if I rotate a sphere, it stays the same. Our bodies have approximate mirror symmetry, at least on the outside, in that our left side looks sort of like our right side, approximately. Not on the inside. Likewise, a fractal has a symmetry, except it's not a symmetry through a mirror, and it's not a symmetry of rotation. It's a symmetry under magnification. Small things look like the big thing. That is sometimes called the *scale symmetry*, a symmetry under a change of scale. It stays the same if you magnify it.

This type of symmetry really hasn't been studied that much until recently, much less than rotational symmetry, mirror symmetry, or some of the others. That's one reason that fractal geometry is a relatively new subject. We haven't looked much at scale symmetry, certainly in the past century, and even much more; it has become a very active topic in the past 30 years or so.

I've been using the word *scales*, and maybe it's time that we should be a bit more precise about what we mean by *scales*. Let's discuss this concept in some depth now because it's important, especially for distinguishing between fractals and non-fractals. Ordinary objects—again, I don't mean real ones with molecular structure. Think of a perfect circle, say. An ordinary Euclidean circle just has a single scale, given by the size of its smallest feature, which in the case of a circle, there are really no features other than its radius. When you go way below that scale, as I showed in an earlier diagram, you see that just a piece of the circle starts to look like a curved line, and very soon, as you magnify further, it just looks like a line segment. There is nothing to see. So below a certain scale, Euclidean objects start to look featureless and unstructured, bland.

In contrast, fractals vary over a wide range of scales. I've suggested in the case of most real ones, over 2 orders of magnitude, some over many more orders of magnitude. Something that varies over a huge number of orders of magnitude in effect has no scale that defines it. In that way, we could think of it as being *scale-free*. It's free of any inherent scale. Let me give you some examples of this that I'm sure you're familiar with, but maybe you haven't thought of them in quite this way.

Suppose you are a geologist and you're outside doing your fieldwork, looking at rock formations. You find something interesting, and you want to make a photograph of it to show your colleagues, or to publish, or to teach the public. Well, there's something that all geologists do when they make a photograph of a rock formation. You know what I'm thinking, right? You've seen these geological pictures. They always put something in the picture to give you a sense of scale. Typical for a geologist, the person is walking around with a rock hammer, so they will put a rock hammer in the photograph next to the structure of interest. Or they might put a coin, or because they're taking a picture, they've taken the lens cap off their camera, they might put the lens cap in the picture. This sort of thing sets the scale, and now you know what you're looking at. Otherwise, the viewer has trouble; the viewer might be unclear. "Am I looking at the cracked face of some interesting mountain, or am I looking at a pebble that's kind of scratchy-looking, or am I looking at a microscopic speck?" It can be very confusing in geology because rock formations are often scale-free.

For instance, let me show you a picture of a rock formation, and you tell me what the scale is in this picture. There are cracked rocks of some kind. Is that a satellite photograph of some barren landscape, or is that a close-up of a smashed-up rock I could hold in my hands, or is it a microscopic picture of a rock that would just be a little dot on your fingernail if you looked at it in real size? Can you guess the scale? I'm sure you have a guess. I'm not going to tell you the answer. I want you to be a little uncertain about it because that's my point, that it's hard to tell.

Let me give you another example from geology that makes this point in a different way. Here is the conventional way to show the scale in a rock formation. I'm going to show you two rock formations. There they are. I'm underscoring that they're fractal. That is, if you didn't have the lens cap there, you would say, "It's sort of hard to tell what scale those pictures are." Is this the side of a mountain? Is this just a patch of dirt, like on the scale of a human being? It's not a very interesting question because I've given it away. I've shown the lens caps, so you know that these two pictures have basically the same scale, except it's a trick photograph. This is the real picture. Do you see what you're seeing? This is the scale it's supposed to be. That's a human finger, and that's a hand. This is a human being standing next to a gigantic trick lens cap. So this is the side of a mountain, and this is just some dirt. Both look like sedimentary rock formations, some kind of outcropping, but it was very—I hope; maybe it didn't work for you, but if the trick worked, then it makes the point that rock formations have this kind of fractal character.

Let me thank the person who provided these photographs, Steve Wheatcraft, a geologist at the University of Nevada at Reno. Thank you very much, Steve, for sharing this trick picture with all of us.

Likewise, though, you can probably think of other examples where scale is a confusing issue. For instance, if you look at a newspaper, and you are obsessing over your stock portfolio, and you happen to see a chart that catches your eye that lists the stock you're interested in. Is this the minute-by-minute stock report on the day that you're reading the paper, or is it a yearly chart, or is it over a decade? It can be tough to tell just from looking at the curve. The curve looks jagged no matter what it is, whereas if you have scale indicators—like on the chart, it says what the year is or whether it's minute by minute. Or another scale indicator would be to look at the size of the fluctuations on the vertical axis. How much is the stock moving around? Then you can tell from that. Otherwise, it can be tough to tell because stock prices, to a certain extent, are often scale invariant.

This is a bit of an unscientific point I want to make now, but I feel like it should be said. There is something inexpressibly beautiful about being scale-free. There is something that appeals to us about the organic nature of scale-free structure. To show you what I mean, see if you agree with me that there's something just beautiful and moving about the fractal forms all around us, such as clouds. Really look at them with the eyes of a child. Isn't that beautiful to see? Here is a big, big cloud, but if you look closer, you'll see sub-clouds that look like the original cloud, and there are even little pieces of the cloud, tiny puffs that resemble the whole thing. I find that beautiful.

How about beech trees? We talked about fractal fig trees in state space. These are real trees, which obviously—branches look sort of like the main trunk, and they split into smaller twigs, and so on. Does it do it for you? It does for me.

How about frost on glass? There's a main body splitting off into smaller structures with tendrils sticking out and even tinier tendrils sticking off of those. There is something gorgeous about the scale-free quality of the fractals that we see in the world around us.

The scale-free structure of fractals also poses certain challenges. There's a kind of downside to these beautiful structures, at least for the scientists. They can be confusing to analyze, and they can ensnare a naive scientist who doesn't know the right things to measure when trying to measure a scale-free object. I should say [that] all of us were naive before about 30 years ago. Fractals were unfamiliar, and people made certain dubious measurements because they weren't really sure what to be measuring. They were making measurements appropriate in the Euclidean classical context that didn't make sense when we were dealing with these more complex objects.

When there is more than one scale to a system, you have to ask the right question, or you may not get a sensible answer. This is a trivial example, but I think it will make the point before we get to harder ones. Suppose you were assigned the task of measuring the surface area of a football field. It's not hard. You think: "I know. It's 100 yards long; it's 50 yards wide. I can multiply. I get 5000 square yards." You're told, "That's the wrong answer because I meant the surface area as seen from the point of view of an ant." Now an ant would measure a different surface area. If you were reporting to the king ant, you would be punished for that answer, because the king ant would say, "You left out all the interesting structure. The blades of grass have surface area, you moron. Why didn't you measure that?" Well, down there in ant world, the ants would have measured a much larger surface area because they would have included all the surface of all the grass blades. The answer might be 100 times bigger than what you got.

That's to illustrate that in a problem where there are two natural scales, the human scale and the ant scale, or the blade of grass scale, there is no single right answer. It depends on the resolution of your measurement. It's meaningless to report just one answer. You have to say, "At this scale, at the scale of the human being, the answer is 5000. At the scale of an ant, it's whatever it is; it's 100 times bigger than that." That's the only correct way to report the answer; give an answer relative to a resolution of measurement. That's the new idea here.

So we need to quantify *how* our answers depend on the resolution we use. In the next lecture, we're going to be examining that issue more closely, using the length of Norway's amazingly jagged coastline as an example. If you're not so familiar with Scandinavian geography, let me remind you. That's Norway, or just a piece of Norway. It's famous for its fjords. Here is the ocean, and I suppose that's the North Sea. I'm not so familiar with geography as I should be, but there are all kinds of inlets and little bays, and there's a very intricate structure coming in.

If I asked you, measure the total length of the coastline—well, at what scale? Are we talking the scale of this satellite photograph? Are we talking about the scale of a person walking along the beach? We'll discuss that example more in the next lecture, but we will find out that a new concept called the *fractal dimension* is the right way to quantify this sort of infinitely rugged shape.

As the examples of broccoli, coastlines, stock prices, mountains—as all of those should suggest to you by now, fractals have real-world applications far beyond the strange attractors and other abstractions that we were busy visualizing when we were focusing on the world of chaotic dynamics. The ideas of fractals have now been incorporated into many branches of science, as we will see in subsequent lectures. But the subject was largely unknown before the mid-1970s, when it was synthesized, popularized, and christened by Benoit Mandelbrot, a polymath then working at IBM and now associated with Yale. I recently got a phone call from Benoit, and it reminded me of something that I had always known. That he is so urbane and funny, with a dry sense of humor and a gift for language, so I would heavily recommend you read any of his work. He's not always the clearest person, but he is amazingly brilliant, and you will learn a lot. I hope you will enjoy him, but don't expect to follow him all the time. It's worth it, though.

[Mandelbrot] has made pioneering contributions to more subjects than anyone else I can think of. His interests over the years have ranged from cotton price fluctuations, to computer graphics, to the structure of galaxies, to the firing patterns of nerve cells, and I could go on. He is literally a polymath. He's amazing. Besides his own original contributions, though, he's a scholar, and he drew on older, scattered work from a wide range of disciplines, everything from pure math to architecture, music, all of it. He thereby created the new field of "fractal geometry," which he named from the Latin word *fractus*, which refers to all that is fragmented, fractured, broken, and irregular.

There is an irony here because earlier mathematicians had created fractals. These are some of the people that Mandelbrot drew on. They had created fractals deliberately as pathological counterexamples, as mathematical monsters, to try to chasten their intuition about what a curve or a surface could be like. For example, in the late 1800s, early 1900s, this was a period when math was turning inward. It didn't want to have very much to do with the natural sciences because there was a crisis of rigor. Things that seemed to make common sense, things that were good intuitive ideas weren't working and were giving weird results mathematically. So they felt they needed to get more rigorous. The question at the time was: What do we really mean by a curve? We think we know what we mean by a curve. Something you can draw with your pencil, or if you were a mathematician, you might say "a continuous mapping of an interval into a space," something like that. But those turned out not to be good definitions because under that definition, that fancy second one I just gave you, a curve could include something—get ready for this—it could include something that goes through every point in a square region and fills the entire 2-dimensional space even though it's an infinitely thin curve. There are objects called *space-filling curves*. And not just 2-dimensional space. You could fill this whole room, literally going through every point in the room with a curve. Would you call such a thing a curve if it fills a solid, 3-dimensional region? You wouldn't. So what do we mean by a curve?

It turns out this thing that I just described, this space-filling curve, is a construction of a fractal. It was that type of thing that led mathematicians to invent fractals long ago. But this is the irony. Those examples were regarded as weird and unnatural. The mathematicians who created them were proud of that. They didn't like nature, and they thought nature was misleading, and [they didn't] deal with intuition. Let's just be rigorous. The irony is, of course, that nature had been using fractals all along. Now, with Mandelbrot's framework, we can see that. We can finally see what we always saw. It's that same psychological thing that has happened in chaos theory, too, where people see things, but they can't see them until they have a conceptual framework for seeing them.

I'm going to be talking about fractals for a few more lectures, and you may wonder, "Why? I thought this was a course on chaos." I tried in the last lecture to show you that fractals and chaos are intimately related, and you might be wondering, "How exactly? Is fractal geometry part of chaos theory, or vice versa?"

You could think of them as two parallel developments. Neither is really a part of the other. They are two parallel developments with tremendous overlap. Both deal with irregularity, but fractals are mainly about irregularity in *space*. Chaos is mainly about irregularity in *time*. That's a crude distinction, but basically that's the difference. One is about time and dynamics; one is about space and patterns in space. Each developed for its own reason, largely independently, but also with important intersections, such as the fractal shapes that we've been seeing earlier in the course, in the visualization of chaotic systems. But that's not really why the public associates fractals and chaos. You always hear them mentioned in the same breath, I think. Why? I would say it's a historical accident that both subjects became media darlings, pop sensations, at the same time in the mid-1980s. By coincidence, chaos theory was being confirmed experimentally at that time just when the first spectacular, multicolored images of fractals began to appear, the most famous of which is one called the Mandelbrot set. You have probably seen that one, and I will be discussing it later in the course.

When people first got a look at the Mandelbrot set, and when they were hearing about chaos, they were getting it from both sides, and we just think of them as being the same subject. They're really not. They're close, but they're not quite the same. Still, perhaps their simultaneous appearance was a kind of a reflection of a zeitgeist in the world of science, a readiness for humanity to finally confront the disorderly and jagged side of nature that it had neglected for so long and to search for hidden patterns within it.

Next time, we'll take a look at one of the most amusing and counterintuitive things about fractals, which is that they can have a dimension that's not a whole number. You can have a fractal that's 1.26-dimensional or something like that, a number in between two whole numbers. It does. I promise you this will make sense, and I'm eager to tell you about it, but I'm hungry now. Excuse me. I see this Romanesco broccoli. I'll see you next time.

A New Concept of Dimension
Lecture 15

First of all, scientists use fractal dimensions all the time to characterize the roughness of highly irregular objects of all sorts: coastlines, tumors, galaxies, even the folded surface of the human brain.

Perhaps the strangest thing about a fractal is its dimensionality. Common sense tells us that a line is 1-dimensional, a surface is 2-dimensional, and a solid is 3-dimensional. But for the highly corrugated shapes typical of fractals, our usual ideas about dimensionality break down. In this lecture we'll explain how to generalize the concept of dimension so that it can handle fractals as well as ordinary, Euclidean shapes. The weird but inescapable conclusion is that fractals are so convoluted that they're not 1-, 2-, or 3-dimensional but somewhere in between, such as 1.26-dimensional! This might seem like an esoteric subject, but it's worth studying for several reasons. Scientists use fractal dimensions to characterize the roughness of highly irregular objects. Before this concept was developed, there was no way to quantify different degrees of jaggedness. We need to understand how fractal dimensions are calculated because they'll come up again later, for instance when we discuss the controversy about the authentication of some recently discovered paintings alleged to be by Jackson Pollock (Lecture 18). And it's fascinating to think about something as basic as dimension in a new way.

The easiest place to start is with objects that are perfectly self-similar. Such objects are made up of copies of themselves, down to arbitrarily small scales. For example, consider a square patch of carpet. It can be divided into many smaller squares, each a tiny copy of the original. For instance, we can chop the carpet into 4 smaller squares by halving each of its sides. Or 9 smaller squares, by dividing each side by 3. Now we're getting close to seeing how dimension enters into it. Notice that $4 = 2 \times 2$, and $9 = 3 \times 3$. Writing this as a number raised to a power, we'd say $4 = 2^2$, and $9 = 3^2$. That power of 2 is no accident. It occurs because the square patch is 2-dimensional. Likewise, slicing a solid cube in half on each side yields 8 smaller cubes. Here, $8 = 2 \times 2 \times 2 = 2^3$, and the power of 3 occurs because the solid cube is 3-dimensional. The power (or exponent) in such an equation tells us the object's dimensionality. If we use the symbol m for the number of smaller copies, r for the reduction factor in each direction, and d for the dimension, we can summarize all this with the equation $m = r^d$.

Now play the same game with a self-similar fractal. We'll find that the exponent—which we still interpret as the object's dimension—is no longer a whole number. Consider the Koch curve, an exquisitely thin and crinkly filament that looks somewhat like the perimeter of a snowflake. It's constructed by iterating a simple geometrical growth rule. Start with a straight line segment. Then, whenever you see a line segment, take out its middle third and replace it with a V shape, provided by the top 2 legs of an equilateral triangle. Keep doing that ad infinitum. The limiting shape is the Koch curve.

The Koch curve refers to the edge of the snowflake, not the area inside. Here's the amazing thing: You can split the Koch curve into 4 perfect copies of itself, each shrunken 3-fold, as if reduced by a photocopier. Incredibly, when the reduction factor is $r = 3$, we get $m = 4$ copies. So, from the equation $m = r^d$ we mentioned earlier, the dimension of the Koch curve satisfies $4 = 3^d$. We can solve for d by using logarithms (don't worry if you've forgotten them). The result is $d = \log 4 / \log 3 = 1.26\ldots$.

Thus the Koch curve is more than 1-dimensional, but less than 2! Like a line, it is infinitely thin and covers no area. But unlike a line, the path between any 2 of its points is infinitely long. You could never walk along it. It's so crinkly that you couldn't even get started. So the Koch curve lives in a bizarre netherworld—more than a line, less than an area. Its dimension of 1.26 reflects that in-between status.

Scientists have generalized the concept of dimension even further, to objects that aren't exactly self-similar (as most real objects are not). For example, suppose we want to measure how long Norway's coastline is. This is difficult, because there are many fjords, each harboring smaller inlets, and so on. The length we measure will depend on how detailed our map is. Thus, as discussed in Lecture 14, there's no one right answer; the answer depends on the resolution we use. The key is to study *how* the apparent length changes with the resolution. The variations often follow a systematic law over a wide range of scales. And that law encodes information about the coastline's fractal dimension. So the right concept here is not the coastline's length but rather its fractal dimension.

To compute the dimension, we overlay a grid on the map and count how many boxes include part of the coastline. Then we change the resolution. As the grid gets finer, more boxes are penetrated. If we graph the number of penetrated boxes against the box size and use a log-log plot, the data fall on a straight line. Its slope tells us the fractal dimension. By measuring Norway's coastline with grids of different sizes, ranging from 0.6 kilometers to 80 kilometers, Norwegian physicist Jens Feder determined that its fractal dimension is about 1.52. Norway's coast is even more rugged than the Koch curve! For comparison, the craggy west coast of England has $d \approx 1.25$, whereas South Africa's exceptionally smooth coast has $d \approx 1$.

Fractal dimensions have now been measured for everything imaginable. For example, consider the convoluted surface of the human brain. If it were a perfectly smooth sphere, its surface would be 2-dimensional. A 1996 study of 30 normal human subjects found that the white matter surface of the brain has a dimension of about 2.3, due to its intricate folding. But don't be too impressed with yourself—the number for cauliflower is 2.8! ■

A 1996 study of 30 normal human subjects found that the white matter surface of the brain has a dimension of about 2.3, due to its intricate folding. But don't be too impressed with yourself—the number for cauliflower is 2.8!

Essential Reading

Gleick, *Chaos*, 83–118.

Mandelbrot, *The Fractal Geometry of Nature*, 25–57.

Schroeder, *Fractals, Chaos, Power Laws*, 1–20.

Feder, *Fractals*, chap. 2.

Liebovitch, *Fractals and Chaos Simplified for the Life Sciences*, 45–72.

Strogatz, *Nonlinear Dynamics and Chaos*, chap. 11.

Questions to Consider

1. What physical effects cause Norway's coastline to be fractal rather than smooth?

2. Consider the Cantor set, a famous fractal constructed as follows. Start with a line segment. Slice it into 3 equal pieces. Throw away the middle third, except for its endpoints. Then repeat this process infinitely often: Every time you see a line segment, remove its middle third. Calculate the fractal dimension of the limiting set.

A New Concept of Dimension
Lecture 15—Transcript

Welcome back to fractals. In the past two lectures, we have been looking at fractals, which you can think of it as being the geometric counterpart of chaos. These fractal shapes have some pretty strange properties, at least until you get used to them. In particular, we have seen that they have infinite amounts of fine structure all the way down to the finest scale, all the way down to the smallest thing that you can see. And the smallest parts of the shape resemble the shape itself, a symmetry property that we called self-similarity in the last lecture. This infinitely repeated structure gives rise to what may be the strangest thing about fractals: their fractional dimensionality. Think about that—fractional dimensionality, a dimension that is not a whole number.

Do you think you know what *dimension* means? If I say something is 1-dimensional or 2-dimensional, do you really know what that means? That's the sort of thing we're going to be talking about today.

Common sense tells us, yes, I do know what *dimension* means. A line, for instance, is 1-dimensional. A smooth surface, like the top of this table, that's 2-dimensional. A solid object would be 3-dimensional. What's the big mystery? For the highly corrugated shapes typical of fractals, our usual intuitive ideas about dimensionality break down. In this lecture, I'll explain how to generalize the concept of dimension so that it can handle fractals as well as the more familiar, ordinary Euclidean shapes. When we do that, the weird but inescapable conclusion is that fractals are so convoluted that they're not 1-, 2-, or 3-dimensional, like we're used to, but somewhere in between. For example, we'll find a fractal today that's 1.26-dimensional. This might seem like an esoteric subject, but it's worth studying for several reasons.

First of all, scientists use fractal dimensions all the time to characterize the roughness of highly irregular objects of all sorts: coastlines, tumors, galaxies, even the folded surface of the human brain. Before this concept was developed, there really wasn't a good way to quantify different degrees of roughness.

We also need to understand how fractal dimensions are calculated because they will come up later when we, for example, look at the drip paintings of Jackson Pollock. In Lecture Eighteen, we will discuss the controversy about some recently discovered drip paintings alleged to be by Pollock, where fractal analysis has weighed in with the verdict. Is it plausible that they are Pollock's or not?

Maybe the best reason to learn about this weird concept of dimension is because it's fun. It is just fun and fascinating to think about something as fundamental and basic as dimension in a brand new way.

The easiest place to start is with objects that are perfectly self-similar. Remember, I discussed in the previous lecture that self-similarity can occur through degrees. That is, we can speak of things that are perfectly self-similar, where tiny pieces are exact copies of the whole, or we can have approximate self-similarity, where the resemblance is less perfect, maybe only statistical. So we're going to use the simplest case of perfect self-similarity to give us a kind of idealized test bed for building up our intuition.

Think about a square group of blocks or a square patch of carpet. I have some cubes here, dice. I want you to ignore their colors and also ignore the number of pips on the faces of the die. That's not what we're interested in. Just look at their shape. They are cubes, but let's pretend that they're squares; that is, flat. There are 2, 4. If you can see what I'm showing you there, I have 4 squares making a bigger square. If I were to cut the square in half in both directions, then I would reveal that a square is composed of 4 smaller squares, each with half the linear dimension of the original square. Clear enough. Let's suppose we did it with 9, because there is a certain point I want to make. I don't mean to insult you with this example, but there is a certain easy point in this context that is going to be a slightly mysterious point in a harder context, so it's worth mastering the easy case.

Let's make 9 of these into a square. Now you see the square has 3 times the linear dimension; that is, it's 3 times as long on each side as any of the individual squares. In fact, what I'm showing here, you could say in words something like—let's look at this picture. I'm thinking of it as a carpet to imagine it's really flat, ignoring the cubicle aspect. We would say that a square carpet is self-similar because it's built up of little squares and it has a dimension of 2. We think we know why a square is 2-dimensional, a square patch, but here's why mathematically we would say it's 2-dimensional in the way that will generalize nicely to fractals. Maybe I should say before I do this, you may think, "I do know what dimension is. The reason that a square is 2-dimensional is because there is this way and there is that way. There's width and height. That's why it's 2-dimensional, and that's why the room is 3-dimensional. I can go left or right, up or down, front or back." So what you're doing when you think of dimension that way is you're telling me that the dimension of an object is how many coordinates I need to specify a point in the object.

That works very well for Euclidean shapes. The problem is, that way of calculating dimension doesn't work for fractals. I'll show you that later when we look at a fractal example, and I'll show you two plausible ways of thinking about the dimension of that fractal along the lines we just did, by counting coordinates, and you'll see it doesn't work. It will give a number that's either 1 or 2, and neither one will seem right. The way I'm arguing now is actually the way that we will generalize some fractals, and that's why I'm belaboring this idea about the square carpet. I'm trying to lead you down the right path.

Back to the square carpet, then. We have the square patch, and we're making this pretty clear point that 4 smaller squares make up the bigger square. In algebra, we would say, "Four equals 2 times itself," 2×2, or using a power, I would say $4 = 2^2$, which is funny because we use the word *squared*, [but] this is algebra. Why are we using the word *squared* that comes from geometry? It's because of this little picture that I've just been showing; $4 = 2^2$, by 9 when I use the chopping into thirds, I'm saying $9 = 3^2$. What you should be noticing is that I keep saying *squared*, but the squared is represented by the number 2, the exponent, the power that 2 is being raised to or 3 is being raised to. The thing that's common between these two equations, $4 = 2^2$, $9 = 3^2$, is this 2, the exponent. The exponent is what signifies the dimension. The exponent is what tells us this is a 2-dimensional object. So it is no accident.

Likewise, let's do this with a cube. Let's see if I can build a cube. I used to know how to do this. Now we're thinking 3-dimensionally in the commonsense way. There we go. I've successfully made a 3-dimensional cube out of smaller cubes. It is a 3-dimensional object, and the question is, if I start playing this game again, chopping it into smaller cubes—which I can do because it's a self-similar structure—how many smaller cubes make up the bigger cube? If I chop it in halves in all 3 directions—voilà!—I get, as you can see, 8 cubes. You knew that was coming. Very good. The reason is that $8 = 2^3$, where that 3 is no accident. The 3 in the third power is because it's a 3-dimensional object that we're talking about. Of course, the English word reflects that. We say *two cubed*, 2^3. So the point of these hopefully not-too-insulting examples is that exponents are the key to understanding dimension—the exponent 2 for 2-dimensional things, the exponent 3 for 3-dimensional things. That tells us the dimensionality of the original object.

Take a look at the square carpet picture. In general—now we're doing algebra—this is the pattern. I've just given you three examples: the two chopped-up versions of the square and the one chopped-up version of the cube. All three of those examples satisfy this pattern: $m = r^d$. Let's see what that means. [The variable] m refers to the number of small copies. When I did the case of the cube, I had eight of them, so m would be 8. [The variable] r is the reduction factor on each side; that is, like in the case of the cube, I was cutting things in half, so that number would be 2, 2 for the reduction factor. I'm taking the big cube and shrinking it down by a factor of 2 to make a little cube. And d is the dimension of the object. So instances of this formula would be like we said, m equals 8 small cubes is the same as 2^3, r being the reduction factor of 2. Or when we had the square, as shown here in the upper right panel, I had 9 small copies, but here I was chopping the side into thirds, so r would be 3, and so I have $9 = 3^2$. I hope that's clear. So this formula, $m = r^d$, is the key to defining the dimension of a fractal. That's what generalizes nicely.

Let's see that by playing the same game that we just played, except with a fractal shape. This is where it gets interesting. So we'll play the same game with a perfectly self-similar fractal. We'll find that the exponent, which we will interpret still as the natural definition of the dimension of the object— it's no longer a whole number. Let's do an example like that. The example we will consider is called the *Koch curve*, named after the mathematician, and I'm a little puzzled about how I should say his name because it's von

Koch. Some books refer to it as the "von Koch curve" and some say the "Koch curve." My instinct is that I should say "von" because I say "von Neumann." Why shouldn't I say "von Koch"? I probably should. Well, I'm not going to because I've practiced by saying "Koch."

The Koch curve is an exquisitely thin and crinkly filament. It's basically all corners, as you will see. It looks like the edge of a snowflake. So if we don't want to fuss over this Germanic name, we could just call it the snowflake curve. Here's what it looks like. The Koch curve is built through an iterative process. We talked earlier in the course about iterated maps using numbers. Fractals also involve iterations, but they tend to involve iterated shapes. We saw that with the Romanesco broccoli, that there was a basic motif, the shape of the broccoli, that was iterated at smaller and smaller scales to make little florets. Here, we're going to iterate a mathematical object to make a fractal that's an idealized mathematical one. So take a look at this illustration.

Here is the Koch curve. It's built up by starting with an ordinary segment of a line. That is stage S_0 in the construction, but don't pay any attention to that symbol. The point is, we start with a line, then we start building out equilateral triangles, like so. What I did was, I took the line above and broke it into 3 equal pieces, a middle third and a final third. Then I deleted the middle third, so you see it's empty down here, and I replaced that missing middle third with 2 legs of an equilateral triangle built on top, creating this flat shape with a hat sitting on it. The rule of building the Koch curve is that we're going to do that obsessively, from now on, infinitely. Every time we see a line segment anywhere, we're going to replace it with this shape composed of 4 line segments, each shrunk down by a factor of 3. That is, every line segment is going to be replaced by a thing that looks like this and then with a little equilateral triangle sticking on the top of it. So S_1, this shape, is sort of the generator of the Koch curve. Look at how that plays out when we go to the next stage in the construction, just to make sure you're getting it. I see a line segment, so obsessively, I go in and put on a little equilateral triangle. There's another line segment, this one here. That needs to be dealt with, so we build out an equilateral triangle, not the whole triangle—just the 2 sides of it, right there. This one needs tending, so it has its own little triangle knobbing out there, and this flat part needs its triangle erected on it. Now, hopefully, you get the pattern, and stages S_3 and S_4 show what happens as we continue building out these 2 legs of the triangle onto successive iterations of the shape.

Now where is the Koch curve in this picture? You can't see it because it's all the way down at infinity. It's what would happen if we did this infinitely often. But that's not as mindboggling as it might seem. You can do that because what will happen is that at some point, you're adding little triangles that are smaller than a quark and you don't have to think about it any more. Or if you want to do it in your head mathematically, you've gone all the way down to infinity as a kind of limiting process, and the 2 shapes become indistinguishable after a certain point, for all practical purposes.

The truth is, if you want to define it precisely, rigorously mathematically, it might be a little bit tricky, but let's not worry about it. Basically, the Koch curve looks a lot like this thing I'm showing at the bottom. If I added more triangles, you probably couldn't really see the difference.

We now want to talk about the properties of this interesting construction. By the way, where did it come from? Why was Koch doing it? We can talk about that at some length in a few minutes, but this is another instance of mathematicians looking inside their own subject, trying to understand what they mean by curves. Is this thing a curve or not? He was constructing it for that reason, to try to clean up the intuition about what could be a curve or not.

Here's the amazing thing relevant to what we're doing: This is a self-similar shape. You can split the Koch curve into 4 perfect, smaller copies of itself. Maybe you can see that. I'll show you in a second. There are these 4 microscopic, or at least reduced, copies of the original. How much do they have to be reduced? Only 3-fold—watch—as if reduced on a photocopier. When I say 3-fold, I mean in both directions in the plane. So imagine going to a reducing photocopier machine, setting it on 33%, putting the Koch curve on there, and hit[ting] the button. It's now 33% of its original size, which means that it was shrunk by that factor of 3 in both directions, sideways and top to bottom. Let's see what that would do for us.

Here's the Koch curve, approximately. What I'm trying to show you is that this region indicated between the arrows is a little Koch curve. If you imagine drawing a conceptual box around this little piece of it, that looks just like the entire shape. I hope that's clear. If you want, try putting your hand over the rest of the picture, and you'll see that this is just a tiny, complete Koch curve right here. I hope that's clear.

Now the claim is that there are 4 such objects in the original. Here's one of them. Let me show you the other 3. There's one; here's another one. It's on this side. It's all of this; it's tilted. Here's another one. You could imagine doing it this way: Suppose you cut out this small shape with scissors. You could then make 4 copies of it, and lay it down, and make the entire Koch curve. So here's one of them, here's 2, this is 3, and this is the 4th. Clear, I hope?

Now, get ready. We're going to calculate the dimension implied by this using our magic formula, $m = r^d$. What do those symbols mean again? The reduction factor is r. That would be 3 in this case because I have to shrink the shape by a factor of 3. I'm putting it on this photocopier, shrinking it by 3 in all of its directions. Shrink by 3; r is 3. How many copies do I get? That's called m. I got 4 copies; m is 4. So I have m equals 4 copies when r is a 3-fold reduction, which means my equation is $4 = 3^d$. And now you're back in high school; how can we solve for d? Well, at one time, you may have known how. You need logarithms to do it. I don't want you to get panicky about this. We don't really need to say anything more about logarithms in the rest of the course, but they are useful for picking out exponents in equations like this.

I am reminded, though, of the TV show *Moonlighting*. That was one where there were two characters named David Addison and Maddie Hayes, and they were detectives. It was the thing that made Bruce Willis a star, and Cybill Shepherd is the woman character in it, Maddie Hayes. They were, one time, trying to solve a very confusing mystery, and she said to him, "David, do you know what I don't understand?" And he said, "Logarithms?" So I guess it's a problem for everybody.

Let's assume you do remember what logarithms are, and taking logarithms of both sides here, it turns out it brings the d down, it puts a log in front of the 4 and the 3, and we get $d = \frac{\log 4}{\log 3}$. If you're really good, you're wondering what base of logarithm. Is it log base 10, or log base e, or log base 2? It turns out it doesn't matter. They would all give the same answer. So $d = \frac{\log 4}{\log 3}$, and when you compute that, you get 1.26. That is the dimension of the Koch curve. It's between 1 and 2. The Koch curve is more than 1-dimensional but less than 2.

Let's think about that. It's a bit strange. You might think it's sort of like a line. It was built from a line. Like a line, it's infinitely thin. No question about that. It's infinitely thin. It certainly covers no area. When I say that, I don't refer to the area underneath the curve. I'm talking about if you were a bug trying to walk along the curve itself, along the perimeter here, this edge—I'm only referring to the edge—the dimension of the edge is this thing that is higher than 1-dimensional. In fact, you could never walk along it, [nor could] a bug or anything because, as it turns out, it's infinitely long. That's an interesting, bizarre point. That makes it different from a line, not an infinite line. Remember, we started with a finite line segment and started crinkling it by adding these triangles. I'm claiming that it's now infinitely long. Why? Because we started with something of a certain length, and then we made it $\frac{4}{3}$ as long. Remember, we divided it into thirds, and in place of the middle third, we erected 2. It became 4 of these thirds. It's now $\frac{4}{3}$ its original length, and if we do that another time, it will be $\frac{4}{3}$ of that, and as we go down in the construction of the Koch curve, the original length is getting multiplied by $\frac{4}{3}$ each time, which is a number greater than 1. It will blow up to infinity.

The Koch curve has infinite arc length, and here is what's even more surprising than that: not just between its endpoints. It has infinite length between any 2 points on the curve. Pick 2 points that look very close together to your eye. They're still infinitely far apart if you try to walk along the curve. Think about that; it's true. There's an infinite amount of arc length packed into any small space of the Koch curve. So that is pretty bizarre. The Koch curve is so crinkly you could never even get started walking along it. Even to take the first step would be an infinity of distance. It's hard to think about. It's counterintuitive; I don't deny that.

The Koch curve lives in a bizarre netherworld. It's much more than a line, but it's less than an area. It certainly contains no area, yet it has got this bizarre, infinitely long character to it. In that sense, it deserves to be between 1- and 2-dimensional, and its dimension of 1.26 reflects that in-between status.

Scientists have generalized the concept of dimension even further than this. This shape that we've been discussing is perfectly self-similar. We would like to generalize to shapes that we would encounter in the real world, which are not going to be perfectly self-similar for reasons that we've discussed.

For example, suppose we wanted to measure the length of Norway's coastline. I mentioned this challenge in the previous lecture. There is Norway's coastline. It would be tremendously difficult because there are so many fjords, each harboring smaller inlets and so on, each having its own little bays. The length we would measure will depend on how detailed our map is. Even if we zoomed in to the finest scale by walking along the beach ourselves with a magnifying glass or something and making our own careful measurements, we would still find endless new features.

As we discussed in Lecture Fourteen, there really is no single right answer. The answer would depend on the resolution that we choose to use. That is actually the key to the problem. The key to the problem is to study *how* the apparent length depends on the resolution. The variations, linked with resolution, often follow a systematic law over a wide range of scales, and that law is going to be the secret to unlocking the dimension. That law encodes information about the coastline's fractal dimension. So the right concept here is not the length. That's the old-fashioned, wrong concept. The right concept is the fractal dimension. That's what we should be measuring. To compute the dimension, what we're going to do is use a concept called *box counting*. We're going to overlay a grid on top of the map of Norway, a grid made of boxes—small boxes, conceptual boxes, just squares—and we're going to watch how those boxes sit in relation to the coastline of Norway. Here's what it would look like. Here's a patch of the coast shown from a schematic drawing. Or we could do this on a satellite photograph. Either way would be fine. Then we overlay these boxes, and now we're asking: Is there a piece of the coastline in a given box or not? So every box that's being shown seems to touch the coastline somewhere. You can imagine a bigger grid, for instance, pieces out here that are in the ocean. Those would count as zeros. They don't include any of Norway's coast. To do this, you take a giant square grid, put it right on the map. Many of the boxes are empty. Some are occupied by a little bit of coast. You just count how many boxes are occupied.

Now that doesn't tell you much, but the game is, we're now going to change the resolution. We're going to imagine doing it again, making the boxes half as big, much finer mesh, and then we can play the same game. That's sort of like zooming in. We're now looking at a finer resolution. So we'll do this finer mesh and then count again how many boxes are penetrated by the coastline. Again, we can just keep shrinking the boxes. This is now sort of like doing that football field problem I mentioned in the last lecture, except now we've gone down to the ant scale, and then we can go to sub-ant scale, down to molecular, all the while looking at the coastline. So by counting the number of boxes that are penetrated by the coast as a function of the size of the boxes, we can extract the fractal dimension, and here's what it looks like.

If you make a graph of the number of boxes penetrated as a function of the box size—so this axis is the box size shown in kilometers. [These are] real measurements made by the Norwegian physicist Jens Feder. Then we count the number of boxes penetrated on this axis. You see that if the boxes are big, like 80 kilometers, you don't need many of them to cover. That's a big chunky thing you're laying down on the coastline. You don't need many of them, so the number is down here.

If we make very tiny boxes, less than a kilometer, say .6 kilometers or something like that, we need a lot of them. Now we're really following all the wiggles. We need more than 10,000 boxes at that resolution. What's miraculous is that the systematic trend here, if we look at the number of boxes as a function of the resolution, the data lie on a straight line. That straight line has a certain slope, which because of the nature of this plot—I'm using what's called a *log-log plot*, using logarithms on both axes. Another way to say it is, each successive increment is a constant factor, not a constant linear separation. It's not a sum. These numbers are going 1–10, not 1–2; 1–10–100. This is going 10–100–1000. On a log-log plot, when I make a graph like this, the slope is a way of revealing the dimension, similar to that logarithm calculation we did just a couple minutes ago.

The answer comes out to be 1.52-dimensional, from Feder's work. Interesting. Norway's coast has a 1.52 dimensionality, somewhere in between a line and a full, filled area. This is a rugged coast. It's much more rugged than the Koch curve, which was 1.26-dimensional.

For comparison, the craggy west coast of England, also pretty convoluted, has a dimension of about 1.25, curiously similar to the Koch curve. And South Africa, which has an exceptionally smooth coast, has a dimension close to 1. Fractal dimensions have now been measured for just about everything conceivable. For example, think about the convoluted surface of the human brain. If it were a perfectly smooth sphere with no convolutions, its surface would just be 2-dimensional. A 1996 study of 30 normal human subjects found that the white matter surface of the human brain has a dimension of about 2.3, due to its intricate folding. But don't be too impressed with yourself, because the number for cauliflower is 2.8.

Next time, we'll see how fractals are changing the way we think about volatility and risk in all its forms. There are fractals in time as well as in space. We'll be looking at fractal processes in the volatility in the stock market, in Internet traffic, and so on, and we'll see how they're changing our conception of risk, volatility—and it's just a beautiful new application of fractals that I can't wait to tell you about. So I'll see you then.

Fractals Around Us
Lecture 16

The new idea in these lectures, especially in this one, is that fractals are not just static geometric shapes. They can also be erratic processes in *time*, such as price fluctuations in the stock market, bursts of data traffic on the Internet, or the unexpected rumbling of an earthquake.

Having established what fractals are and how they can be created by chaos, we devote the next three lectures to a survey of their importance in science, commerce, and the arts. The goal of this lecture is to open your eyes to fractal processes in nature and in patterns of human activity and to explain why their presence changes everything we thought we knew about risk and volatility.

How can we think of a temporal process as akin to a fractal shape? The connection is that stock prices and other erratic processes vary over a wide range of scales, both in time and in the size of their fluctuations, and in that sense they can be regarded as "scale-free" (see Lecture 14). To reinforce what we mean by scale-free structure, let's look a bit more deeply at how fractal shapes differ from non-fractal ones. To do so, we're going to draw on concepts from probability and statistics for the first time in this course. Non-fractals display features at one predominant scale. Variations may occur, but they tend to follow a bell curve distributed about some average size. Typically, the sizes of the features are "normally distributed" about a mean value.

In contrast, fractal shapes display features over a much broader range, with the small features resembling the larger ones in some way. They follow a power-law distribution, with an exponent corresponding to the shape's fractal dimension (see Lecture 15). Power laws are the algebraic expression of the scale-free structure of fractals. We focus on power laws for the rest of the lecture because they are the link between shapes and processes. They occur for all sorts of things besides geometric shapes. The idea is that events (earthquakes, stock price changes, and so on) can vary dramatically in size. Typically, large events are much rarer than small ones. In a surprising number of different settings, the relation between the size of events and their frequency follows a power law, just as the sizes of features do in fractal shapes.

Power laws have very counterintuitive properties, totally different from those taught in traditional statistics courses. That's because a power-law distribution has what's called a *heavy tail* (also known as a *fat tail* or a *long tail*). Extremely large "outliers," though still rare, are much more common for power-law distributions than for bell curves. They can have a stunning impact on a system's average behavior.

For example, suppose the world's richest man, Bill Gates, walks into a room of 100 people. His presence changes the average income in the room dramatically. But if the world's *tallest* man walks in, he won't change the average height by much. The difference is that heights are normally distributed, so the tail of the distribution drops off extremely fast (exponentially fast, which is the fastest you can get, in practice). Whereas incomes follow a power law in the long tail of the distribution (as the Italian economist Vilfredo Pareto reported in 1896, in the first scientific use of a power-law distribution). Why the difference? Heights are constrained by biology, resulting in a characteristic scale of around 5 or 6 feet, whereas incomes are unconstrained (and hence scale-free) in capitalist societies.

Likewise, data traffic on the Internet obeys fractal statistics.

Or consider some examples from finance. On October 19, 1987 (now known as "Black Monday"), the Dow Jones Industrial Average dropped by 22% in a single day. Compared to the usual level of volatility, this was a drop of more than 20 standard deviations. Such an event is essentially impossible, according to traditional bell-curve statistics. (Its probability is less than 1 in 10^{50}.) The explanation is that changes in stock prices and other financial markets (such as exchange rates) don't follow normal distributions. They are better described by fractal statistics, as Mandelbrot and his disciples have emphasized.

Likewise, data traffic on the Internet obeys fractal statistics. File sizes vary widely, from tiny e-mails to bigger photographs to enormous movies, leading to violent bursts in overall traffic. This came as a surprise in the early days of the Internet. Engineers were used to handling voice traffic over the phone network, where the statistics obey tame bell curves. The "burstiness" of Internet traffic makes it much harder to ensure reliability of the system. (Compare how often you're frustrated by delays on the Web versus how often you fail to get a dial tone.) A related complication is that the traffic fluctuates on a wide range of time scales, from hours to fractions of a second. Bell-curve models developed for voice communications don't capture this, but fractal models do a better job.

Earthquakes, wildfires, floods, and other natural hazards are also governed by fractal statistics. These processes gyrate much more wildly and frequently than one would expect on conventional statistical grounds, complicating the task of risk management in the financial and insurance industries. For example, earthquakes obey a power law, called the Gutenberg-Richter law, which relates their frequency to their size (measured by the amount of destructive energy released). The data of payouts from the insurance industry are correspondingly "bursty." Although we may never be able to predict where or when such catastrophes will occur, the hope is that a better understanding of fractal processes will provide a more rational basis for assessing their overall risk and for guarding against them. ■

Essential Reading

Schroeder, *Fractals, Chaos, Power Laws*, chap. 4.

Suggested Reading

Ball, *The Self-Made Tapestry*, 210–16.

Buchanan, *Ubiquity: Why Catastrophes Happen.*

Liebovitch, *Fractals and Chaos Simplified for the Life Sciences*, 73–105.

Mandelbrot and Hudson, *The (Mis)behavior of Markets.*

Taleb, *The Black Swan.*

Turcotte, *Fractals and Chaos in Geology and Geophysics.*

Questions to Consider

1. Give some examples of what Nicholas Taleb has christened "Black Swan" events—extremely unlikely events in history, economics, or politics that occurred nonetheless, and that changed the world.

2. In what ways would life be different if stock prices, personal incomes, earthquakes, and everything else obeyed normal (bell-curve) statistics rather than fractal (power-law) statistics?

Fractals Around Us
Lecture 16—Transcript

Welcome back to fractals. In the past three lectures, we've been building up the theory of fractals. Up until now, we've defined fractals as shapes with infinitely rich, fine structure, in which the small parts are just scaled-down versions of the original whole, or at least in some way reminiscent of the structure of the whole. We've seen why fractals are so intimately related to chaos, and we've learned how to calculate their dimension, which turned out, surprisingly enough, sometimes to be a number that is not a whole number—it's in between two whole numbers—a fractional dimension.

Now we devote the next three lectures to a survey of the intriguing ways that fractals are being applied in science, commerce, and the arts. The goal of this lecture is to open your eyes to fractal processes in nature—processes in time, not just shapes—and in patterns of human activity, and to explain why their presence changes everything we thought we knew about risk and volatility.

The new idea in these lectures, especially in this one, is that fractals are not just static geometric shapes. They can also be erratic processes in *time*, such as price fluctuations in the stock market, bursts of data traffic on the Internet, or the unexpected rumbling of an earthquake. How can we think of a temporal process as akin to a fractal shape? The connection is that stock prices and other erratic processes vary over a wide range of scales, both in time and in the size of their fluctuations. In that sense, they can be regarded as "scale-free," a concept that we introduced in Lecture Fourteen.

To reinforce what we mean by scale-free structure, let's look a bit more deeply at how fractal shapes differ from non-fractal ones. To do so, we're going to draw on concepts from probability and statistics for the first time in this course. We don't need much. We just need the idea of a probability distribution, which you may remember from seeing the curve after taking a test in school where all the different grades are plotted, and [it shows] how many people got how many grades. That's the kind of thing we'll be showing—distributions of some property.

Our first point, then, is: How are fractal shapes different from non-fractals ones? Take a look at this. We're showing on the left an example—this is a schematic thing. I'm not talking about time yet. I'm still just in the world of shapes to get across this idea of scales and how many scales are involved. That's the distinguishing thing we want to focus on here, that a non-fractal process has one typical scale. Here, I'm trying to convey that by showing a series of disks, little circles of different sizes, but they're basically all the same size. There is one typical scale such that if I made a graph showing the number of features of a particular size as a function of the size of that feature, most of them are centered around some typical value.

There's the classic curve that you hear about in connection with test scores, or heights of people, or anything else. It's commonly called a *normal distribution*, and it's just often called the *bell-shaped curve*, especially when people talk about intelligence testing or whatever. This is the bell curve that you've heard about so much. Statisticians speak of the *Gaussian distribution* or the *normal distribution*; that's jargon for what we're talking about. I'll probably tend to call it the bell curve. But I really do mean the classic bell curve that's talked about in statistics courses—the Gaussian, or the normal, distribution.

In contrast to something distributed like that, where basically everything has the same size except for little fluctuations around that mean value, fractal shapes or processes have features over a wide range of scales. Their features can vary over a much broader range, with small features often resembling the larger ones in some way. In the illustration shown here, the features are signified by these little circles, or there are some big ones. Here's a big monster one, a humongous one. So we have a bunch of different scales, and that's what I'm showing in the corresponding distribution picture. If you look at the number of features as a function of the size of the features, you see that there are a lot—that is, the curve is high corresponding to small features—that is, there are a lot of little ones. There is only one big one in this picture, but in general, these big events, or big features, are somewhat quite rare.

What is really important is the way that the distribution behaves as a function of size. We see it dropping down according to some kind of decaying curve, and that curve takes a particular form, algebraically speaking, called the *power law*. It's a very important concept that we will be emphasizing in this lecture.

A power-law distribution is one of the key ideas in thinking about the statistics of fractals. Here is an expression, then, of what we mean by a power law and why it's called a power law. It's to show the number of events or features of a certain size as a function of that size. In the case of a temporal process, like the frequency of earthquakes or stock market price changes, this is the sort of event or feature we're interested in. So what we're showing in a picture like this is that there are many events of small size, not so many events of large size, and the relationship is that the number of events, which is y, is proportional to x, the size, to some power, $-d$. The negative sign means it's really more like $\frac{1}{x^d}$. It's a decaying function as x increases, as you can see from the picture. This d is the exponent that we'll be talking about, the power that x is being raised to.

Now, power laws are the algebraic expression of the scale-free structure of fractals, a concept we also talked about previously. There's no inherent scale here. There are small scales, large scales, everything in between. They're all there in the fractal. Power laws are what we want to focus on for the rest of the lecture because they are the link between shapes that we've been thinking about and processes in time. They occur for all sorts of things besides simply shapes.

The idea is that events occur, like earthquakes, stock price changes, and so on, and the large ones are much rarer than the small ones, as I've just shown. In a surprising number of different settings, that relationship between frequency of events and their size follows a relationship of the type we've just been looking at—a power law, just like the geometric features of a geometric fractal. We saw power laws in connection with thinking about the coastline of Norway. I don't know if you remember, but it was in the last lecture. I was making a lot of noise about $m = r^d$. If you think about that, m referred to the number of small copies in the whole; m was sort of like, number of features. [The variable] r was a scale factor, a reduction factor. That's sort of like a feature size. And d is that dimension, an exponent. So that relationship, $m = r^d$, was exactly of the type we're talking about here, where the number of features is being related to their size. Here, it's number of events of a particular size that we're looking at, but the basic idea is the same, and the power-law structure is the same.

Now here is an extremely important idea and a really interesting one that will help you understand our modern world much better if you can assimilate this. It's one of the contributions that this course will make to your thinking about everyday life. I think it may change the way you look at things in the everyday world. It has to do with understanding power laws. They have very counterintuitive properties, totally different from those taught in traditional statistics courses and possibly very different from what you think you understand by common sense. That's because a power-law distribution has what's called a *heavy tail*, sometimes called—it's not the most polite expression—a *fat tail* or a *long tail*. There was a bestselling book recently called *The Long Tail*, which used that term, but statisticians usually speak of heavy tails. I'll probably use all those words.

What's interesting about the heavy tail is that it tells us that extremely large "outliers"—events of whopping size, though still rare—they're in the tail of the distribution. The *tail* refers to the part far out on the extreme right, the very rare event. Although rare for a power law, they are much less rare than they would be for a normal distribution. In other words, enormous, potentially cataclysmic events are predicted to occur much more frequently than you would expect based on normal statistics. Such outliers can have a stunning, dramatic impact on a system's average behavior.

Let me give you a simple illustration of that. Think about money. We all have an intuition about money. Suppose the world's richest man, Bill Gates, walks into this room right now, or a room of 100 people, let's say. His presence would change the average income in the room very significantly. It's just one person, but his contribution—whether we talk about his annual income, which I don't know. He's estimated to have a wealth of something like $50 billion just if he keeps the money in the bank and it's getting interest. I'm sure he's smarter than that, but if he were just getting interest, he would be making on the order of several billion a year just from the interest alone. So his income is out there in the billions. I don't know about you other guys, but speaking for myself, this would change the average income in the room very significantly, just one person.

In contrast, suppose the world's *tallest* man walks in—tallest, not richest—he wouldn't change the average height in the room by very much. That's an interesting thing. Why the big difference? It's because heights and incomes are distributed in very different ways. Heights are normally distributed. The tail of the distribution drops off extremely fast, in fact, exponentially fast, which is about the fastest that something can drop off in math, in practice. So what that means is you really don't ever see anyone with a height, let's say, 10 times the average. There are no people 50 feet tall. You don't really see anyone even 2 times the average. There has never been any recorded human being 10 feet tall. The point is made, that the distribution is quite tight, although it has outliers. There are, after all, 8-foot tall people occasionally. But they're not as extreme as they would be for incomes, where you have people with incomes of $10,000.00 a year, $100,000.00 a year, and I can go many factors of 10 out to Bill Gates's several billion a year.

The point is incomes don't follow a normal distribution; they follow a power-law distribution. They have this enormous, heavy tail going way, way out there. You can keep going out past millions out to billions. So they drop off very slowly. There are people out there with Bill Gates. There's Warren Buffett, and various Arab sheiks, and other people—so there are lots of people out in the tail, though still rare. That power-law distribution of the tail of incomes was discovered in the late 1800s by Vilfredo Pareto. It was the first scientific use of a power-law distribution, so this is not really a contrived example, to be talking about incomes. That's where power laws were first noticed in regard to their statistical significance.

Why the big difference in heights and incomes? It's sort of obvious, if you think about it. Heights are constrained by the laws of biology and physics, resulting in a characteristic scale for the process of human height. There are biological reasons that people are not 50 feet tall. But incomes are not constrained in the same way by any laws of economics or politics—at least in a capitalist society, where you can make as much money as you can make, subject to taxation, of course, but other than that, you really have basically unconstrained income—so there is no inherent scale for income. The absence of the scale makes the distribution scale-free, and then power laws pop in there automatically.

Consider some other examples from finance; again, to underscore the difference, the really dramatic difference, between things that are normally distributed and things that are power-law distributed.

On October 19, 1987, now known as "Black Monday," the Dow Jones Industrial [Average] dropped by 22% in a single day. Compared to the usual level of volatility, the typical fluctuations in the stock market, this was a drop of more than 20 standard deviations. I said I wouldn't be using much statistics here, but I'm sure you have some feeling for what I mean by standard deviations. Roughly speaking, a *standard deviation* is the measure of the width of a bell curve. When I say something is 1 standard deviation away, that's a deviation, but not much. There are plenty of things that fall within 1 standard deviation. About two-thirds of the time, things will fall within 1 standard deviation. Two standard deviations away, you would expect to see 95% of your observations will be in there. Five percent would not fall in. Those will be outliers.

You have probably heard about Six Sigma used in thinking about business practices and so on. But this is now talking 20 sigma, 20 standard deviations. That is all but theoretically impossible, according to traditional statistics. If you calculate the probability of a 20-sigma event using a normal distribution, which—notice the word *normal*. There's a normative sense there. This is the way things are supposed to be. Normally, that would occur with a probability of less than 1 in 10^{50}, 1 followed by 50 zeros. That's essentially impossible. That cannot happen in the lifetime of the universe, and yet it happened on Black Monday.

We see events like that happening more than you'd think, a lot more than you'd think. The reason is that the normal distribution doesn't work for many of the things in the modern world. It does work for heights; it works for sizes of ball bearings; it works for certain things constrained by laws of physics or chemistry. But in many other situations, the normal distribution is not just wrong, it's really misleading and will give you the wrong intuition about the way the world works. That's really the big point of this lecture.

You need to be aware of power-law distributions and their fractal properties because much of our world depends on them.

Fractal statistics are the really better description for many of these events in the financial and commercial world, as Benoit Mandelbrot himself and his disciples have emphasized. One more example from this world, let me show you four charts, sometimes called *fever charts*, almost like a fever as a function of time, except these are fever charts for the price of certain commodities or stocks. Two of the charts I'm going to show you are real, and two are fake. I want you to see if you can pick out which ones are produced by mathematical models and which ones are real. There will be pictures of prices versus time. Try to pick the fakes.

There are four of them, and you notice that I've deliberately taken off any scale indicators. I'm not showing you the length of time, or the numbers on the vertical axis for what the price jumps are, or anything else. Do you have a feeling in your head? It's not easy to tell.

Let's try a different approach. Rather than showing you price as a function of time, let's look at the "day-to-day" changes in price. In saying "day," I should put that in quotes because I haven't told you if [these are] data measured over a microsecond, or over a year, or whatever. I just mean the times between successive events. Let's look at the price changes from one "day," so to speak, to the next—again, the same data, but now plotted a different way.

So here are the daily changes in price. Now I have a feeling that you can guess something about which one is bogus—that is, produced by a mathematical model—and which two are real. Maybe you don't know that. Clearly, like when the police do a lineup, sometimes you can tell this is not the guy that committed the crime. There's one that sticks out of this lineup, right? Number 2.

Number 2 is the prediction of the traditional statistical model known as a *random walk model* generated from bell curve data. That is, we matched a standard deviation and a mean to the data and just assumed that events occur at random, independently—price changes going up and down according to the random walk—and this is what the picture would look like graphed that way. You see fluctuations. There are some outliers; that is, there are spikes up and down. But overall, the picture doesn't look nearly spiky enough. It's just not plausible. It's clearly not real. It doesn't look like two of the others. There are three here, two of which are real, the remaining ones. But now it's a little bit harder. Which one of the remaining three is the imposter? I don't know if you have a feeling. This one is thicker, clearly, than the other two. This one is quite "bursty," right? You see that there are bursts of volatility here and then quiet periods. Notice, it's sort of quiet, without much going on; then some more big bursts; then quiet. Likewise, this one is very bursty here and then sort of quiet. Forget about number 2. This one has really big spikes in it here and there. Any guess?

Let me show you what the right answer is. The first one is relative changes in IBM stock prices over a period of those years from 1959 to 1996. This one is a deutsche mark–dollar exchange rate, and this one is from Mandelbrot's recent work on applying fractals to finance. If you find this sort of thing interesting, I would encourage you to take a look at Mandelbrot's work. His book *The Misbehavior of Markets* makes for interesting reading.

Let's move along now to talk about the Internet. Data traffic on the Internet, another feature of our modern world, also obeys fractal statistics, and this has confounded the engineers responsible for the Internet, at least in its early days. At that time, people were thinking of the Internet as analogous to the phone network. Engineers had a lot of experience with the phone network, but it was different in that the phone network, remember, has limitations on it based on human biology again. How long can a person talk; how fast can we talk; how much can we tolerate hearing from somebody else? There are sort of intrinsic limitations based on it being human-voiced. Whereas on the Internet, I can send you any size file; I can send you a movie. I can send it as fast as I want, so it's sort of scale-free. It's unconstrained in the same way that incomes are unconstrained, whereas voice is constrained in the same way that heights are constrained.

What we see on the Internet is that file sizes vary widely, from tiny e-mails to bigger photographs to enormous movies, leading to violent bursts in the overall traffic on the system. As I say, that came as a surprise in the early days of the Internet because engineers were used to handling voice traffic.

In the case of voice traffic, the statistical description of bell curves works beautifully. The whole theory of voice communications and network engineering for the phone network was based on the bell curve, and it worked very, very nicely. But the burstiness of Internet traffic makes it much harder to ensure reliability of the overall system. That's a significant consequence of this power-law effect. Think about how often you're frustrated by delays on the Web, congestion on the Web, compared to how often you would fail to get a dial tone using a landline. You really don't have any reliability problem in the phone network compared to the Web.

A related complication is that the traffic on the Web fluctuates on a wide range of time scales. It's not just that the sizes of the files vary a lot. The time scales vary a lot, too, from hours to fractions of a second. Let me show you some data about that. First I'll be showing you data having to do with bell-curve modeling compared to Internet traffic, and you'll see what a poor job it does. Here is a bell-curve model versus reality. I'm going to show you a series of paired pictures, the ones on the left being a model prediction from the bell-curve way of thinking, and the one on the right being measured data from traffic coming into a certain big corporation, reported by the engineers who published this study, Walter Willinger and Vernon Paxton. I have a feeling it might be AT&T since they used to work for them. They don't name the company.

The "packets arriving every 0.1 second" is a measure of the traffic coming into the corporation. That's a very short time scale, one-tenth of a second. How many packets are arriving in that time? What they find is that it's quite noisy. It's jumping around, up and down, depending on when these packets come in. Remember, packets are a measure of Internet traffic. It's jumping around. This is the model. This model is to be compared to what it really was, which is this curve, which is also jumping. It's not point by point the same, but it has the same overall statistical character.

To make it a fair comparison, they took the actual process, looked at its mean and standard deviation—standard bell-curve statistics—and used the same mean and standard deviation for a random process called the *Poisson process* and then generated random numbers with that to simulate what they were seeing on the right.

These two looked reasonably comparable, at least overall. Then, what's interesting, and this is like analogous to picking out the imposter in that Mandelbrot finance study, suppose we zoom out. Instead of looking at this very short time scale of one-tenth of a second, look at traffic coming in over a longer time period—10 times longer—1 second. How many packets do you see then?

Now we go to two more slides. Here's the model. Here's reality, and the black regions indicate the zoom in the sense that this black region would have been expanded to produce the figures that we looked at a minute ago. That is, it's all squished down because earlier we had been looking at a shorter time interval, and that shorter time interval is this region here shown in black.

Let me quickly go through these because what I'm going to show is—I keep zooming out to time scales of 10 seconds. Now we're starting to see that the bell-curve model is looking way too flat compared to the real data. Finally, at a time scale of [the number of] packets arriving every 60 seconds, the bell curve has produced a very flat picture, much different from what is seen in reality. The reason for it being so flat is that if you think about how many packets arrive every minute, things kind of average out. So even though it's noisy on a time scale of one-tenth of a second, if you wait as long as a minute, things average out so that typically you can say how many packets will arrive in that minute, and you will be pretty sure of it. This is the kind of thing insurance companies make big use of. They know very well how many traffic accidents they have to deal with in a given year because even though they don't know if you're going to have an accident, they have an averaging because they've insured thousands or hundreds of thousands of people, and things average out. So insurance relies on this kind of averaging that the normal distribution implies. But it doesn't work for the Internet. The average behavior being so flat here doesn't correspond to reality.

In contrast, if we look at a fractal model, as I'll show here, then when we make a match across the short time scale of 0.1 second to get the thing started, to get the model calibrated to reality—here is the model on the left, reality on the right. When we zoom back to larger time scales of 1 second, 10 seconds, or 60 seconds, the overall structure of the graphs is similar throughout. The fractal model is capturing the burstiness and the different time scales of Internet traffic over a wide range of scales from one-tenth of a second up to a minute. The fractal models do a much better job.

Finally, let me turn to earthquakes and other natural hazards, things like wildfires, floods, other disastrous events in nature. They turn out to also be described by fractal statistics. It's very important, as I say, for the insurance companies to know about this, as well as it should be a concern to us, like people living in Southern California need to understand this for reasons we will be talking about in a second.

Processes like this—earthquakes and so on—gyrate much more wildly and frequently than you would expect on conventional statistical grounds, which complicates the task of managing both the financial industries and the insurance industries. Here is an example of the difficulty. Let's look at the distribution of earthquakes. By that I mean, not their distribution in space. I'm just going to fix my attention on one part of space, this being California. We'll ask: How frequently do earthquakes occur of a particular size? How often do we see magnitude 8, which would be a tremendous earthquake, a really devastating earthquake? How often do we see magnitude 4? A rumbler but not a devastating one. Well, here's what the data look like. This [graph] is earthquakes showing a power law in their frequency as a function of their size; that is, the data are this line here. I'm showing the number of earthquakes per year occurring in southern California having certain magnitudes greater than a number m, magnitude m being shown down here, running from 4 up to an estimate for magnitudes greater than 8. These magnitude-8 earthquakes have not been directly observed, but they have been inferred from the geological record. You can see the signature of those events back into the past through thousands of years. They basically melt some of the rocks, and you can see some of that. They let you calculate when these earthquakes occurred, going back hundreds of years. From that, we can estimate the frequency of these enormous, really terrible earthquakes.

Notice that that dot corresponding to the terrible ones lies really right on the extrapolation of the line for the moderate to big ones. There is one pattern extending all the way from the largest earthquakes to the smallest, and that pattern is a power law. To be clear, it's a power law if you graph the frequency versus the amount of destructive energy released in the earthquake. It's not a power law with respect to magnitude. It turns out, magnitude involves logarithms again, and it's the logarithm of energy that is related to magnitude. That's just a technical detail, but if we graph things as a function of energy rather than magnitude, then you find that power laws govern the distribution of earthquakes. Remember the bottom line of that. It means that big earthquakes occur much more often than you might think, certainly much more than they would if they were normally distributed.

That complicates the task of insurance for earthquakes. If we look, for instance, at earthquake insurance payouts, here's what it looks like. This is the insurance payout data for California earthquakes, showing what's called the *loss ratio*. You can think of it as the amount of money put in, in the form of claims. This is what people would like to be paid versus the premiums that have been paid in to the insurance company. So that's a measure of how bad a year it was in terms of earthquakes, how much had to be paid out.

Here are some years: 1971 was not terrible. They paid out a loss ratio of less than 20. Here are some good years for the insurance companies, plus people didn't have earthquakes, so they're happy. This was bad year. There's a big spike there, but you expect some variations from time to time. Look at all these data over these years, from 1971 to 1993, I think is that last one shown, and try to get a sense, using statistics, for what is a typical scale. If you were an insurer, what would you typically expect to have to pay in a given year? You might say a number like 70 or something, dominating the data using this big one, or maybe a number like 40.

Here's what happened in 1994, the year not shown. The loss ratio for 1994 was 2273, which is off the chart. This is 140. If I were to graph 2000 it would be way off the top of the page, more than 10 times the data of what's shown there.

I hope that brings home the point to you that you can't really extrapolate trends very well when the process you're trying to extrapolate is governed by a power law. This, of course, has to do with the Northridge earthquake in 1994, which caused some $10 to $30 billion dollars of damage, depending on estimates.

Although we may never be able to predict where or when such catastrophes occur—fractals are not going to solve that for us—but the hope is that a better understanding of the fractal statistics involved here will provide a more rational basis for assessing overall risks of these events and for guarding against them in a more rational and sensible way.

Next time, we'll turn our attention from the fractals around us to the fractals inside us. I'll see you then.

Fractals Inside Us
Lecture 17

Whenever your body needs to send something from place to place—signals propagating along a nerve, hormones, blood, oxygen, anything that needs to be carried throughout your body to service yourself—those sorts of things are transported on a superhighway system based on fractal branching networks.

Why might evolution have favored the architecture of the fractal? If you want a plumbing system that can reach every cell in a 3-dimensional body, this is a great way to do it. Fractals also create an enormous amount of surface area in a confined space, crucial for such processes as gas exchange in the lung or absorption of food in the intestine. Finally, the blueprint for a fractal tree is simple and hence easy to encode genetically: Grow a tube, split it into several smaller branches, repeat.

The fractals inside us might also hold the key to some of life's greatest mysteries. Why do we live for about 80 years, rather than 80 seconds or 80 centuries? And why does a mouse live only a year or two, even though it's made of the same molecules and genes as we are? Despite all that biologists have learned about the mechanisms of aging, no one knows how to calculate an animal's typical life span from first principles. The rest of this lecture discusses a theory that offers insight into questions like these. It's based on the fractal architecture of all living things. The theory is provocative because it challenges us to look at life with a physicist's eyes, seeking unifying principles and ignoring the rich diversity that biologists cherish. Proposed in 1997 by Geoffrey West, a physicist at the Santa Fe Institute, and his biologist colleagues Jim Brown and Brian Enquist, the theory has already explained some of biology's most far-reaching laws, mathematical regularities that *all* creatures seem to obey, from the smallest microbes up to elephants and whales.

For example, what are the daily energy needs of a mouse, a dog, or an elephant? To measure these needs, you can look at how much an animal eats, how much oxygen it uses, or how much heat it produces. However you make the measurement, the metabolic rates for diverse mammals always fall neatly on a curve when they're graphed versus their average body mass. Pound for pound, the little guys burn a lot more energy than we do. It's more revealing to plot *total* caloric intake versus mass (so now we're comparing different species in absolute terms, not on a pound-for-pound basis). When we use logarithmic axes, the data fall on a beautifully straight line, implying that the underlying law is a power law—just as we saw in fractals (in Lecture 16). The line rises with a slope of 3/4. Hence, bigger animals need to eat more than little ones, but only in proportion to their mass raised to the 3/4 power. When this law was reported by the veterinary scientist Max Kleiber in 1932, some biologists were upset (and some still are today). They had expected to find a 2/3-power law, based on simple considerations of surface area and its expected role in respiration, heat loss, and other physiological processes.

Kleiber's 3/4-power law has since been shown to hold over 27 orders of magnitude in body mass, extending from subcellular molecules to elephants. It testifies to the unity of life on Earth. It is one of the most comprehensive laws in all of science. What is the explanation for it? The mystery goes even deeper. There are many other scaling laws in biology, also with 1/4 powers of mass in them. The lifetime of a mammal increases in proportion to $M^{1/4}$, where M is its mass. The time intervals between heartbeats and breaths are also proportional to $M^{1/4}$. These results have the strange implication that a mammal of any size, from mouse to elephant, can expect to live for about 1.5 billion heartbeats or 300,000 breath cycles! Is this a coincidence or something fundamental about life? No one knows.

The theory of West et al. takes a step toward answering these questions by predicting the exponents in the scaling laws for metabolism, lifetimes, etc. They asked themselves: What is the *optimal* design of a network of branching tubes carrying nutrients, oxygen, or other resources, such that the resources are delivered to all parts of a 3-dimensional body as fast as possible and with minimum energy? Invoking some simplifying assumptions and the laws of fluid mechanics, they found that the network had to have a self-similar branching pattern. In other words, it had to be a fractal. Then they imposed the realistic constraint that the network's terminal units are invariant for creatures of different sizes. For the circulatory system, the terminal units are the capillaries (which are known to be roughly the same size for an elephant or a mouse). With these assumptions, the 3/4 power fell out of the math, as did the other known scaling laws. The same argument also explained the 3/4-power law seen at the cellular and subcellular level. The theory also correctly predicts that the size of the smallest possible mammal is a few grams—consistent with the mass of a shrew—and that mammalian cells in culture always have the same metabolic rate, regardless of what mammal they come from.

The optimality assumption rankles some biologists, since natural selection need not be optimal.

The theory is controversial. There are technical disputes about the calculation itself. The optimality assumption rankles some biologists, since natural selection need not be optimal. But it's intriguing that the observed behavior is close to optimal. In any case, optimality is a natural benchmark.

The modeling philosophy irritates some people. Drastic simplifications are used to expose broad trends—a common strategy in physics but anathema to biologists. Interesting details are missed. For example, humans live longer than the overall trend for mammals would predict, given our weight. West et al. would retort: That's the point! By highlighting the trend, one can tease out what requires biological explanation versus what follows more generically. ∎

Essential Reading

For the most accessible introduction to the theory of West, Brown and Enquist, see: Whitfield, *In the Beat of a Heart*.

Supplementary Reading

Ball, *The Self-Made Tapestry*, chap. 5.

Liebovitch, *Fractals and Chaos Simplified for the Life Sciences*, 14–25.

McMahon and Bonner, *On Size and Life*.

1. What causes aging?

2. How much should you eat? Kleiber's law states that a warm-blooded animal of mass M, measured in kilograms, typically needs about $70 M^{3/4}$ calories per day. Translated into pounds, this becomes $38 W^{3/4}$ for an animal weighing W pounds. Apply this formula to yourself. (Hint: If you need help calculating the 3/4 power, multiply $W \times W \times W$ to get the third power (W^3), and then take the square root of that number, and the square root again, to get $\sqrt{\sqrt{W^3}} = \left((W^3)^{1/2} \right)^{1/2} = W^{3/4}$. Then multiply by 38 to estimate your daily caloric needs.)

Fractals Inside Us
Lecture 17—Transcript

Welcome back. Let's remember where we are. We're in the midst of discussing fractals. At first, we were discussing them in the context of the abstract world of state space. You might think of these as fractals in our minds, that is, fractals in the mathematical world. Then, we saw that fractals were abundant in the real world around us, in everything from the shapes of coastlines and clouds to processes in time, like earthquakes and Internet traffic.

In this lecture, we will turn our attention inward to the fractals inside us, in our very bodies, and actually in the bodies of all living things. Whenever your body needs to send something from place to place—signals propagating along a nerve, hormones, blood, oxygen, anything that needs to be carried throughout your body to service yourself—those sorts of things are transported on a superhighway system based on fractal branching networks.

For example, think about how our blood vessels work. Starting from the heart to the biggest vessel, the aorta, and then subsequent branchings, our blood vessels branch about 15 times in a self-similar, fractal pattern all the way from the heart down to the smallest capillaries. You've probably seen pictures of that, but they're really quite dramatic, and you may not have thought about them as fractals before, so let's take a look. Here is what the blood vessels in your lungs look like. Look at that structure. These are called the pulmonary arteries. They're the only arteries that carry dark, deoxygenated blood. You're used to thinking of arteries carrying the fresh, red blood full of oxygen, but the pulmonary arteries carry dark blood that has given up its oxygen. They take it from the heart to the lung, where it gets reoxygenated.

What you're seeing in this picture is this incredible branching system of a big vessel branching into smaller ones, and then smaller and smaller. The whole purpose of this, of course, is that these blood vessels are splitting so many times so that they can place blood in close proximity with the air that you have inhaled, allowing oxygen to diffuse across the tiniest vessels, through the vessel wall, into the pulmonary veins, where the now-oxygenated blood can be carried back to the heart and pumped to the rest of the body.

Meanwhile, carbon dioxide, which is produced during metabolism, is diffusing out of the blood and into the lungs, where it can get exhaled. As this suggests, it's not just the blood that we have to worry about but also air: gas exchange. If you look at the structure of the airways of the lung, they have a similar fractal structure, as you can see in this next picture. There are the airways, the bronchial tree. It's very similar, a big vessel that branches and branches and branches. This one, too, shows this sort of self-similar branching structure.

It's not just animals like us that use this sort of geometry to allow for gas exchange for transport of food. Plants do it, too. If you were to look at the veins of a leaf, there is another case of a fractal network with a big vessel branching into many smaller ones. In this case, what the plant is doing is using a fractal network to try to carry water and nutrients around the leaves and also to bring photosynthesized food that was produced in the leaves down to the rest of the plant.

My first point is simply this: that all living things rely on fractal networks. I could give you more examples than the three I've given you, but I think you can believe this. We could ask the question, "Why has evolution favored this particular kind of design, this architecture?" There are really several reasons, and I'm not sure any one of them is dominant, but here are a few that come to mind.

First of all, fractal networks are clearly a very efficient way to provide resources or communication links throughout a 3-dimensional, solid body, using only 1-dimensional tubes. Think about that. That's an interesting engineering problem. If you want to reach every point in a 3-dimensional solid, a living thing, but all you have are these vessels, how can you get everywhere, get close to every cell (because every cell needs nourishment or needs communication, in the case of hormones or nerves)? A good design, then, to do that is to make a tree that branches and branches many times, and you can get close to everything that way. So if you wanted a plumbing system that can reach every cell in a 3-dimensional body, a fractal network is a great way to do it.

There's another reason. Fractal architecture also creates an enormous amount of surface area in a confined space, the cavities of the body, and that great amount of surface area is crucial for processes like gas exchange that we already talked about, carbon dioxide and oxygen, or for absorption of food in the intestine, which also has its own version of fractal structure.

Finally, there's a third reason fractals have been chosen or selected by evolution, which is that they're really kind of easy to program. If you think of them as a software problem, the design of a fractal network is really pretty trivial. You just genetically encode it like this. The rule is: Grow a tube, and then split it into several smaller branches and repeat. That's it. Those are the instructions for making a fractal network. Just keep growing a tube, splitting, growing more tubes and splitting, and pretty soon, you'll have a fractal tree. That much I think is probably fairly clear and maybe even obvious—that fractals would have benefit at the architectural level for living things, and that's why they're used.

The second point, and what I will be basically spending the rest of the lecture on, is I think quite a bit deeper and maybe much less familiar to you. We're going to see that fractals may hold the key to explaining some of life's greatest mysteries. I realize that sounds like a big claim, and it is, but these are the kinds of mysteries I'm talking about: Why do we live to be about 80 years old and not 80 seconds or 80 centuries? It might sound like a silly question, but think about that for a second.

A physicist would ask the question this way: What sets the time scale for life? Why is it measured in tens of years, 80 years, and not in units of seconds, or microseconds, or centuries, or millennia? It seems like an easy question. If we understand anything about biology, we should be able to give a ballpark estimate of life span using first principles. And yet we don't know how to do it. It's amazing. You don't even hear that question being asked in biology courses, but it's the first thing that any physicist would want to ask. How do you do a back-of-the-envelope calculation of life span? So we'll see in this lecture that physicists who have approached this problem are starting to make strides toward answering that kind of question.

If you think the question is trivial, or you think you see the answer to it, ask yourself this question: Why does a mouse live only about a year or two, even though it's made of the same molecules and genes as we are? It's a mammal. It's not really different from us at the molecular level, and yet its life span is quite a bit different, by close to a factor of 100. Why is that?

Despite all that biologists have learned about the molecular mechanisms of aging—and there has been great progress in that direction in recent years—no one knows how to calculate an animal's typical life span. That's just one of the kind of mysteries that this new theory I'll be describing takes us toward addressing.

In the rest of this lecture, I want to discuss a recent theory that is, I think, our best hope for answering questions like these. What's so cool about it is that it's based on the fractal architecture of all living things. The theory is provocative. You may not buy it. That's fine. A lot of biologists don't buy it, but I find it very stimulating and very provocative because it challenges us, as I have suggested, to look at life through a physicist's eyes. That means we're going to be seeking broad, unifying principles and ignoring the incredible richness and diversity that biologists are taught to cherish. It's a whole different way of thinking about life. We're going to look at the big picture and ignore what to a physicist are details and what to a biologist might seem like the whole story.

The theory that I will be describing was proposed in 1997 by Geoffrey West. He is currently the president of the Santa Fe Institute, which is devoted to studies of complex systems. He is a former particle physicist, the kind of person who used to work on quarks and collisions that would have taken place in the superconducting super collider if that machine had successfully been built, which it wasn't. The funding was canceled in the '90s. That was important for this story because he was so disgusted by the way high-energy particle physics seemed to be going that he looked for some other branch of science to think about, and because of a brush with a serious health problem, he got to thinking about life span. I think that may have been the genesis of this theory that I will be describing.

West did not have much background in biology, and he sought the help of two biologists, who would become his collaborators on the theory that's our topic. They are Jim Brown and Brian Enquist at the University of New Mexico. So it's Brown, Enquist, and West; that's the story. But I'll tend to refer to it as just Geoff West and his team because I think he was really the driving force in this, at least from the physics side.

This theory has already explained some of biology's most far-reaching laws. This is the great part of the story. I'll be telling you about these laws in a minute. You may not have heard of them. They're not often taught in biology courses. If you have not heard of them, I think they will knock your socks off. They did knock my own socks off when I first heard about them. I did happen to hear about them in a freshman biology course in college, and it was one of the things that made me want to learn more about mathematical biology.

The laws that I'm going to tell you about are mathematical regularities that *all* creatures seem to obey. There are very few laws like this in biology that span all of life, but these laws do. All creatures seem to obey them, from the smallest microbes all the way up to the biggest animals we know, to whales and elephants. These laws have to do with relationships between the size of a creature and things like its average life span, how fast it uses energy, how many calories it burns per day, things like that, many other properties—but all relating to the size of the animal. Laws like this have been known for many years. Actually, they were discovered in the 1930s. But explaining them, explaining where they come from, has been a baffling problem, and it has bothered biologists and other scientists, including engineers who have looked into this, for about the past 70 or 80 years.

Let's begin by discussing this question of calories since I mentioned that. How many calories do we need each day? What determines that? In other words, what are our metabolic needs? And not just us, because the point of view we will be taking, as I mentioned, is that we want to look at this across species. So I ask the same question for a mouse, or a dog, or an elephant. First of all, how are we going to measure energy needs? There are various ways you can measure it. You can look at how much an animal eats, or you can strap on a sort of an oxygen-measuring device to see how much oxygen a horse consumes, or you can even just measure the heat that the animal produces.

However you make the measurement, the pattern always comes out the same, and it's an amazing pattern. The metabolic rates for diverse mammals—let's just focus on mammals for now, and later I'll be talking about other creatures, including plants and microbes—but if you just do this for mammals, as the first studies were done, what you find is that the metabolic rates fall very neatly on a curve when they are graphed as a function of the body mass of the animal. Here is what that relationship looks like.

What I'm showing here on the vertical axis is the number of calories used per hour. Here I actually mean calories, not kilocalories. When you think about calories, typically you really mean kilocalories, thousands of calories. But I just mean here calories in the ordinary physics sense. So this is calories used per hour per gram of body mass. We're looking on a pound-for-pound comparison. What you're seeing are—here is the tiniest mammal, called a shrew. It weighs just a gram or two—tiny. Here is a mouse. Here is a flying squirrel, a bat, a cat, a dog, a sheep, a human, a horse, and an elephant. What is remarkable is that they all lie on this beautiful curving shape. The data are not scattered all over the place. They're in a nice curve. What the curve means is that pound for pound, the little guys use a lot more energy than we do. To put it in terms that are pretty stark, this tiny shrew eats more than its weight in food every day. Think about that. Try doing that. Clearly, we don't do anything like that. We don't eat our whole body weight each day, and we don't need to. The graph shows that there is a kind of economy of scale to being big. There is an advantage to being big. Metabolism becomes more efficient on this pound-for-pound basis.

It's also interesting to plot *total* caloric intake. Not just pound-for-pound comparison but the total number of calories that a mammal needs per day versus the mass of that mammal. That's what I'll show you in the next slide, but let me forewarn you that I'm about to pull out logarithms again because we don't like to look at curved lines as much as straight lines. Straight lines are easier to interpret, and if I use logarithmic axes—that is, instead of showing a constant increment along each axis, I'm going to show a constant factor of 10 between successive tick marks—then you will see this picture. The curved line becomes a beautifully straight line.

Now showing again, mouse, going up through rat and guinea pig, cat, goat, chimpanzee, a steer, a cow, and an elephant. All of them fall on a certain line. The line I'm showing here is a solid line. I'll talk about the dashed line in a minute, the one that doesn't seem to go through the data, but first let's focus on the solid line. It's a straight line. That means that the underlying relationship being shown here has the algebraic form of a power law. We talked about power laws earlier in our discussions of fractals, and we said that they were a signifier of self-similarity, the most basic aspect of fractals, their self-similar structure—meaning that one variable is related to another variable raised to some power. In this case, the relationship in the straight line shows that the metabolic rate—in this case, measured as heat production, kilocalories per day—grows like body mass but not in direct proportion to body mass. It grows less than that, like body mass to the 3/4 power. That's the interesting thing. The line goes through the data with a slope of 3/4. That means that bigger animals do need to eat more than little ones in absolute terms but not in direct proportion to their mass—less than that—only in proportion to their mass raised to the 3/4 power. That's another way of looking at this economy of scale that I mentioned.

When this law, the 3/4-power law, was discovered and announced by a veterinary scientist named Max Kleiber in 1932, some biologists were very upset about it. Some still are today because they had expected a 2/3 power. That's the dashed line shown in the picture. You see it doesn't go through the data.

I don't want to go into too many details about why people thought it would be a 2/3 power, but let me try to give you the gist of it.

The thinking was that if all mammals were basically just replicas of each other except for scale, like if one was just a shrunken-down or an expanded version of another, you would get the 2/3 power. The reason is that if the creatures are made of the same kind of tissue, with the same density, and their shapes were basically the same Euclidean shape, like thinking of an animal as being made of a bunch of cylinders—I've got the cylinder of my body, the cylinders of my arms, cylinders of my legs, and so on, maybe a big sphere on my head—if you think of a creature as being a conglomeration of Euclidean shapes, for Euclidean shapes like cylinders and spheres, the volume of a shape like that is proportional to its linear dimension (say its width) cubed, whereas its surface area only grows like the square of the typical side or linear dimension. So when you think about important physiological processes, like gas exchange, or respiration, heat loss, whatever, those things happen through surfaces. Something is passing through the surface of your skin or surface of other parts, of maybe internal surfaces, like your intestines. The thinking was that since these effects are surface-dominated and you've got this volume growing like length cubed—that's the body, all the cells—but heat losses controlled through surfaces—that something squared—when you put all the squares and cubes together, you end up making a prediction of a 2/3 power. But it's wrong. The 2/3 slope does not go through the data points, whereas Kleiber's 3/4 power does, and that's the mystery. Why 3/4, and why not 2/3 ?

This 3/4 power has since been shown to extend over 27 orders of magnitude. The picture here showing from mouse to elephant, I'm going over something like 6 orders of magnitude. But if we go over all creatures known, even below creatures, down to subcellular molecules used in respiration, 27 orders of magnitude. That is an astonishing number. That's why I called it one of the most far-reaching laws in all of science. That's a factor of 1000 trillion trillion that this law holds—incredible.

Here is the picture of it. The metabolic scaling shows 3/4 power. It's a complicated diagram, and I'm going to slowly work my way through it, but what I want you to see for now is just that there's a straight line here with a slope of 3/4. This is the mammals that we talked about earlier. What we're showing is logarithm of metabolic power, now measured in watts, versus the log of the mass of the animal, measured in grams. So here's the familiar Kleiber's law for mammals that we talked about.

Let me ignore what's going on down here for now—we've been talking about cells in culture—and instead mention that we again have the same slope of 3/4 power if we look at the level of a cell; down to an organelle inside the cell called the mitochondrion, the factory for making energy in the cell; down to molecules inside the mitochondrion, the enzymes in the respiratory complex.

The point is that this law with the 3/4 power holds very, very broadly and is a testament to the unity of life on Earth. It's really one of the most comprehensive laws in all of science. So what's the explanation for it? Before I tell you what we think the explanation might be, I want to underscore one more thing about these 1/4 powers. I've been focusing on the 3/4 power and how mysterious it is. There are other 1/4 powers also, very mysterious if we look at other things, like life span. If we look at the lifetime of a mammal [and] ask how much longer does an elephant live than a mouse or a person, it turns out it does not increase in proportion to our mass. It goes like mass to the 1/4 power. It increases but slower than mass. One-quarter power: the time interval, not just of a whole life, but of the events within a life, the time interval between heart beats, between consecutive breaths. All of those things grow like mass to the 1/4 power. Physiological time is somehow related to mass to the 1/4 power, and that has a really cool and weird implication, which is that any mammal of any size, from a mouse to an elephant, including us, can expect to live the same number of heartbeats, no matter how big it is—about 1.5 billion heartbeats. That's what you get. If you want to measure it in breath cycles, you get about 300,000 of them and you're done, on average—whether you're a mouse, or an elephant, or a person, or a horse. It's amazing. We all take the same number of breaths, on average, but we don't know why. Is it a coincidence, or is it something very fundamental about life? Nobody knows. The theory of West, Brown, and Enquist takes us a step toward answering these questions. It predicts these 1/4-power exponents for metabolism and lifetime and so on.

Here's how they got it: They asked themselves—we recognize that all these things have these fractal networks inside them or at least branching networks, at this point, just assuming branching, like we see, not assuming the fractal part yet. If you have these branching networks in all living things, what's the best way to make one? What is the *optimal* design? That's the question they asked themselves. What is the *optimal* design of a branching network that will be carrying nutrients, or oxygen, or other resources in such a way that a few things happen? First of all, this network has to reach every point in a 3-dimensional body. It has to be space filling. And because this is constrained by the laws of physics, you might imagine that evolution has chosen a design that would be most efficient, that would carry the resources with minimum energy wasted and ideally as fast as possible to get to the cells at the end.

Under some reasonable simplifying assumptions and using the laws of fluid mechanics, West and company found that under these assumptions and satisfying this optimality criterion, the network had to have a self-similar branching pattern. In other words, it had to be a fractal. The fractal structure was dictated by this minimization-of-energy criterion.

Then they imposed one further constraint, but known to be realistic, that at the very ends of the network—this is not like a mathematical fractal that's going to branch all the way out to infinity, getting down to infinitely small scales—this is a real network, so it has to end. It has to terminate. The terminal units—in the case of the bloodstream, we would be talking about capillaries—are basically the same size for all mammals. That's an interesting known fact. The capillaries in an elephant or a mouse are just capillaries, and they are about the same either way. So assuming that, that we have these invariant capillaries, invariant terminal units, and the other things that we have hypothesized, with those assumptions, West and company—through a lot of technical work that I can't really go into here, but you can certainly read about it if you're interested—through a technical calculation, they showed that the 3/4 power popped out. It's a consequence of those simple things that we assumed. So did the other scaling laws that have been measured about breath cycles and heartbeat cycles and so on, and others. They actually have a list of 20 or 30 of these scaling laws, and they were predicting the exponent bang on for all of them. They could explain everything that's been measured so far and also make predictions of things that hadn't been measured, and those were coming out right. So there is something that smells very good about this theory.

For instance, the same argument also explained this 3/4 power that I mentioned down here in the level of cells and subcellular molecules. You might wonder what network we're talking about there. There's no bloodstream inside the cell. There are networks, and it's unclear whether we're talking about real networks of actual tubes inside the cell or some kind of virtual network involving chemical reactions that branch in their structure, sort of an abstract but effective branching network that's in there. We don't quite know about that yet, but West and company hypothesized that there is some sort of network that is conveying oxygen from the cell surface down to the mitochondria, where the oxygen is processed, and yet another network that goes the farther step from the mitochondrial surface down to the inner membranes of the mitochondria, where the respiratory enzymes do their job. That's the end of the line. By making the same assumptions that they had about efficiency, they again explained the 3/4 power.

Other interesting testable implications that came out of the theory were that mammals cannot be arbitrarily small. There is a smallest mammal, which we know is a shrew. But they calculated from first principles what the mass of that smallest mammal should be, and it came out right, a few grams. The reason was that if you had a hypothetical mammal smaller than that, the pulsatile blood flow coming from its heart (the pulse—blood doesn't flow steadily; it pulses because of the heartbeat) would become impossible at the scale of anything smaller than a shrew. It would have to be replaced by a kind of oozing, steady flow, like what you see in the smallest capillaries, and it seems that that would be incompatible with life of mammals as we know it.

The theory also predicts something that is shown on this graph, that cells taken from any mammal and put in culture—so it is taken out of the body and just kept alive in a dish—they should all have the same intrinsic metabolic rate. It's not that an elephant's cells at the cell level are using energy slower than a mouse. It's something about the geometry of the elephant that makes them slower. That was a prediction, and it's right. When you look at the cells in culture all the way across mammals, they all lie on this flat line. They all have the same metabolic rate, whereas in vivo, in the animal, then the elephant starts to show its advantage. The explanation seems to be that in the bigger animals each capillary has to supply more cells, so there is less capillary source, less nutrient supply, to go around per cell, and so the metabolic rate for each cell has to slow down. In culture, though, those differences would be rendered irrelevant because there is no supply network. All the cells are getting the nutrients they need in culture, and so this prediction, as I say, was borne out.

It's just a great theory. I happen to love it, but to be honest, I have to emphasize that it's still very controversial. There are people who hate it, who think it's very misleading and not to be trusted. So there are all kinds of disputes about it. There are technical disputes about the calculation itself. Some say the fluid mechanics was done wrong. There are disputes about the data analysis, saying that it's not really a 3/4-power straight line that fits the best. It's really a series of lines that are all 2/3, but they happen to just lie on a line that looks like 3/4, but that's an artifact of comparing different species. That's what some folks would argue, that it's silly to compare across species. On the other hand, you can look at the data and see [that they do] lie very nicely on the 3/4 power, so I think there is something to explain there.

What really bugs some people is this optimality assumption. That rankles biologists. Especially if you've read books of Stephen J. Gould, you know [that] he is always making the argument that natural selection does not need to be optimal, and that's true. Selection works with what's available based on the evolutionary history of an organism. And even if you do optimize something, like the energy expended or the travel time, you might not be optimizing something else. There could be conflicting costs, so optimality criteria can be very confusing.

A good counterargument to that is, "Yes, fine, but optimality is still a natural benchmark." Let's just see how well organisms do in terms of coming close to optimality. Even if they don't have to be strictly optimal, how close can they get?

The whole modeling philosophy here is also bothersome to many people trained in biology because what we're doing and what West et al. insisted on doing was take this physicist approach of making drastic simplifications to try to expose the broad trends. That's a common strategy. We're taught that in physics, but it's anathema to biologists, who come from a different intellectual culture, because you miss all kinds of interesting details when you do that. For example, it turns out that human beings live longer than the overall trend line for mammals would predict given our weight. But West and company would retort, "That's the point exactly!" By highlighting the trend, you can tease out what requires biological explanation. Why do people live longer than you'd expect? You isolate what needs explanation versus what follows generically from deeper chemical or physical principles.

In the same spirit, some biologists have complained, "We know that metabolic rate depends on more than a creature's weight. If you look at a plant or a microbe, its metabolism is 200 times slower than a mammal, even if you make the adjustments for body weight." Right. But don't forget that they have different body temperatures. Those creatures are not warm-blooded, and so if you take body temperature into account in addition to weight and recognize that chemical reactions go slower at lower temperature, you can reduce that variance down from a factor of 200 to a factor of 20.

So, again, what West and company are saying is, "Look at basic physics about mass, about temperature, and figure out what variance remains after you have taken out those things, and that's what biology should be about. That's what needs explaining." Personally, I think they're onto something wonderful, and I think it's groundbreaking work—fantastic.

In the next lecture, we'll complete our survey of fractals and their applications, now moving away from science and looking at the interplay between fractals and the visual arts. I'll see you then.

Fractal Art
Lecture 18

This lecture examines two of the ways that fractals connect with art.

We begin with the computer-generated fractal images that adorned T-shirts, postcards, and coffee mugs in the 1980s. The most famous and dazzling of these is the Mandelbrot set. Mandelbrot was exploring the transition to chaos, as others had done earlier (see Lectures 8 to 11). But he studied simple iterations of points on a plane instead of a number line. The Mandelbrot set encodes a fantastic amount of information about the dynamics of these maps (much like the orbit diagram did for the logistic map; see Lecture 10). The most mild-mannered maps are shown as black points. The more explosive ones are color-coded according to how hot-tempered they are. The real pleasure comes when we zoom in for a closer look. Successive magnifications reveal phantasmagoric structures that look like sea horses and tendrils, lightning and kelp. Burrowing even deeper down, we are astonished to find the Mandelbrot set itself, reappearing in miniature, eerily repeating like variations on a theme. The pattern is infinitely rich, even though it was made by repeating a simple rule. It's also subtly self-similar—not blatantly and artificially so like the Koch curve (see Lecture 15) or other simple fractals. This combination of simplicity, richness, and subtle repetition is what makes these images so appealing. They're neither too regular nor too random—just like good art. These images captivated the public when they first came out around 1985. Personal computers had only just become available around 1980. Fractals were easy to program and compute, and they allowed many people—for the first time in their lives—to make their own mathematical discoveries and to feel how exhilarating math could be.

A more controversial application of fractals to art involves a recent analysis of the drip paintings of Jackson Pollock (1912–1956). Pollock famously said, "I am nature." His paintings are often described as "organic." He created them by channeling chaos, flinging and dripping paint from a stick onto huge canvases laid on the floor of his barn, all the while moving around, leaning off-balance in a controlled way. Are his paintings just random splatterings, as some cynics would say? Or are they fractals, full of hidden structure reflecting their chaotic genesis?

In 1999, a team led by physicist Richard Taylor reported that the drip paintings are fractals. They scanned photographs of the paintings and calculated their fractal dimensions, using the box-counting method we used for Norway's coastline (see Lecture 15). They found one fractal dimension for small scales (between 1 millimeter and around 2 or 3 centimeters). They attributed this to the chaos introduced by the fluid mechanics of the dripping process. They found a different dimension for the larger scales (between around 5 centimeters and 2.5 meters, the size of the canvas). This, they say, was determined by Pollock's technique and body movements.

Finally, they claim that as Pollock refined his technique over the years, the large-scale fractal dimension of his paintings increased. It started at around 1.1 in 1945, corresponding to a loose web of paint trajectories. Then it grew to around 1.5 by 1948, a value more like that of coastlines and other natural fractals. Finally, it reached 1.72 in a densely interlaced 1952 painting, *Blue Poles*, now valued at more than $30 million.

Pollock's alleged fractal signature was invoked as a damning line of evidence in a recent art authentication dispute. In 2005, Alex Matter, the son of a longtime friend of Pollock's, announced that he had found a stash of 32 drip paintings among his parents' belongings. If authentic, they would be worth many millions of dollars. Art historians had conflicting opinions about their authenticity. The Pollock-Krasner Foundation asked Taylor to analyze six of the paintings. He found none of them had Pollock's characteristic fractal geometry.

But in 2006, Taylor's methodology was ridiculed by another team of physicists. They showed that childish pictures of doodles would, when subjected to fractal analysis, give similar results to what Taylor found for Pollock's paintings. Taylor, Mandelbrot, and others dismissed this challenge, saying that the doodles are obviously not fractals, so computing their fractal dimension is meaningless. I agree with the critics that fractal dimension, by itself, is too blunt a tool to be convincing, either in art or science.

But I also believe that the paintings are indeed fractals, in part because of other, supporting lines of evidence. Taylor has produced his own drip paintings and demonstrated that they needn't be fractal. In failed fractals, small portions of the painting don't have the same general appearance as the whole. Thus, it was remarkable that Pollock somehow managed to make his drip paintings self-similar down to very small scales. This does *not* occur automatically. Apparently that's how Pollock instinctively knew when he was done painting, even if a casual observer couldn't see the logic.

But in 2006, Taylor's methodology was ridiculed by another team of physicists. They showed that childish pictures of doodles would, when subjected to fractal analysis, give similar results to what Taylor found for Pollock's paintings.

Taylor has also provided evidence that chaos was crucial to Pollock's process. He built a mechanical device he dubbed a "Pollockizer"—a pendulum that swings over a canvas while dripping paint onto it continuously. The pendulum normally swings regularly but can be made chaotic by perturbing it with a periodic sequence of electromagnetic kicks as it swings. The chaotic drip trajectories look like Pollock's. The non-chaotic ones don't. As of this writing, the jury is still out on the question of the new paintings' authenticity, although the evidence—fractal, chemical, and human expert opinion—weighs against them. ∎

Essential Reading

Gleick, *Chaos*, 215–40.

Taylor, "Order in Pollock's Chaos."

Supplementary Reading

Peak and Frame, *Chaos Under Control*, chap. 7.

Peitgen and Richter, *The Beauty of Fractals*.

Rehmeyer, "Fractal or Fake."

Questions to Consider

1. Can images of the Mandelbrot set really be considered art? What about Pollock's drip paintings?

2. To what extent can artistic style be quantified mathematically? Will computers someday replace human connoisseurs in art authenticity disputes?

Fractal Art
Lecture 18—Transcript

Welcome back. In this lecture, we'll examine two of the ways that fractals connect with the visual arts. This is our first excursion into art and our last, and I hope you will enjoy it. I'm certainly looking forward to it.

This lecture is, of course, part of our ongoing journey across the disciplines. We've spent the past five lectures focusing on fractals, which remember, at least in our course, grew out of our thinking about chaos. Now fractals will return the favor. We'll see chaos reemerging from fractals, first in connection with an incredible fractal called the Mandelbrot set and then later in the art of Jackson Pollock, the great abstract expressionist painter.

Let's begin with computer-generated art. You've seen these, I'm sure. These are the fractal images that adorned T-shirts, coffee mugs, and screen savers. You basically couldn't avoid them in the 1980s. They seemed to be everywhere. The most famous and dazzling of these is called the Mandelbrot set. It's a fractal that you've seen. Here's what it looks like. It's great.

By the way, before I start, I want to thank Wolfgang Beyer, who made this image and all the other gorgeous images I'll be showing you of the Mandelbrot set.

Take a look at it. You see a sort of a strange-looking shape until you come to love it. Some people think it looks like a gingerbread man lying on his side. Some people see something more insect-like, maybe some kind of beetle. It's a particularly weird beetle, if it is that, because it seems to have many little copies of itself all around its body. Here's one; here's one. In fact, there are infinitely many of them, some big, some small. They're all over the place. And more than that, it seems to have bolts of lightning shooting off of its little sub-beetles. So it's quite an exotic-looking structure and kind of beautiful when you get used to it.

But what does it mean? What is the math behind this picture? Where does it come from? Why do we care about it scientifically? Well, it arose when Benoit Mandelbrot was himself exploring the transition to chaos. That was a topic that occupied us quite a bit in Lectures Ten and Eleven, and into Lecture Twelve. At that time, we were focusing on the iterated map that was called the logistic map that arose in population biology, and we used the iterated map to explore the transition to chaos in biology. But then we also saw that it served as a broader paradigm for thinking about the transition to chaos in all kinds of systems, including chemical oscillations and fluid mechanics. We saw that many of the predictions based on the logistic map were later confirmed in real experiments.

At around this time, Mandelbrot himself wanted to think about something analogous to the logistic maps. He wanted to explore the transition to chaos mathematically in his own way. He had the idea that instead of using the kind of numbers that we're used to, like populations—a number that is a real number that would sit on the number line, in other words—he had the idea to study iterated maps on a plane instead of on a line. Specifically, he worked with what mathematicians call *complex numbers*. Now *complex* sounds intimidating, but it doesn't mean complicated. I just mean complex in the sense of a complex of two things stuck together, in this case, a real number and a so-called *imaginary number* welded together into a complex to form a complex number.

I just mentioned imaginary numbers, and I suspect you've heard about them. You may even remember them from high school. Do you remember $\sqrt{-1}$? You called it *i* for "imaginary" because from an ordinary way of thinking, there is no number that when you square it gives -1. But mathematicians defined a number that does that. They call it *i*, $\sqrt{-1}$, and a complex number, then, is something that is going to be formed out of a real number attached to one of these strange (at first) imaginary numbers. You can think of it like this: Think of a complex number, instead of being on a number line, it's a generalization of that idea to a point in a plane where one axis will be the ordinary real numbers, the ordinary real number line, and then the vertical axis will plot these imaginary numbers, whatever they mean, and then a typical point in the plane has both aspects. It has some real part and some imaginary part.

In daily life, we don't use complex numbers, but mathematicians and scientists use them all the time. They turn out to be very natural in higher math. They come up in our studies of electricity and magnetism, and you really can't be a professional engineer, or mathematician, or scientist unless you can cope with complex numbers. They seem quite comfortable, and we don't really think of them as imaginary or complex any more; they're just our friends, like the ordinary real numbers.

Mandelbrot thought, "Why not play the same games as we played with the logistic map for real numbers? Let's play them with a complex version of the logistic map for complex numbers." That was what his idea was. The thinking, as I say, being that complex numbers are, in a certain sense, a more natural generalization of real numbers. Maybe something beautiful would come out from this point of view. And something beautiful did come out—the Mandelbrot set, which encodes a fantastic amount of information about the dynamics of these maps as they undergo their transition to chaos, much like the orbit diagram did for the logistic map in Lecture Ten.

You may recall at that time, when I was talking about the orbit diagram, I gave the analogy of it being a kind of family portrait, showing different character that the logistic map would have as we changed a knob, a parameter in it. It was the growth rate in that model.

There is a similar knob in the problem that Mandelbrot was studying, and the Mandelbrot set shows us the behavior of the maps he was looking at by graphing something about their behavior as a function of the setting of this knob. It's not really a knob in the usual sense because it's a complex number, so it's like two knobs, a real part and an imaginary part.

Looking again at the picture, each point in the plane has a different color. There are black points. Those turn out to correspond to very mild-mannered maps, in a sense I will explain in a minute. They are very tame. They don't do anything too wild. On the other hand, there are members of the family shown in yellowish or blue or white. Those are more explosive, hot-tempered maps in a certain sense. The colors denote how volatile or hot-tempered these different mathematical maps are.

Let me try to say that more precisely. This is going to be a slide with some complex numbers on it and some heavy math. If you don't want to do algebra, you can space out for a minute or two here. It's not going to be terribly important. But for those of you who do like math, here is what the Mandelbrot set really means.

We're going to choose a complex number c, and now here's the map. We're going to take some number and define—that we'll consider the old number, whatever it is—and then we'll square it. You can square complex numbers just like you can square real numbers. Multiply them by themselves. It makes sense. You can do that. So you get the square of some number, add this magic number c to it, and that defines the new number. Then we're going to keep doing that.

For example, if we start with 0—0 is the old number—then 0^2 is 0; I add c; I get c ($0^2 = 0 + c = c$). So the new number is c. In that sense, 0 maps to c. Then I feed c into this machine. So I get c as the old number; I square it; I add c. So c produces $c^2 + c$. I keep doing that. I feed this into the maw of the machine, and I get $(c^2 + c)^2 + c$, and so on, and so on. It starts to get tedious to pronounce, but mathematically, a computer doesn't have any problem just grinding forward for millions of iterations.

Here is the issue now. When we do this, we generate a sequence of complex numbers, which you can think of as a point hopping around on the complex plane, like we saw when we were looking at the Hénon map, if you remember that, where there was a point popping around, making a boomerang shape eventually. Here, the points are going to hop around, and the questions will be: Do those points stay bounded? Do they live inside a certain square around the origin in the complex plane, or do they eventually pop off all the way out to infinity?

If the sequence stays inside a certain square—you pick a square that's length 4 on each side in the complex plane, sitting around the origin. [If] you stay in that square forever—if you're trapped in there—then you're defined as being in the Mandelbrot set. The original point c was in the Mandelbrot set and gets colored black. Whereas if you hop out to infinity, now we want to color-code you—that is, this point c—according to how fast you hopped out to infinity. That's quantified by the number of steps you took until you escaped from that certain square. That's the gist of it.

Having done all that, we can then color-code the Mandelbrot set, and we get an image that looks like this. The real pleasure comes when we zoom in for a closer look at the Mandelbrot set. That's what we're going to do now in a sequence of slides. We'll first look at some area, the area shown here in the white rectangle, and we'll blow that up to be full screen. Notice all kinds of interesting structure is revealed. Now you see all these little beetles marching down, an infinite procession of them marching into this crack. It's beautiful. Now let's zoom in on this part. You see new structures. Each successive magnification reveals beautiful, phantasmagoric structures. Some of them look like sea horses; some look like tendrils, or lightning, or kelp. I'm going to burrow in to reveal all that's in here. It's incredible. Now let's go in this box. Gorgeous. What's in here? That looks sort of like a sea horse tail. When you zoom in closer, you see it has little substructures, eyes to it. Now we're starting to see something curious. Keep your eye on what's in the middle of this box. Do you see what's there? It's an old friend of ours. It's a tiny copy of the Mandelbrot set itself, reappearing in miniature, eerily repeating, like variations on a theme.

Now let me show you the whole sequence in a continuous zoom, but watch out. You might want to take a Dramamine first because it could make you dizzy. Here we go. We start with the Mandelbrot set and just zoom in. So this Mandelbrot set is infinitely rich, even though it was made by just repeating a simple rule in the complex plane. It's also subtly self-similar. It's not blatantly self-similar or artificial-looking, like the Koch snowflake curve was or the other simple fractals that we discussed.

This combination of simplicity and richness and subtle repetition is what makes these images so appealing. They're neither too regular nor too random. Just like good art, I would say. Something that is too random or too regular is boring. This is neither, and it's part of what makes it so aesthetically appealing.

These images captivated the public when they came out around the mid-1980s. The big break was when the Mandelbrot set made it to the cover of *Scientific American* in 1985. That set off a tidal wave of experiments, where ordinary people, who had just gotten personal computers only a few years earlier, could now recreate the Mandelbrot set right there in the living rooms, and this allowed people to participate in math for the first time in their lives, in many cases; to make their own mathematical discoveries; and to feel the exhilaration of seeing something for the first time mathematically that no one had ever seen before.

A more controversial application of fractals to art, moving to the second half of the lecture, involves a recent analysis of the drip paintings of Jackson Pollock. If you're not familiar with Pollock, let me show you an example of what his work looks like. This is from a 1950 painting, *Autumn Rhythm*. If you're not used to it, it looks like a lot of paint thrown on the canvas, and that's pretty much what it is. But it's made to remind you of nature. Does it look like something in autumn to you? Pollock himself said famously, "I am nature." His paintings are often described as being "organic." He created them by deliberating channeling chaos. Interesting. He used the dynamics of nature, the chaos inherent in the natural world, and he tried to use that in his own painting. He would fling paint from a stick. He would dip it into the paint can and then fling it onto a big canvas that he had rolled out on the floor of his barn. Sometimes he would pour it onto the canvas or drip it. All the time he was doing the flinging or dripping, he would purposely move around the canvas in an off-balance way, but it was a controlled off-balance. There is a great movie about this where you can see Ed Harris playing Pollock, and it will show you what his motions looked like.

The question, then, for us is: Are these drip paintings of Jackson Pollock just random splatterings that any child could make, as some cynics would say, or are they fractals? I'll leave the question of whether they're beautiful to you, but are they fractals? Is there structure in them? Are they not random? Are they full of hidden structure, reflecting their chaotic genesis, remember chaos having this mixture of order and randomness in it?

In 1999, a team led by physicist Richard Taylor reported that the drip paintings are, indeed, fractals. Richard Taylor is unusual in that he has a Ph.D. in Physics, but he also has a Master's degree in Art Theory. In fact, his Master's work specifically focused on Jackson Pollock.

To address this question—are these fractals?—Taylor and his team took very high resolution photographs of Pollock's paintings and scanned them and then calculated their fractal dimension as if they were a coastline of England or Norway or something like that, just treating these as natural objects. Let me give you some examples to give you an intuition about what the different dimensions would mean.

First, let me show you how we would calculate a dimension with box counting. We actually discussed this when we talked about Norway. In this case, we're not showing the coast of Norway but just a little excerpt of one of Pollock's paintings. We cover it with a mesh. There's a whole mesh of squares here, little squares. Some of them, you see, are white. That means that no part of the painting is contained in those squares. So these are empty, whereas the blue squares do have a little bit of paint in them. What we do, then, is ask: "How many of the squares are occupied, have some paint in them?" We count them. That's the game. We count the number of occupied squares that have been penetrated by the painting. That number in itself is not interesting. What's interesting is how that number changes if we refine the mesh, make the mesh have smaller boxes. So we'll do that again. You see, in the image below, we have now made the mesh smaller. These boxes are much smaller, and there are now more empty ones, as you would expect. The question is: How does the number of occupied boxes change as we change the size of the mesh? We make a graph of the number of boxes occupied versus length of the little squares. If that graph turns out to be a good straight line when we graph it on logarithmic axes, on a log-log plot, that's an indication of self-similarity. It would mean that these images are fractal, and the slope of that line, as we discussed when we were talking about the coastline of Norway, will give you the fractal dimension of the original painting.

Taylor and his team did this sort of data analysis, and they found that the data didn't fall on a simple straight line. In fact, [the data] fell on 2 lines joined at an elbow. So it looks something like this. There would be one straight line for part of the picture, corresponding to box sizes going up to maybe 2 or 3 centimeters, and then there's a different slope for bigger scales for between a few centimeters and meters, maybe 2 meters—the size of the whole canvas. So they attributed this broken line, 2 lines with the elbow, as being due to two different physical processes. At the lower scale, there is chaos introduced by the dripping process itself, the way that paint fractionates into a fractal splatter when he was throwing it or flinging it. It's almost like we saw those fractal drops in that Hokusai painting, *The Great Wave off Kanagawa.* Remember, as the wave comes in, it sprays drops of all different sizes. That's actually known from measurements of paint, that it will splatter by making drops of many different sizes. So this lower dimension, the drip dimension, for small scales, Taylor and company interpret as due to the physics of paint itself, whereas this larger-scale dimension, the dimension having to do with scales on the order of meters or tens of centimeters, they interpret as having to do with Pollock's body English as he was moving around, the way he moved around the canvas and moved his body position, which as I say, he deliberately went off balance.

[Taylor and his team] claimed, further, that as Pollock refined his technique, working on these drip paintings for about 10 years, from a period from about 1943 to about 1952, the drip dimension, the one associated with the paint, systematically changed over the years. In fact, they say that this is the hand of the artist, that you can tell by calculating a dimension, plus or minus one year, when Pollock would have painted it.

Here's what different dimensions would look like. Early in his development, Pollock was painting things with dimension close to 1. A dimension of 1 itself would just be a straight line. That's not a fractal. Something that's loose like this would be 1.1. As it gets denser, 1.6; 1.9 is almost filling the whole canvas; and 2 would not be a fractal. Again, that's just a filled square. So as we look at Pollock's style over the years between 1943 and 1952 or so, you see that the data showing the different dimensions—there's not much in here, but these data points are then followed by ones way up here, and it just keeps going systematically higher. Here is one point that is off the curve. It's interesting. This one has a dimension close to 1.9, and Pollock didn't like this painting and later destroyed it. He was somehow trying to get things. He felt it was too dense. It's as if he intuitively wanted to make something that matched this pattern as a function of time.

The interesting thing, though, is that if you look at some of the numbers—like these paintings in here have dimensions around 1.5 around 1948—that was a value close to the value seen in coastlines. If you remember, Norway had a coastline dimension of 1.52. So [Pollock] is now painting things like coastlines.

Finally, here is a 1952 painting with a dimension of 1.72, which is very densely interlaced. It's a famous painting, considered by many to be Pollock's masterpiece. *Blue Poles*, it's called. It took him 6 months to paint it, and it's now valued at more than $30 million. Here is *Blue Poles*. Many different colors of paint, reds and yellows; the poles, these black images, going through. So that's Jackson Pollock's drip painting with its systematic increase in fractal dimension.

The plot thickens because this alleged fractal signature of Pollock was recently invoked as a damning line of evidence in an art authentication dispute. In 2005, Alex Matter, the son of a longtime friend of Pollock's, photographer Herbert Matter, announced that he had found a stash of 32 drip paintings among his parents' belongings in a storage locker. They were labeled in his father's handwriting as being done by Pollock in the 1940s.

Pollock had a turbulent life. He was an alcoholic. He died in a car crash. He possibly had some kind of mental illness; it's unclear. In any case, he had a very up-and-down life. When he was down, he didn't even have money for groceries, and he would give away his paintings or sell them very cheap just to get money for food. So we don't really know how [Herbert Matter] came to have all these paintings. They may have been a gift—we don't know—but in any case, [Alex] Matter said he found these paintings in the storage locker. He claims that they are real Pollocks. You have to appreciate that the discovery of any new Pollocks—now worth tens of millions of dollars if Matter's claim is true—if authentic, these paintings would individually be worth probably tens of millions.

Art historians had conflicting opinions about the authenticity of the paintings, just for an issue of how they were found and what they looked like. The question came up: From the point of view of Richard Taylor's fractal analysis, what would he have to say about it? The Pollock-Krasner Foundation asked him to analyze six of these newfound paintings. Applying his mathematical method, Taylor found that none of them had Pollock's characteristic fractal geometry.

The plot thickens more because in 2006, Taylor's methodology was ridiculed by another team of physicists. They showed that if they took very childish doodles—an example that they made was just drawing a bunch of little stars the way I'm indicating here. They filled up the canvas with a bunch of stars on top of each other, different sizes, all over the place, and that was it. When they subjected these childish paintings of doodles to the same kind of analysis that Richard Taylor had used, they claimed it gave similar results to what Taylor had found when he looked at Pollock's paintings. In other words, they said [that] the test for fractals is not convincing, that this kind of silly painting also looks like a fractal. Are we serious? Is this really how we're going to analyze these paintings?

Richard Taylor replied, and so did Benoit Mandelbrot, the father of fractals, that it's a silly challenge because the doodles are obviously not fractals. You can just look at them and see that they're not, so computing their fractal dimension is meaningless. You have to first establish that the image is self-similar by looking at it. The fractal dimension alone isn't going to compute that in a reliable way. If it's self-similar, then the fractal dimension is meaningful.

In any case, the team criticizing Taylor is also unhappy that Taylor's data extended over only about 2 or, at most, 3 orders of magnitude in the size of the boxes, saying that isn't enough to make a convincing fractal, plus the fact that the line wasn't even straight. It has an elbow in it. So they alleged there is no real good evidence for self-similarity or fractal structure in the Pollock paintings.

Taylor said, "Hold on; that's true of most of the fractal literature, that 2 or 3 orders of magnitude is typically what we see; usually 2 orders of magnitude." The critics replied, "Yes, but that literature is worthless, too." So it's quite a controversy going on and on. I would weigh in myself and say that I agree that fractal dimension, by itself, is too blunt a tool to be convincing either in art or science. But I also believe that the paintings are probably fractals, in part because of other supporting lines of evidence that Taylor has given. For instance, he has produced his own drip paintings. He himself has some artistic ability and tried to make his own drip paintings. He demonstrated that the drip process alone doesn't guarantee that it will result in a fractal. They need not be fractal. That is, Pollock was doing something special.

Here is a drip painting by Pollock. In the bottom panel, we take one of these regions and blow it up to make a full size, and then blow up a piece of that, and blow that up to be full size. You're seeing that the painting preserves its structure under successive blow-ups. Each image sort of looks like the previous ones statistically, whereas if you have an amateur, so to speak, making a drip painting, then the structure is lost under successive magnification. You can see that this top image doesn't really have the same statistical character as the bottom one. So in failed fractals, the small portions of the painting don't have the same general appearance as the whole.

It was remarkable that Pollock somehow managed to make his drip paintings self-similar down to very small scales, down to millimeters. This does *not* occur automatically. Apparently, that's how [Pollock] knew when he was done, instinctively. He would be painting for months, and to anyone else it just looked like splattering. But there were things he was trying to do, and when he was done, he knew he was done, even if a casual observer couldn't see the logic.

Taylor has also given other evidence that chaos itself, in the literal sense that we've been studying in this course, was crucial to Pollock's process. Taylor built a mechanical device that he called the "Pollockizer," which is a pendulum that has paint in it that swings over a canvas while dripping paint onto it continuously. The pendulum normally swings regularly, like an ordinary pendulum, but you can make it chaotic by perturbing it with a periodic sequence of electromagnetic kicks as it swings. Here are the results of those paintings by pendulum. If the pendulum is set to be non-chaotic—that is, you don't give it a kick—it produces a drip painting that looks like this. That certainly does not look like a Pollock. If it's set in chaos mode, it produces a painting that looks like that, which is sort of like Pollock's paintings. Here's Pollock's painting *Number 14* from 1948. It's not identical to the chaotic pendulum, but you can see it has the right character.

As of this writing, the jury is still out on the question of those disputed paintings and their authenticity. The evidence—fractal, chemical evidence about pigments, human expert opinion about style—seems to weigh against these new paintings being authentic, but who am I to say? That is something that will be settled in the future. In any case, I think it's interesting that the mathematics of fractals has played a part in this authentication dispute.

In the next lecture, we'll see how scientists have come to embrace chaos, almost in the same way that Jackson Pollock did—that is, deliberately using chaos to try to do something interesting and find practical advantages to chaos. We'll see in the next lecture in particular how chaos is being used to design low-cost, fuel-efficient travel through outer space.

Outer space. It's interesting. The chaos in the solar system is now being used for space mission design. It seems fitting, in a way, or maybe ironic, that after starting the whole subject—the three-body problem in astronomy, which we saw in Lecture Four, started the subject of chaos theory—after being regarded as an enemy for these 100 years, the chaos in the three-body problem is now coming back as our friend.

I'll see you next time.

Embracing Chaos—From Tao to Space Travel
Lecture 19

> The idea of chaos as a source of creativity also occurs in the writings of
> the philosopher Friedrich Nietzsche, who wrote, famously—you have
> probably heard this quote, but I like it—he said, "I say unto you, one
> must have chaos in oneself to be able to give birth to a dancing star. I
> say unto you, you still have chaos in yourselves."

In this lecture and the next, we'll take a look at how chaos can be used for practical purposes in engineering and technology. Artists and philosophers have long embraced chaos as a source of creative energy. In Lecture 18 we saw how Jackson Pollock channeled natural chaotic processes while creating his drip paintings. The idea of creative chaos also appears in the writings of philosopher Friedrich Nietzsche (1844–1900): "One must have chaos in oneself to be able to give birth to a dancing star." Perhaps the earliest expression of the fecundity of chaos is in the ancient Chinese philosophy of Taoism, from the writings of Lao Tzu in the 6th century B.C.E. Taoism is a philosophy based on nature, and Tao is the natural order of all things. Here, *hun-tun* (chaos) is the primal state, the mother of the universe. But unlike other conceptions of chaos (the abyss of the Greeks, the void of the Hebrews, or the malevolent disorder of Christian traditions), *hun-tun* signifies a primordial confusion that is fertile, pregnant with possibility, a source of life and creativity. In Taoism, nature is self-organizing. Order emerges spontaneously from chaos, with no need for a purposeful God to wrest order from disorder. There are parallels here with the modern scientific conception of chaos. Chaos contains hidden order, as we've seen in our discussions of strange attractors, iterated maps, orbit diagrams, and fractals. Nonlinear systems can organize themselves spontaneously (see Lecture 23).

Yet it's taken a long time for most scientists and engineers to see chaos as an ally. They traditionally avoided chaos, and where it was unavoidable, they squelched it. A new approach emerged around 1990. Close in spirit to Taoist teachings, the notion was to embrace chaos and to make use of its unique properties. An influential publication along these lines was titled "Controlling Chaos." By its very nature, chaos might seem uncontrollable. Yet in some ways the opposite is true. Precisely because tiny nudges to a chaotic system can have such potent effects, these systems are highly responsive. Edward Ott, Celso Grebogi, and James Yorke, theorists at the University of Maryland, showed mathematically that gently tapping on a chaotic system in the right way could coax it into periodic behavior. Rajarshi Roy, a physicist then at Georgia Tech, demonstrated this experimentally. He forced a laser to emit chaotically pulsating light and then nudged the chaos into various forms of periodicity. Operating in this way, the laser produced much higher power than usual, while remaining under control.

Aerospace engineers exploited a similar idea years ago. They purposely designed the F-16 jet fighter plane to be unstable, to enhance its maneuverability. The F-16 is almost impossible for a human pilot to control, but its computer automatically makes split-second corrections to keep the plane stable.

The F-16 is almost impossible for a human pilot to control, but its computer automatically makes split-second corrections to keep the plane stable.

Perhaps the most striking use of chaos has been in the design of fuel-efficient trajectories for space travel. Space flight is expensive. The costs of flying freight to the Moon are about $250,000 a pound. The conventional route is fast (3 days for a one-way trip), but it relies on a heavy-handed maneuver called a *Hohmann transfer*. A spacecraft parked in orbit around the Earth fires its engines until it reaches almost bullet speed, then coasts for 250,000 miles. Finally, to settle into lunar orbit, it pivots around and blasts its retro-rockets until it slows down enough for the Moon's gravity to capture it. The slow-down maneuver alone costs more than $130 million in fuel!

A new approach, invented by mathematician and former NASA scientist Edward Belbruno, exploits the chaos inherent in the three-body system of Earth, Moon, and spacecraft. He thinks of it as "surfing the gravitational field." Just as a big ocean wave carries a surfer along with it, Belbruno devised trajectories that "go with the flow" from the Earth to the Moon. The spacecraft *coasts* into lunar orbit without using any retro-rockets to slow down. There are only two catches:

NASA/JPL, Cici Koenig.

Interplanetary Superhighway Makes Space Travel Simpler.

- The trajectories become chaotically tangled when they reach a "fuzzy boundary," a gravitational tug-of-war region between the Moon and Earth.

- They take 2 years to get there!

Belbruno's superiors at NASA's Jet Propulsion Lab scoffed at this—and didn't trust chaos anyway—so they eventually let him go. But NASA came back asking for help in 1990. Could Belbruno salvage a Japanese space mission gone awry? Japan, aspiring to be the third country to reach the Moon, had launched two probes. One had headed toward the Moon but lost radio contact. Could Belbruno find a way to send the other spacecraft—a communications relay station the size of a desk—to the Moon, even though it was never intended for that purpose and had practically no fuel? The route had to carry the spacecraft to the Moon with just the right speed. Too fast and it'd sail past the Moon; too slow and it'd crash.

Belbruno (who was then collaborating with Jim Miller, the scientist from NASA who'd initially approached him) worked backward from the desired speed and altitude required. They found an amazing solution: a trajectory that first wandered a million miles away, out to where the Sun's gravity also became significant. This detour saved the day. Belbruno and Miller guided the spacecraft out to the Earth-Sun fuzzy boundary, then quickly pushed it through the chaotic tangle by firing the craft's rockets, after which it coasted back to the Earth-Moon fuzzy boundary. The Japanese spacecraft—named *Hiten*, meaning a Buddhist angel that dances in heaven—safely arrived at the Moon on October 2, 1991, after a 5-month trajectory.

Belbruno's ideas have now become mainstream. Extensions of his method, called *low-energy transfers* along the *Interplanetary Transport Network*, were used in NASA's unmanned *Genesis* mission and in the European Space Agency's *SMART-1*. Other ingenious uses of chaos may be crucial for the future of space travel, especially as NASA seeks inexpensive ways to achieve its mission of getting back to the Moon, and possibly to Mars by 2020. Even though low-energy transfers are too slow for people, they might be fine for delivering the supplies the astronauts would need on the Moon or Mars. ■

Essential Reading

Belbruno, *Fly Me to the Moon.*

Supplementary Reading

Frank, "Gravity's Rim."

Girardot, *Myth and Meaning in Early Taoism.*

Klarreich, "Navigating Celestial Currents."

Ott, Sauer, and Yorke, *Coping with Chaos*, chaps. 12–14.

1. When you're trying to be creative, do you prefer to make a mess first, or do you favor a more carefully controlled process? What are the pros and cons of each strategy?

2. Why might the ancient Chinese have arrived at such a different view of chaos from Western cultures?

Embracing Chaos—From Tao to Space Travel
Lecture 19—Transcript

Welcome back. In this lecture and the next one, we'll take a look at how chaos can be used for practical purposes in engineering and technology. The utility of chaos would come as no surprise to those artists and philosophers who embraced it as a source of creative energy. For example, in Lecture Eighteen, we saw how the abstract expressionist painter Jackson Pollock deliberately channeled natural chaotic processes in the creation of his drip paintings as he was flinging paint and moving his body in a controlled, off-balance way.

The idea of chaos as a source of creativity also occurs in the writings of the philosopher Friedrich Nietzsche, who wrote, famously—you have probably heard this quote, but I like it—he said, "I say unto you, one must have chaos in oneself to be able to give birth to a dancing star. I say unto you, you still have chaos in yourselves." I love that; chaos giving birth to a dancing star.

Perhaps the earliest expression of the fecundity of chaos is in the ancient Chinese philosophy of Taoism, from the writings of Lao Tzu in the 6th century B.C.E. Taoism, which I certainly am not an expert on, but as I understand it from others, is a philosophy based on nature, and Tao is the natural order of things. In this philosophy, there is no particular clear translation of *chaos*, but the closest would be the concept of *hun-tun*, the primal state, the mother of the universe, a state of undifferentiated potential and infinite possibility.

Let me read you a quote from Lao Tzu from his book the *Tao te Ching*, chapter 25. He wrote:

> There was something chaotic yet complete, which existed before the creation of heaven and earth. Without sound and formless, it stands alone and does not change. It pervades all and is free from danger; it can be regarded then as the mother of the world. I do not know its proper name but will call it Tao.

Unlike other conceptions of chaos in the creation myths—for instance, the abyss of the Greeks, the void of the Hebrews, or the malevolent disorder in

Christian traditions—*hun-tun* signifies a primordial confusion that is fertile, pregnant with possibility, a source of life and creativity.

Hun-tun also refers to a creation myth where the Earth sits inside the heavenly sphere like the yolk in an egg. A related modern word in that context is *won-ton*, as in the soup with dumplings floating in a clear broth, like the Earth floating in the celestial sphere. In fact, the words are so close that if we were to look at the Chinese words for *won-ton* and *hun-tun*, they really look the same. They have the same characters, except that a food radical is replaced by a water radical.

In the philosophy of Taoism, nature is viewed very differently from the way we're used to thinking of it in the West. Nature is self-organizing. Order emerges spontaneously out of chaos, with no need for a purposeful God to wrest order from disorder. There are modern parallels to this view of nature in the modern scientific conception of chaos theory. As we have seen throughout this course, chaos contains hidden order. We saw that in our discussions of strange attractors when we were looking at the Lorenz system. There is hidden order in the form of iterated maps, orbit diagrams, and fractals. This view of order inside of chaos is very much in the spirit of Tao.

From our point of view, nonlinear systems can show the self-organization that was mentioned as one of the features of Tao philosophy; that is, they can organize themselves spontaneously. We haven't talked about that much yet in the course, but we will in Lecture Twenty-Three. We'll look at the issue of self-organization, in particular, how systems synchronize themselves.

It has taken a long time for scientists and engineers to see chaos in a positive light—as a potential friend, an ally. Traditionally, engineers especially avoided chaos. They didn't want anything to do with chaos, and where it was unavoidable, they would squelch it. A new approach emerged around 1990. Close in spirit to Taoist teachings, the notion was to embrace chaos, to welcome it, and to see if some use could be made of its unique properties. That's what we will be focusing on for the rest of this lecture.

An influential publication along these lines is called "Controlling Chaos." By its very nature, chaos might seem uncontrollable, but in some ways, the opposite is true. Precisely because tiny nudges to a chaotic system can produce such potent and disproportionate effects, these systems are highly responsive. That's the big idea: that chaotic systems can amplify small disturbances, and you can use that.

A strategy for controlling chaos was developed by three researchers at the University of Maryland—Ed Ott, Celso Grebogi, and Jim Yorke, the authors of the paper I just mentioned, "Controlling Chaos." They pointed out that in a chaotic system that has a strange attractor, like the Lorenz attractor, in addition to the chaotic motion around the strange attractor—that is, the darting from wing to wing in an unpredictable way—there are also periodic motions, repetitive motions embedded in the strange attractor, periodic orbits that exactly repeat.

We didn't talk about those because you can't see them easily. They're unstable. They're like a pencil balancing on its point. In principle, a pencil could balance on its point. That would be an equilibrium state, but the slightest nudge will cause it to tip. The same is true of these periodic orbits in the Lorenz attractor. They're in there, but any little numerical error or any puff of wind or anything will kick you off of those onto some other type of behavior. We know they're in there. We know there's a kind of periodicity built into the attractor, but it's hard to see it because it's so unstable.

Still, these orbits carry a lesson for us, which is that a chaotic system has the inherent potential for a kind of order—periodicity—if it could be stabilized. The beauty of this idea is that you wouldn't have to manhandle the system to make it periodic. It wants to be periodic on its own. Its own dynamics would take care of that naturally if you could stabilize a system so it wouldn't fall off these periodic tracks. In fact, there are infinitely many of these periodic orbits embedded in there, like a thicket. That is interesting because it means that a strange attractor has a kind of infinite potential to it. It's an infinite library of periodic motions if you could stabilize them. So if they could be made stable, these periodic orbits would provide an enormous repertoire of possible rhythmic or periodic behavior that might be useful for something.

That was the idea in "Controlling Chaos." They developed a scheme for controlling chaos in this way and showed that the periodicity was, in some sense, inherent and natural—as I say, you didn't have to manhandle the system to force it to be periodic. You could just give it a little tweak from time to time. Sporadic kicks would nudge it into periodicity, which it would follow, and because of instability it might start coming off, but you could flick it back on. So you can be very gentle and control it onto these periodic paths. Their scheme was then tested in a number of real chaotic systems to see if it would work. Could you create periodicity out of chaos with this very delicate control?

One interesting example along those lines was the stabilization of a chaotic laser, which was worked on by a team at Georgia Tech led by Rajarshi Roy. Raj is a physicist and a specialist in lasers. I'll tell you more about him later. He's a friend of mine, and I would love to tell you, but let me move along.

[Raj] and his team were thinking about a chaotic laser. It was chaotic in the sense that it wasn't producing a steady beam of light. The intensity was pulsating fitfully. That sort of chaotic pulsation in intensity in lasers had been known for a long time, since the creation of lasers in the 1960s. From the very beginning, people found that lasers were prone to instabilities, and that bothered them; they didn't want that. They wanted a nice continuous beam. It was regarded as a scourge that if you pumped a laser too hard and tried to make it produce a lot of energy, it would start oscillating and then go chaotic. So over the years, laser scientists have worked hard to develop control schemes, either keeping their lasers at low power or finding ways to control them from going into these chaotic, fitful pulsations. That's what all practical lasers do today in CD players, or checkout scanners, or when an ophthalmologist is doing laser surgery on your eye. Don't worry; those lasers are not chaotic. They behave nice and steadily, producing a nice constant beam.

With the rise of chaos theory, those old instabilities started to seem interesting, like maybe we should look at that chaos in lasers, and maybe we can do something with it. Maybe we shouldn't have just suppressed it all these years. It was worth a second look, and that's what Raj Roy and his team did. They showed that using the Ott, Grebogi, and Yorke scheme— or actually, a slight modification of it—with gentle, sporadic control, they could stabilize a laser that they had driven to be chaotic. They could put it on any kind of periodicity that they wanted almost, as long as it was one of those infinitely many periodic behaviors embedded in the system's natural dynamics. This meant that Raj could drive the laser to much higher performance, much higher output power than normal, and still have a means of controlling it.

This idea of exploiting chaos, using it, harnessing its great responsiveness was an idea that aerospace engineers had many years ago. Something similar to it, maybe not exactly chaos, but in the same spirit, had to do with the design of the jet fighter plane called the F-16. That plane was the first plane to be deliberately designed to be aerodynamically unstable to enhance its maneuverability. It's almost impossible for a human pilot to control it because of the tendency to fly off into an unstable flight. But the F-16 has a computer built into it that makes split-second adjustments to keep the plane flying the way it's supposed to, keeping it stable. In effect, the computer is almost flying the plane, and sometimes the system is referred to as "fly by wire" because rather than flying with a steering wheel, the pilot uses electronic controls—joysticks, in effect—to keep the plane doing what it's supposed to.

In case you're worried that the computers will fail and the pilot can't fly the plane, obviously that was thought of, and the engineers have made the system quadruple redundant. There are four backup systems to make sure that the pilot doesn't lose control. This incredible instability of the F-16 makes it very, very responsive and great at pursuit and evasion in dogfights. It was a tremendous breakthrough in fighter design.

Conversely, systems that are too stable can be difficult to control. I just tried to illustrate with the F-16 that it can be very responsive by being unstable. A system that is too stable can be tough to control, like the old cliché about a battleship, that when it's moving in one direction, it will be tough to turn it around because it's so stable in its course. So in that way, if you want changes, it's tough to deal with systems that are too stable.

You might also think of a horse. If you could tame a thoroughbred, that is a good horse to ride. It has tremendous potential, but you have to tame its chaos.

Perhaps the most striking use of chaos in this regard has been in the design of fuel-efficient trajectories for space travel. You have to realize that space flight is very expensive. The costs of flying freight to the Moon—and not just freight, the cost of flying anything; it could be the fuel itself that you're taking on the spacecraft—the cost of flying freight to the Moon is about $250,000 a pound. That's serious money. So you would like to use as little weight as possible. How do we get to the Moon? There is a conventional way of getting there, and it's fast. It only takes 3 days for a one-way trip. But it relies on a very heavy-handed maneuver called the *Hohmann transfer*, developed in the early days of rocketry.

Here's how it would work: A spacecraft is parked in orbit around the Earth. It's orbiting the Earth. It's already blasted off and gotten into Earth orbit; it's sitting there. Then it fires its engines—for a second time, having blasted off—to bring itself up to close to the speed of a bullet. It's going 2 miles per second. A bullet goes about 3 miles per second, but you get the idea. This rocket is now moving at close to bullet speed, through the tremendous use of fuel, and then it's going to coast at that enormous speed, for the rest of the way mostly, to the Moon, about 250,000 miles, so it's coasting at bullet speed to the Moon. Now, it's approaching the Moon, screaming at bullet speed and, of course, it could easily fly past. So to get into lunar orbit—I'm trying to be a rocket here, not too easy—but if I were the rocket coming this way at bullet speed, I would have to make a maneuver to turn around and blast my retro-rockets in the opposite of the direction of travel to slow myself enough for the Moon's gravity to capture the rocket. The slow-down maneuver alone, the turning around and blasting the rockets yet again, by itself costs about $130 million in fuel.

The advantage of the Hohmann transfer is that it's very well understood mathematically. We have understood it for 80 or 90 years, something like that. All you have to do is solve a pair of two-body problems. Remember when we were talking about Newton's work on astronomy and, in particular, the planets orbiting the Sun, he solved that by pretending it was a two-body problem, just the Sun pulling on the planet. That's well understood now for 350 years, and that's what the Hohmann transfer is based on. You solve a pair of two-body problems. First, the spacecraft is being pulled by the Earth, and then when it gets close enough to the Moon, you ignore the Earth's effect and assume the spacecraft is being pulled by the Moon. It's so simple that it can be worked out in a pocket calculator that NASA engineers would have. They like the Hohmann transfer for that reason. They understand it, and we've been very successful in getting to the Moon.

I want to spend the rest of our time talking about a new approach, which was designed and invented by a mathematician working at the time for the Jet Propulsion Lab, the part of NASA that deals with unmanned space flight. This man's name is Ed Belbruno. He was trained as a mathematician, but he found himself working at JPL, the Jet Propulsion Lab. He knew about the three-body problem. He had studied dynamics—nonlinear dynamics, chaos; he knew about those things—and he wondered, maybe there's a better way than the Hohmann transfer, especially with the concern about the costs of space travel. NASA is trying to do everything much more cheaply than it used to and still stay safe. Can they be more efficient? He wondered if there would be a way to use the chaos inherent in the three-body problem of the Earth, the Moon, and the spacecraft—now treating it as a three-body problem, not two bodies any more. Could he somehow use the chaos that we know is there from Poincaré's work on the three-body problem that we discussed in Lecture Four—can we use that chaos to advantage?

Belbruno thought of his idea as "surfing the gravitational field." That's the metaphor, surfing. Just as a big ocean wave carries a surfer along with it, Belbruno devised trajectories that would, so to speak, "go with the flow," the gravitational flow from the Earth to the Moon.

In this picture [from] *How To Do Things*, the spacecraft coasts into lunar orbit. In other words, when it's getting close to the Moon, it doesn't turn around and blast retro-rockets. It just coasts and finds itself in orbit around the Moon. It still has to blast off the Earth, so you need to use your fuel there. But having left the Earth, you could sit above the Earth, parked in orbit, and then nudge yourself farther and farther away from the Earth by spiraling out gradually, using only a low, continuous thrust. In terms of fuel, the cost is negligible. But it takes a long time. You have to spiral out from the Earth until you start getting a significant gravitational tug from the Moon. Once you're there, you just let the Moon pull you in to orbit. It's almost like the difference between a glider and an airplane. That is, you could just glide along with the currents of the air, letting the natural dynamics of the air take you, or you could try to steer your way through and power your way through using a big jet engine. Those are the differences. The jet engine method is like the Hohmann transfer. This is sort of like gliding on air currents, except we are now talking about gliding on the gravitational field all the way to the Moon.

It's an attractive vision. Maybe a little kooky, but interesting, and Belbruno worked out the details of it. He found that he was able to devise trajectories that would do this. There are only two catches: First of all, there are many possible trajectories of a spacecraft surfing in this way, and when they get close to a certain region near the Moon, these trajectories become very sensitive and chaotic. As you would expect, there is chaos in this three-body problem. Slight differences in the trajectory start to become amplified tremendously in a certain region near the Moon that Belbruno called a "fuzzy boundary." It's a boundary in the sense that on one side of it—it's not actually a region around the Moon. It's a little more complicated than that because there's a geographical region around the Moon, but we also, in speaking about the state of a particle moving in the three-body problem, we would also have to keep track of its velocity; that is, the state space for this problem is more than just position relative to the Moon. We also have to be thinking about velocity. So really, the fuzzy boundary lives in this abstract state space that's higher dimensional than just the 3 dimensions of real space where the Earth and the Moon live and the spacecraft.

Already it's getting a bit complicated, but what Belbruno found is that these trajectories become very chaotic when they get to this region in the fuzzy boundary, on one side of which you're going to be captured by the Moon, on the other side of which you're captured by the Earth's gravity. In the fuzzy boundary, it's like a gravitational tug-of-war, and the spacecraft doesn't know which way it's going exactly, so it's prone to chaos.

That's one catch. You can get to the Moon like this, but your trajectory may be chaotic. The other catch is that it's going to take 2 years to get there. Compared to 3 days for the Hohmann transfer—well-understood math, no chaos; expensive, yes, but we know what we're doing—versus this new idea: 2 years, with chaos, but at least it's cheap.

Belbruno's superiors scoffed at his idea and certainly didn't trust the idea of basing missions on chaos, so you can imagine what they did; they let him go. He was looking for jobs as an assistant professor, and he did find gainful employment as a math professor.

The story takes an interesting twist at this point because NASA came back to Belbruno. They needed his help. Something had happened. A JPL engineer named Jim Miller showed up at Belbruno's door one day asking for his help: Could Belbruno use his ideas to help salvage a Japanese mission that had gone awry?

Japan had aspired to be the third country to reach the Moon. Of course, the United States and the former Soviet Union had already done this, but Japan wanted to be third to show their technical prowess, and so they had launched a pair of unmanned probes. One had been sent on its way to the Moon; the other was supposed to serve as a relay station. The one that was headed toward the Moon got lost; it lost radio contact. They didn't know where it was. It was gone.

So that was the question: Could Belbruno find a way to send the other spacecraft—this communications relay station that was like the size of a desk—to the Moon, even though it basically was never intended for that purpose and had very little fuel, practically none? Could he use his low-energy transfer to get this desk out to the Moon?

Now, the route had to carry the spacecraft to the Moon with just the right speed. If it went too fast, it would sail right past the Moon; too slow, and it would crash into the Moon; just right, and it would be weakly captured by the fuzzy boundary around the Moon. It could just glide into orbit. From there, it would take only a tiny amount of maneuvering and a little expenditure of the fuel that it had left to convert that orbit in the fuzzy boundary into the standard elliptic orbit of the type that is usually desired and that we understand well. So that was the hope. Get to the fuzzy boundary. If you can get me there, I can then just use a little fuel. I'll be in orbit; Japan will have succeeded.

Belbruno was by that time collaborating with Jim Miller, the scientist from JPL who had originally approached him. What they did was calculate [the following]: Suppose this spacecraft was already at the Moon, at the fuzzy boundary where it needs to be. Could we work backwards from there to figure out the right conditions to get it there with the desired speed and at the right altitude above the Moon? They did find a way. They found an amazing solution. It was amazing in that the trajectory they found first wandered a million miles away, out to where the Sun's gravity became significant. So it was now a four-body problem. Their computer solution told them that would work if you could somehow get out to the neighborhood of where the Sun's gravity would become significant, somehow you could end up back at the Moon with just the right velocity.

This detour saved the day. Belbruno and Miller guided the spacecraft out to the Earth-Sun fuzzy boundary. Any pair of objects will have the fuzzy boundary for the spacecraft that's interacting with those two. So there's a fuzzy boundary between the Earth and the Sun. They got out to there. Then they quickly nudged it through that chaotic tangle by firing the craft's rockets, using only a little bit of fuel, after which it coasted back to the Earth-Moon fuzzy boundary, and they had succeeded. The Japanese spacecraft, named *Hiten*, meaning a Buddhist angel that dances in heaven, safely arrived at the Moon on October 2, 1991, after a 5-month trajectory.

It's a great story. You can read this story in Belbruno's recent book called *Fly Me to the Moon*. He will tell you the whole saga of his work. It's quite nice. It's short and quite a colorful story. He's a character. I think you'll enjoy it.

In any case, Belbruno's ideas have now become mainstream, that is, extensions of his method, now called *low-energy transfers* along what is regarded as the *Interplanetary Transport Network*. There is a picture of these routes linking all the different planets through these low-energy chaotic paths. This Interplanetary Transport Network and this kind of routing of space missions were used in NASA's unmanned *Genesis* mission between 2001 and 2004. They used that idea. They also used it in the European Space Agency's mission called *SMART-1*, which arrived at the Moon in 2004.

The math behind all these ideas is still being actively developed and refined. The concept of the fuzzy boundary hasn't quite survived. It turned out it was great intuition, but it was a little too fuzzy. It didn't fit well into existing chaos theory, and it has been replaced by a new conception involving tubes and other structures in state space. This sort of work is being developed by one of the great dynamicists and chaos theorists, Jerrold Marsden at Cal Tech, and his former student Shane Ross, and Martin Lo at JPL, and a number of other people. This has gone mainstream, and it's quite a vindication of the whole chaos approach to engineering. It's a thrill for our whole community.

Other ingenious uses of chaos may be crucial for the future of space travel, especially as NASA seeks inexpensive ways to achieve its mission of getting back to the Moon and also getting to Mars by 2020.

Even though these low-energy transfers would be too slow for transporting people, they might still be useful for transporting equipment or freight or other supplies that the astronauts would need on the Moon or when they finally get to Mars.

In any case, it's interesting to think about who else might have been using these low-energy transfers through the solar system. It seems likely that comets and near-Earth asteroids have been using this superhighway through the solar system for billions of years, possibly even involved in the collision that is thought to have extinguished the dinosaurs.

In the next lecture, we'll discuss another example of the usefulness of chaos, but this one will be closer to home. The theme is privacy: how to use chaos to encrypt secret messages.

I'll see you next time.

Cloaking Messages with Chaos
Lecture 20

Today's digital technology makes it pretty tough to decrypt a cell phone call, but it wasn't always like that. ... Do you remember when reporters intercepted Princess Diana's cell phone conversations with her lover, James Gilbey? It made the news because apparently his pet name for her was "Squidgy."

This lecture continues our examination of the uses of chaos. Now the theme is encryption—how to use chaos to send secret messages. Our story begins around 1990, in the days before cell phone calls were routinely encrypted. Do you remember when reporters intercepted Princess Diana's cell phone conversations with her lover James Gilbey, later publicized as the embarrassing "Squidgy" tapes? Prince Charles's even more salacious conversations with Camilla Parker Bowles became public in the same way. When Newt Gingrich, then Speaker of the House, was under investigation for alleged ethics violations, his calls were intercepted by Democratic loyalists and used to humiliate him. In all these cases, even a minimal level of privacy would have helped—not military- or banking-level encryption, just enough to thwart a casual eavesdropper.

Chaos has two potential uses in this regard. First, it's noisy. If played through a loudspeaker, it sounds like static because it's made of all frequencies, just as white light is made of all colors. So it's good for concealing signals; it cloaks them over a wide band of frequencies. Thus, an eavesdropper might not even realize any information was being sent. And even if he did, he'd have trouble pulling the message out of the chaos. Second, chaos could act as a carrier (not just a concealer) of messages. Traditionally, signals are carried by sine waves (as in AM or FM radio). A sine wave has just one frequency. Chaos, being made of infinitely many frequencies, might be more versatile.

This line of thinking first occurred to Lou Pecora in 1988. Pecora is a playful, light-hearted physicist at the Naval Research Laboratory in Washington, D.C. His background is in solid-state physics, not communications or chaos theory. He became curious about chaos in the mid-1980s, when it was the hottest subject in physics. But to please his employer, he felt he needed to find a practical justification—military or otherwise—for devoting research time to chaos, then newfangled and trendy. As soon as he posed the question, he immediately thought of communications.

This raised the problem of synchronization. All communications require a transmitter and receiver to be synchronized. For instance, the process of tuning a radio to a particular station locks the receiver to the frequency of the broadcast transmission. Once synchrony is established, the song on the radio is extracted by *demodulating* it from the radio wave that carries it. So the first step was to figure out how to synchronize two chaotic systems. "Synchronized chaos" sounds like an oxymoron. The butterfly effect would seem to destroy any hope of achieving synchrony. That's true if the two chaotic systems are independent. But perhaps the butterfly could be dispatched by linking the systems somehow—they're communicating, after all! Pecora and his postdoctoral fellow Tom Carroll tried weeks of computer experiments on various pairs of chaotic systems, without success. One night, while rocking his baby daughter to sleep, Pecora had a brainstorm: "I need to drive chaos with chaos—I need to drive the receiver with a signal that comes from the same kind of system."

That idea worked. In computer simulations and then in experiments on electronic circuits, Pecora and Carroll found a way to synchronize two identical chaotic systems. The communication was one-way, with a chaotic transmitter driving a chaotic receiver. Pecora was asked to explain his work to the National Security Agency, which led to a scene reminiscent of a Monty Python sketch.

In 1993, graduate student Kevin Cuomo and his adviser Alan Oppenheim, electrical engineers at MIT, published the first experimental demonstration of chaotic encryption. Cuomo built a pair of matched circuits that ran the chaotic Lorenz system (see Lecture 7) and showed that it could transmit and mask human speech. When he demonstrated his circuits to my class, he masked a recording of the hit song "Emotions" by Mariah Carey. One student, apparently with different taste in music, asked, "Is that the signal or the noise?" Listening to the hiss of the encrypted music, one had no sense that there was a song buried underneath. Yet when the combined song plus mask was transmitted, the receiver synchronized almost perfectly to the chaotic mask, but not to the song. So after instant electronic subtraction, we heard Mariah Carey again.

The next breakthrough was a 1998 lab study conducted by Gregory VanWiggeren and Rajarshi Roy, physicists then working at the Georgia Institute of Technology. They extended chaotic encryption to lasers and fiber optics, instead of electronic circuits. Messages are converted into light and masked by the wild fluctuations of a chaotic laser. After transmission down the fiber, the optical message is decoded by a matched chaotic laser. This system transmits 150 million bits per second, thousands of times faster than electronic circuits.

Chaotic encryption has since been extended to real-world tests in the fiber-optic network of Athens, Greece, in 2005. This study was the first to demonstrate chaotic encryption outside of a lab.

Chaotic encryption has since been extended to real-world tests in the fiber-optic network of Athens, Greece, in 2005. This study was the first to demonstrate chaotic encryption outside of a lab. It worked with existing infrastructure, thus confirming its feasibility. It was fast and reliable. Transmission speeds reached 1 gigabyte of encrypted information per second—about the same as commercial transmissions—and lost only about 1 byte in every 10 million. Yet questions remain about the security of chaotic encryption. Kevin Short, a mathematician at the University of New Hampshire, has broken nearly every chaotic code proposed to date. Still, chaotic encryption might be useful as another layer of security in the e-commerce transactions that race around the world every day (which are currently encrypted by software using mathematically heavy algorithms). With the growing concerns about cyberterrorism, national security, and cell phone and Internet privacy, this potential application of chaos is worth exploring. ∎

Essential Reading

Strogatz, *Sync*, chap. 7.

Supplementary Reading

Ditto and Pecora, "Mastering Chaos."

Ott, Sauer, and Yorke, *Coping with Chaos*, chap. 15.

Questions to Consider

1. If you were a code-breaker, what techniques would you use to extract a message buried in chaos?

2. How is the chaotic masking method different from other forms of encryption?

Cloaking Messages with Chaos
Lecture 20—Transcript

Welcome back. In the last lecture we saw that in the 1990s a few maverick scientists were starting to embrace chaos, looking for a way to use it to advantage rather than ignoring it or avoiding it or squelching it, as scientists and engineers had traditionally been taught to do. This lecture is about another possible use of chaos. Here, the theme will be encryption—how to use chaos to send secret messages.

Before I get into the details of the story, I have to tell you I'm really excited about this lecture, in part, because I think it's one of the niftiest developments in the whole subject in the past decade or two. Also because it happens that some of the work was done by people who are friends of mine, so it gives me pleasure to be able to tell you about their research.

Encryption is something that's on our minds all the time these days. We expect to do online banking or shopping without any fear of eavesdroppers or thieves stealing our passwords or our bank account numbers or other important information, although as you know, these sorts of things do happen. But not because of the encryption. A lot of people forget to turn on the encryption at their home wireless network, and pretty much all online banking does use encryption. That's the only way it can work. If you use proper encryption, those encryption algorithms are based on methods that come from a different branch of math than we've talked about. It's not from chaos theory. It's from the theory of numbers. We call it *number theory*, or sometimes you'll hear it called *prime number theory* because it especially has to do with properties of prime numbers and the difficulty of factoring big composite numbers into their prime factors.

The math behind this tells us that there's a provable level of high security to these methods. We know that they're very, very tough to break. We're sure of it. So that level of encryption that comes from number theory is, with good reason, thought to be very safe and strong. Similarly, in addition to banking and other things that we do through our computers, we expect our cell phone calls to be private, even though, in principle, the information is going out over the air. You would think that a determined eavesdropper could intercept our phone calls and possibly use them to embarrass us or whatever, potentially decoding them after pulling them out of the air.

Today's digital technology makes it pretty tough to decrypt a cell phone call, but it wasn't always like that. I don't know if you remember this, but in the early 1990s, when the research I'm about to describe started, there were some big stories about people having their cell phone calls intercepted. Do you remember when reporters intercepted Princess Diana's cell phone conversations with her lover, James Gilbey? It made the news because apparently his pet name for her was "Squidgy," which is a term of affection that I'm not so familiar with. I think in British English it may mean something like soft and squishy in a good way. People made a lot of fun of Princess Diana for being Squidgy.

There was an even more embarrassing incident where Prince Charles had a cell phone call to his lover, now wife, Camilla Parker Bowles, that became public in the same way. The stuff that he was saying to her was so nasty that I don't think I'm going to remind you. You may want to look that up if you don't remember what I'm talking about. It was bad.

Newt Gingrich had his own problems. He was Speaker of the House around that time, and he was under investigation for alleged ethics violations, and his cell phone calls were intercepted and made public by Democratic loyalists who wanted to humiliate him.

In all these cases, even a minimal level of privacy would have been very welcome. I'm not talking about military- or banking-level encryption, just anything that would have been enough to thwart a casual eavesdropper, like these reporters. They don't really know how to decode messages, and they didn't need to. They could just pull them out of the air and hear what was being said. That's what you want to make sure you can avoid, at the very least.

Chaos has two potential uses in regard to private communications. First of all, chaos sounds noisy. We haven't talked about the sound of chaos. I've been mostly focusing about the shape of chaos in state space, the strange attractors and so on, or graphed as a time series. We see some erratic wiggle.

But chaos also has a sound, an auditory dimension to it. You can hear it by playing a wave of chaos through a loudspeaker. If you made a voltage signal that was erratic and drove a loudspeaker with it, you would hear chaos. What does it sound like? It sounds like static. It sounds like noise because it's made of all frequencies.

Noise, something called *white noise*, like white light being made of all colors of the spectrum, is made of all frequencies. Chaos has some characters like that. It's made of all these different frequencies, not like a sine wave, which is a single frequency, a pure tone, like a musical note that would come if you hit a tuning fork, a very simple sound. Chaos is an admixture of infinitely many different frequencies, so it has a noisy sound. Why is that good? It makes it good for concealing information. It can blanket a message over its whole frequency band. All the different frequencies in whatever the intended message are can be covered by the chaos like a mask, like a cloak. So an eavesdropper hearing a combined message with chaos on top of it might not even realize the message was being sent because it was shrouded in chaos. Even if the eavesdropper did realize there was a message buried under there, it would be very hard to pull the message out of the chaos. That's the hope.

There was a movie along these lines that I found very haunting. Maybe you saw it. It didn't have to do with chaos theory. It predated chaos theory, but did you ever see the movie called *The Conversation*? It's a Francis Ford Coppola movie. Gene Hackman is the star, and he plays a professional spy, an eavesdropper, who is a sound engineer, and Cindy Williams, who you may remember from *Laverne and Shirley*, is walking around with her boyfriend in a street fair. I think it takes place in San Francisco, if I remember. They're on a very noisy street with street musicians pounding on bongos and a lot of sound of people. It's the sort of place you would go if you wanted to have a private conversation. If you thought someone might be spying on you or trying to listen in, you would talk in a noisy place like that with the hope that even if they did spy, they would have trouble pulling your conversation out of the noise. That, of course, is what the movie is about. Gene Hackman has a directional microphone. He records all the noise. He records what he hopes is the underlying message, and I won't tell you the rest of the story. It's quite a movie.

So this old idea that you can make something secret by burying it under noise, that's the first possible use of chaos, to use chaos as a mask.

The second possible use is to use chaos as a carrier, a novel carrier. Normally, like with radio, you use sine waves to carry songs or whatever else is being sent over the radio. That's the traditional way of doing things in either AM or FM radio. A sine wave has just one frequency. Chaos, being made of infinitely many frequencies, you would think might have some advantages. It might be more versatile. That's another possible idea. At this point, we're just brainstorming. Is there anything we could do with chaos with regard to communications?

This line of thinking first occurred to someone named Lou Pecora around 1988. Lou, who is one of these friends of mine, is a very playful, lighthearted physicist—good-natured, self-effacing guy, a joker—who works at the Naval Research Lab in Washington. His background has nothing to do with communications. Lou is trained as a solid-state physicist. He's supposed to work on things like spin waves and magnets and positron annihilation in solids, that kind of thing. So he has a background in solid-state. In the late 1980s—you remember what was happening then—James Gleick's book had come out about chaos, and chaos theory was the hottest thing there was in all of science. Everybody was talking about chaos, and certainly any physicist was very aware of chaos theory. What's it all about?

Lou was curious, and I think it's possible he might have been less than excited about some of his research. I'm not sure about that, but in any case, he was very excited to learn about chaos, and curious. So he thought, "Is there something I can do to please my employer, the Naval Research Lab? If I'm going to get into chaos, I have to justify why I'm looking at this thing." He had to find a practical justification. It could be military or otherwise. The Navy was willing to consider things not just purely military, but at least it should be practical. Otherwise, why should he devote his research time to chaos? [It] seemed not only newfangled [but] maybe trendy in a bad way, at least to some people.

Lou posed the question to himself this way: "What could I do with chaos?" That was a new question. People hadn't been asking that question. He immediately thought of communications for the reasons I just suggested. That made him instantly think about synchronization because all communications depend on synchronization. They require a transmitter and a receiver to be able to synchronize so that messages can be conveyed. They have to be locked in to the right frequency. For instance, when you listen to a song on the radio, what's going on? You have to tune the radio to a particular station. That locks your receiver, your radio, onto the frequency of the broadcast transmission. Once synchrony is established, then the song on the radio is extracted. You can hear it by a process electrical engineers call *demodulation*. They have to demodulate the song from the radio wave that's carrying it.

So the first step in Lou's vision would be: Can we figure out how to synchronize, not sine waves, but two chaotic systems? That also was a question that people hadn't really thought about at that point.

I hope you realize that this already starts to sound a bit strange because "synchronized chaos" sounds like an oxymoron. How could you have chaos in sync? In fact, if you think about it a little more, you might say, "Wait a second. That's not even possible because if I have two systems that are both chaotic, one a transmitter and one a receiver, that are both supposedly chaotic, how could they stay in sync? The butterfly effect is the most basic thing about chaos, so any little difference between them will snowball over time and not be able to stay in sync." So people thought—to the extent that they even thought about a question like this—that the butterfly effect would destroy any hope of synchrony. That would be true if the chaotic systems are independent. Yes, it's true. Two independent chaotic systems, not communicating in any way, could not stay in sync. Any little disturbance would cause them to drift apart quite fast. But perhaps the butterfly could be dispatched by linking the systems together somehow. They're not independent, and of course they're not. They're communicating, after all. That's the whole point. We're trying to do communication. So this butterfly argument is wrong. There's nothing that forbids two systems from being both separately chaotic and yet perfectly in sync with each other.

It would be like if you were to watch a modern dancer moving in some weird, chaotic way. There could be another modern dancer who is very sensitive to what the first modern dancer is doing, and they're both moving chaotically in the same way. They're both chaotic, but they're synchronized. There's nothing in principle impossible about that.

Pecora had to try to figure out a way to do this now, and he and his postdoctoral fellow, Tom Carroll, tried for weeks in computer simulation to try to get various pairs of chaotic systems to synchronize with each other, moving in lockstep, but it didn't work. It was without success. They kept trying everything they could think of. They tried the Lorenz system with a logistic map or maybe the Rössler attractor we talked about. They were trying all kinds of different combinations, different parameter values, changing the knobs. Could they get anything to synchronize? No, they could not. It's typical of research. It's frustrating. You spend a lot of your time being stuck, and that was the way Lou Pecora felt at this point. He was stuck. He was frustrated.

[Lou] went to a conference and was listening to the talks. He was in a bad mood because he wanted to solve his problem. He couldn't enjoy the talks or learn anything because he kept thinking about his problem, but he couldn't solve it. The conference ended, and he went home, and it was late at night; he had just caught a long flight. He got home around midnight or 1:00. He went in the house, and his baby daughter was crying. His wife said, "It's okay, Lou. I'll give her a bottle or whatever she needs. You just go to sleep. You take it easy." But he said, "No, I don't think so. I can do it. Maybe it will help. I'm just going to sit here and unwind with her." So he's holding the baby in his arms, giving her the bottle, rocking her to sleep, and then he figured it out. Then he solved his problem. He had a brainstorm. He thought, "I need to drive the chaos with chaos that comes from exactly the same kind of system. I need to drive the receiver with a signal that comes from a matched chaotic system. They have to be chaotic in the same way. I'm going to drive chaos with the same chaos, and then maybe I can get it to synchronize."

Pecora and Carroll tried it, and the idea worked. It worked in computer simulations, and then Tom Carroll, who is very good at building things, built electronic circuits, and they experimented on real electronic circuits. They found a way to synchronize two identical chaotic systems—or in the case of real circuits, nearly identical, as identical as they could make them. The communication was a one-way street, as you would expect. It was a chaotic transmitter driving a chaotic receiver. So they were able to establish synchrony. Remember now, that's step one. Once we get two synchronized chaotic systems, now we're ready to try to do communications.

Lou, being very naive about communications, thinks to himself, "I don't know. What should I send as a signal? How about a sine wave?" That's a physicist's first impulse: "I'll send a sine wave as my signal. Can I send a tuning fork sound to the receiver?" So he sent the signal, and he buries it under chaos, like we talked about, and figured nobody would know that there is a little tuning fork sound in there.

He starts talking it up around the Naval Research Lab, and his superiors thought it sounded interesting, so they told him one day, "Someone is going to come to your lab and see your system, and you're not to talk to the man or ask him any questions. Just let him do his work. Don't bother him. Don't ask him anything." Lou said, "Okay, but how do I refer to him? What do I call him?" They said, "Don't call him anything, or call him Bill." So "Bill" shows up one day, a young guy with a lot of equipment. He asked to see the system, and Lou said, "I dare you to figure out what's hidden under the chaos." This young engineer, mysterious "Dr. X" that Pecora and Carroll started calling him, set up a few little experiments and within about maybe 10 or 15 minutes said, "You've got a sine wave hiding under there." Lou, of course, was flabbergasted. So this was not a secure way of doing encryption at all. It was elementary for any code-breaker to break.

Still, there was some interest in any code. The National Security Agency is interested in these things, and they arranged for Lou to come to this ultra-secretive organization to tell them about his ideas. So he went to the National Security Agency, and he told me that for him it was like speaking into a black hole, that information goes in, but none comes out. You can say stuff to them, but they won't react, and they don't say anything to you.

This led to a scene that reminds me of Monty Python. For whatever reason, Lou forgot something at the National Security Agency, and he needed to get back in touch with his contact there, whom we'll call "Colonel Y." Lou couldn't remember how to reach Colonel Y, and so he thought, "National Security Agency. How am I going to get to Colonel Y? Maybe I'll look in the phone book for the National Security Agency." As it turns out, it's in the phone book, which is a surprise in itself. Lou called the information desk at the National Security Agency. An operator picks up. "May I speak to Colonel Y?" The operator says, "I can't confirm or deny that any person named Colonel Y works here." Lou said, "I understand that, but I came there a few weeks ago and really need to speak to him. I understand why you can't tell me, but could you just send me the number, or is there some way? He will want to talk to me. Can you give me the number for Colonel Y?" "I cannot confirm or deny that Colonel Y works here." So Lou said, "Is this the information desk?" The operator said, "Yes, what information would you like?"

Pecora and Carroll's work on synchronized chaos was published in 1990. The part about synchronized chaos was published, not the part about communications. It became one of the most highly cited papers in the history of chaos theory. It was cited thousands of times. Their further work on using chaos for encryption was not published for quite a while for security reasons and maybe because of a patent possibility.

It was left to others to show the first experimental demonstration of chaotic encryption. This was reported in 1993. It was done by a graduate student named Kevin Cuomo, working at MIT, with his advisor, Al Oppenheim, in electrical engineering. I was at MIT at that time and had the fortune to know these two gentlemen, and even worked with them a little bit. Cuomo was a student in my chaos class. I remember when he came to my attention. There were maybe 60 or 70 people in the class, but he stood out because in preparation for the final exam—where I told the students they were allowed to bring one page of notes of anything that they wanted to write, handwritten formulas or whatever—He very proudly showed me his handwritten page. He said, "I have the whole course right here on this one side of one sheet." All the formulas were written inside little boxes with very precise, machine-like handwriting, like as if it was typed.

I had learned over the years in grading students that when a student writes like that, they are usually good. It doesn't always work, but it's a pretty good

319

rule of thumb. By the way, if you have messy handwriting, don't worry about that. Messy handwriting signifies nothing. With messy handwriting, you might be a genius or you might not. But if you write like a little typewriter, you are probably smart. At least, that's my rule of thumb. Other teachers may disagree.

Back to Cuomo, though. What did he do in his thesis? He built a pair of matched electronic circuits that ran the Lorenz system. Basically, each little voltage at different points in the circuit was one of those variables in the three-variable Lorenz system that we talked about in Lectures Seven and Eight or thereabout. This was already a feat in itself. No one had been able to build a Lorenz circuit, a circuit that implemented the Lorenz system. It was, in effect, an analogue computer for the Lorenz system. It turns out [that] it has a very large dynamic range, and there are electronics difficulties. People had tried, but no one succeeded until Cuomo.

So he built the first Lorenz circuit, but then he built a second one. One was going to be a transmitter, one a receiver, and then he showed that you could use this to transmit and mask—that is, conceal—human speech. It was a dramatic thing. He came back to my class a year later as a visitor, and he demonstrated his circuit to the class. First he showed that he could synchronize the two circuits. He demonstrated that with an oscilloscope. Then he really brought the house down when he showed how you could use it to mask human speech.

The speech signal that he chose was a recording of a hit song at that time by Mariah Carey called "Emotions." I'm not even going to try to sing it for you. One student in the class, apparently with different taste in music, said, "Is that the signal or the noise?" It was the signal, but then Cuomo buried it under the hiss of chaos. When he played the combined signal plus noise, you had no idea that there was Mariah Carey buried under there. You just heard this hiss. Yet when the combined song plus mask was transmitted to the receiver, the receiver synchronized almost perfectly to regenerate the mask, only the mask. In other words, the mask, which is Lorenz chaos, was sent across the wire, and when the receiver received the mask plus Mariah Carey, it regenerated just the mask. That's the nifty idea. It just regenerated the mask so you could sort of electronically subtract the mask, and there underneath it is the Mariah Carey song.

When he played that version, the combined signal with the mask taken off, it was clearly Mariah Carey. It was instantly recognizable, though a little bit fuzzy. It wasn't perfect synchronization, and of course, it couldn't be because remember, he wasn't sending a perfect, clean Lorenz mask to the receiver. There was a little bit of, so to speak, contamination, with all due respect to Mariah Carey. There was contamination of the chaos, if you want to think of it that way, with human speech, and so there was no way a Lorenz circuit could regenerate all of that. It was a sort of corrupted signal, but it did a very, very good job of regenerating the chaos, enough that it could be subtracted and unmask the message.

The next breakthrough came in 1998. It was a laboratory study done by Greg VanWiggeren and Rajarshi Roy, whom I mentioned in the last lecture. Raj was the advisor. He's a physicist; at that time, they were working at Georgia Tech. I would like to tell you for a second about Raj. He's one of my best friends, a very sweet, gentle, kind person. I always feel a little ashamed of myself when I'm with him at a conference because no matter how bad a lecture we've just heard, he always finds something positive to say about it, something that he could learn. He can learn from anybody. He sees the good in everything. But I don't want you to think he's a pushover because he's not. He's a very incisive, creative, bold scientist with a daring creative streak and with great hands, which is really the highest compliment you can give to an experimentalist. He has the magic touch, the golden touch, and his experiments seem to yield gold every time.

In this case, what he and his student, Greg VanWiggeren, were doing was to try and extend the earlier work of Pecora and Carroll and Cuomo and Oppenheim, who had all been working on electronic circuits. They were going to extend chaotic encryption to lasers communicating with each other through fiber optics instead of electronic circuits. In their approach, messages are converted into light. Instead of an electronic signal, you make an optical version of the message, and then they're masked by the wild fluctuations of a chaotic laser. I mentioned in the last lecture that lasers can be driven into chaos. That was one of Roy's specialties, to study chaos in lasers. So he made a chaotic laser that was then used to mask the message. Then the combined message plus mask was transmitted down the optical fiber and decoded by a matched chaotic laser. This system had the advantage, compared to electronic circuits, that it transmitted 150 million bits per second, thousands of times faster than you could do with electronic circuits. So there was hope of tremendous speeds using optical communications.

This idea of chaotic encryption through optical fibers has now been extended to the real world. It has been tested in the real-world network of Athens, Greece, in the optical fiber network that's already installed there. What was striking about this study is that it was the first to demonstrate that chaotic encryption could work outside the controlled environment of a lab. It worked with the existing infrastructure. Nothing had to be tweaked; it just worked, confirming its feasibility. The method was fast and reliable. By now, the engineers that did this study were able to get the transmission speeds up to 1 gigabyte of encrypted information per second, which is about the same as commercial transmissions right now. But this was now encrypted, and they lost only about 1 byte in every 10 million. So it's fast and reliable.

Still, questions remain about the security of this kind of encryption, about chaotic encryption. Remember Lou Pecora's unfortunate attempt to conceal that sine wave. That was easily unmasked. It has turned out that other forms of chaotic encryption have also not been that tough to unmask. Kevin Short, a mathematician at University of New Hampshire, showed that he could break essentially every chaotic code that has been proposed to date. I'm not sure if he has done that yet with the most recent studies from the Athens optical fiber network, but he was able to unmask Raj Roy and Greg VanWiggeren's earlier optical fiber work on chaos.

So it seems likely that this method is not very secure. That was something I had been warned about from the very beginning. When I used to talk to Al Oppenheim and Kevin Cuomo, Al admonished me in my excitement about all of this from the very beginning. He said, "You must never call this secure. This is not secure. Secure is a very special thing. Secure means that if a spy comes and kills someone, and takes the encoder and brings it back to their base, and tries to decode the message, they still can't decode it. That's secure. That's what secure would mean. This is maybe minimal privacy at best. We don't know how good it is; we're just trying to learn." So I remember that admonishment very well to this day, and Oppenheim's intuition was right. These methods have not proved to be secure.

But that doesn't mean they will always be insecure. This is a newborn baby. We don't quite know what its use will be, and if its security can be strengthened, it might be useful either on its own or as another layer of security in the e-commerce transactions that race around the world every day, which are currently encrypted by these number theory algorithms that are quite mathematically heavy and in some ways not so elegant or fast. So this could have a role for chaotic encryption.

In any case, with the growing concerns about cyberterrorism, national security, cell phone privacy, and Internet privacy, this potential application of chaos is still certainly worth exploring.

In the next lecture, we'll turn our attention to another role for chaos: chaos in health and disease in the human body. Is chaos good for you or bad for you? The answer might surprise you.

I'll see you next time.

Chaos in Health and Disease
Lecture 21

> Other researchers, as I mentioned, have their own promising algorithms, but nobody has fully solved the problem of epilepsy seizure prediction yet. But I think we're on the right track, and if we are, maybe Iasemidis's dream will come true. He dreams that some day an early warning system can be programmed on an implantable computer chip, planted in the brain, and then when the beast appears, the chip will trigger a kind of "brain defibrillator" to fire a preventive electrical volley or maybe release an anti-seizure drug.

Our next stop along the frontier of chaos research lies at the crossroads of mathematics and medicine. The topic is biological rhythms, from the electrical chatter of brain waves to the pulse of a beating heart. Building on decades of biological research about the function and mechanism of such rhythms, chaos theorists have been asking questions about their dynamics. Can the mathematics of chaos be used to predict epileptic seizures? The problem is important. Epilepsy afflicts 50 million people worldwide. During a seizure, an epileptic might convulse, black out, collapse, or stop breathing. Seizures attack suddenly, so epileptics dare not drive or even give their baby a bath. Just a few minutes of forewarning would help the patient get to a safe place and possibly take preventive medicine.

But is there any hope of predicting seizures? Clinicians traditionally viewed them as random, yet there do seem to be precursors. Some patients experience *auras*. They see strange lights, feel sick, or smell odors. Others are alerted by their dogs that a seizure is coming. Recently, bioengineer Leon Iasemidis and neurologist Chris Sackellares, and a few other teams of researchers worldwide, have been trying to predict seizures with the methods of chaos theory.

They seek clues in EEG (electroencephalogram) tracings taken before, during, and after a seizure. Such records come from patients who were being monitored in preparation for surgery to remove an epileptic focus from their brains. The data reveal that the EEG becomes *less* chaotic, not more, during a seizure. The EEG's fractal dimension and Lyapunov exponent (which measures the strength of the butterfly effect) both drop abruptly when a seizure starts.

Even more exciting, though, is what happens several minutes *before* the seizure. The time series of Lyapunov exponents measured from different electrodes start to synchronize, lining up in amplitude and phase. According to Iasemidis, this is "the mark of the beast," a clue that a seizure is imminent. His interpretation is that different brain regions are falling into pernicious lockstep.

What's still unknown is how often this pattern is followed by a seizure. In a preliminary study in 2003, Iasemidis found his algorithm correctly predicted 82% of seizures, with about 75 minutes of advance warning (on average). But it also made false predictions 3 or 4 times per day, which is about as many seizures as actually occurred. Several other researchers have their own promising prediction algorithms, but nobody has fully solved the problem yet. Iasemidis dreams that an early warning system can be programmed onto an implantable computer chip. When the beast appears, the chip will trigger a "brain defibrillator" to fire a preventive electrical volley or to release an anti-seizure drug.

Next we consider cardiac arrhythmias. Arrhythmias are irregularities of the heart's electrical activity. In the worst kind, *ventricular fibrillation*, the heart quivers and fails to pump blood, causing death within minutes if untreated. Sudden cardiac arrest kills over 300,000 Americans each year. Is arrhythmia a form of chaos? If so, can it be controlled with methods developed for other chaotic systems (see Lecture 19)? An elegant experiment along these lines was reported in 1992 by a team led by Alan Garfinkel, a physiologist and chaos theorist at UCLA.

They induced arrhythmia in rabbit heart tissue by applying drugs to it. The action of the drugs was gradual. The tissue beat periodically at first, and then progressed through two period-doubling bifurcations that later gave way to chaos. This is similar to the scenario we studied in the logistic map (in Lectures 9 and 10). The appearance of the interbeat interval plot confirmed that the arrhythmia was genuine deterministic chaos, not randomness.

Next the researchers controlled the chaos by harnessing the order hidden within it. They nudged the tissue into periodic behavior by stimulating it electrically, with sporadic but carefully timed pulses dictated by the theory. In contrast, simple pacing of the tissue didn't stop the arrhythmia.

What are the larger implications? Unfortunately, many arrhythmias are far more complex than this. In fibrillation, spiral waves of abnormal electrical excitation circulate on the heart, sometimes degenerating into multiple smaller wavelets. The point is that real arrhythmias are complex in both space *and* time. Many cardiologists and mathematicians are working hard to decipher their dynamics. With the insights that emerge, we may be able to develop gentler defibrillators, or perhaps more effective anti-arrhythmia drugs.

Goldberger believes that a bit of chaos is good for you. It allows you to adapt, to avoid rigidity in your physiology.

Finally, and most controversially, when chaos occurs in the human body, might it ever be a sign of health rather than sickness? This paradoxical idea was proposed by cardiologist Ary Goldberger of Beth Israel Hospital in Boston. In his lectures, he likes to show a graph of four patients' heart rate variability, and he asks the audience: Which of these is the sickest? The one with the most regular rhythm is! That person is suffering from congestive heart failure. The healthiest person has a surprisingly variable rhythm, with dance-like chaotic variations from beat to beat. The pattern is self-similar, at least in a statistical sense. It's a fractal process in time (see Lecture 16), displaying the same kind of variability at time scales of 300 minutes, 30 minutes, or 3 minutes. That variability reflects the normal workings of the involuntary nervous system. Goldberger believes that a bit of chaos is good for you. It allows you to adapt, to avoid rigidity in your physiology. He has also shown that variability decreases as we age. So stay young, stay chaotic! ∎

Essential Reading

Gleick, *Chaos*, 275–300.

Supplementary Reading

Garfinkel et al., "Controlling Cardiac Chaos."

Glass and Mackey, *From Clocks to Chaos*, chaps. 1, 8–9.

Goldberger et al., "Chaos and Fractals in Human Physiology."

Iasemidis et al., "Adaptive Epileptic Seizure Prediction System."

Liebovitch, *Fractals and Chaos Simplified for the Life Sciences*, 225–35.

Winfree, *When Time Breaks Down*, chaps. 2, 5.

Questions to Consider

1. What diseases might be caused by too much chaos in the body? Which ones might be caused by too little?

2. In a similar spirit, name some diseases in which some part of the body becomes too rhythmic, or not rhythmic enough.

Chaos in Health and Disease
Lecture 21—Transcript

Welcome back. Our next stop along the frontier of chaos research lies at the crossroads of math and medicine. The topic is biological rhythms, from the electrical chatter of brain waves to the pulse of a beating heart. Building on decades of biological research about the function of these rhythms—that is, what they are for, and also the mechanism at the molecular level, how they work biochemically—chaos theorists have been asking new questions about their dynamics, how these rhythms behave in time.

Let's look at three examples today. I'll begin with the question of epileptic seizures. Can the mathematics of chaos be used to help predict epileptic seizures? This is a really important problem. Epilepsy affects something like 50 million people worldwide. It's a pretty awful disease. During a seizure, an epileptic might convulse. You've probably seen this sort of thing, at least in movies or maybe in real life, these rhythmic convulsions as a person is having a seizure. It's very frightening. They may black out. They could collapse. In some cases, they will stop breathing.

Seizures sometimes come on very suddenly, so epileptics have to be extremely careful. They can't really drive. It's not safe for them to drive, or even something like giving their baby a bath is dangerous for that reason.

Just a few minutes of warning would be a tremendous help for these people. It would mean that an epileptic patient could get to a safe place, lie down, maybe even have time to take some sort of medicine to help prevent the seizure or reduce its severity.

But is there any hope of predicting seizures? Clinicians have traditionally regarded them as pretty random and unpredictable. On the other hand, they know, and you probably know from what you have heard, that there are precursors. For instance, some patients experience what are called *auras*. These could take various forms. They might see strange lights, or smell something disgusting, or feel sick, or have a weird taste in their mouth. These auras suggest that something neurological is there before the seizure and the patient is sensing it. You may also know that anecdotally, and it seems to be true, some patients are alerted by their dog that a seizure is coming. So the dog is picking up some very subtle cue, some precursor to the seizure.

Recently, a bioengineer at Arizona State, Leon Iasemidis; and his collaborator, neurologist Chris Sackellares; and a few other teams of researchers worldwide have been trying to predict seizures with the methods of chaos theory. They seek clues in brain wave recordings, which you know as EEG, the electroencephalogram. These are tracings that are taken before, during, and after a seizure. That might raise a question in your mind. If seizures are so hard to predict, how are we able to hook up the patient to an EEG so we can record everything before, during, and after? That's the curious thing. These EEG records are taken from a special population. These are patients who are known to have epilepsy and who, in fact, are about to go in for surgery to have an epileptic focus removed from their brains. These people are being monitored continuously for something like 3 to 5 days, and they actually have electrodes surgically implanted in their brain. So they are in there continuously recording for 5 days. The electrodes are in there, of course, to help localize the diseased tissue so that when the surgeons do the surgery, they will have a much better idea exactly what needs to be removed.

Also, these patients are a gift to science. Because of the multi-electrode placements in their brains, they provide all the information that we're able to use to judge the time course of seizures before, during, and after. In fact, they give us the best information available about how seizures develop and unfold.

What do the data show? Let me show you now a fairly complicated series of tracings, and I'll try to walk you through this. The upshot will be that the EEG recordings show that the normal chaos in the brain—by the way, we haven't really discussed that, but normally your brain, as funny as it sounds, likes to have a state of chaotic electrical activity. The rhythmic convulsions that we just talked about are associated with a pathological periodicity in the brain. Too much rhythmicity is bad. That's associated with epilepsy. So you would want a normal state of independent regions of the brain acting, effectively, chaotically. What the EEG recordings on these patients show is that their brain waves become *less* chaotic, not more, during a seizure. In other words, a seizure is not more chaos; it's actually less.

What you'll see in the diagram is some measure of how strong the chaos is. There are various ways we could measure it. We could look at the fractal dimension of the strange attractor associated with the chaos. We talked about strange attractors when we were looking at the Lorenz system in Lecture Seven. Or we could measure something that we haven't talked about yet called the *Lyapunov exponent*. It's quite a mouthful, but here is what it means: It's a measure of the butterfly effect. When you have two nearby trajectories in state space—remember, in a chaotic system, they will diverge from each other exponentially—that's the butterfly effect, the sensitive dependence on initial conditions. What the Lyapunov exponent measures is the rate of exponential growth in this separation. So you can try to measure that and look at that as a proxy for how strong the chaos is at different sites in the patient's brain. What you find is that this Lyapunov exponent starts to drop abruptly when a seizure starts.

In the chart I'm about to show, you'll see five tracings of Lyapunov exponents measured over a short term, over 10 seconds or so, measuring how strong the chaos is at five different electrodes placed at various locations in the patient's brain. You'll see the chaos dropping in each of them, bottoming out when a seizure is ignited. Then, after the seizure starts, the brain becomes much more chaotic, as it's supposed to. So let's take a look.

Here is the chart. You see several different-colored lines; these are the measures of the Lyapunov exponents at the five sites. They're humming along with—the actual number is not so significant. It's measured in bits per second, but the thing to notice is that all the curves are dropping. When the seizure starts, they stay low. By the time the seizure is over, they all go back up. Even more interesting than that, though, is what happens several minutes *before* the seizure. Not just that the seizure makes things more chaotic. Can you see something striking happening before the seizure? These curves look like they're coming together right about here. In fact, they're all starting to be more or less on top of each other. That is, the time series of these short-term Lyapunov exponents, measured from the different electrodes, start to synchronize. They're lining up in amplitude and phase. You can see that they're getting in sync about 50 minutes or so before the seizure.

Or we could see it another way on the next graph, which is a statistical measure of the amount of synchrony. Here's that graph. If we look at something called the *T-index*, measuring how much synchronization there is between the chaos at the different electrodes, you see it's high at first, and then it comes down, and it crosses below a certain line. This line is a measure of when the synchrony is statistically significant. So somewhere about, as I say, 50 minutes, which our eyes saw, this statistical test is seeing it, too. Now there is a significant amount of synchronization, and that was a predictor of the seizure that was about to occur 50 minutes later. The point of this statistical test is that you don't need human pattern recognition to see the synchrony. A computer could see the synchrony by measuring that statistic and so could automatically determine that it has reached dangerous levels, and possibly in the future, if there were an implantable device in the patient's brain, maybe the computer could trigger some medication to be released or something like that. In any case, this statistical test is another way of seeing that synchrony occurs 50 minutes before the seizure in this particular patient.

According to Leon Iasemidis, this synchronized chaos is what he calls "the mark of the beast," a clue that a seizure is imminent. His interpretation is that different brain regions that are normally supposed to be acting independently are starting to fall into a pernicious lockstep. An interesting thing here, though, is that it's not that the brain waves themselves are lining up. If you looked at the EEG in the usual way, brain wave recordings raw, you wouldn't see this synchrony. So it's not that the brain waves are lining up. It's that the strength of the chaos at the different electrodes is lining up. In other words, you can't see this phenomenon by looking at the EEG in the conventional way. You only see it if you look at it through the lens of chaos theory, by thinking about Lyapunov exponents, measures of chaos.

So it's a very exciting thing. We're now starting to have an idea about what electrical signals might be precursors to seizures. But what's unknown is how often this particular pattern that we just talked about is truly followed by a seizure. Is it perfectly predicted? Does it sometimes make incorrect predictions by crying wolf, false positives? In a preliminary study in 2003, Iasemidis found that his algorithm, based on the T-index that I just showed, correctly predicts 82% of seizures, with about 75 minutes of advance warning time on average. That's very good. But it also made false predictions about

3 or 4 times a day, which turns out to be about as many times a day as these poor patients have seizures. So it is not a perfect predictor yet, and there's a lot of room for improvement. Iasemidis realizes this. He is the first to admit it, and he has been working on that, trying to improve the statistical methods, and possibly tuning them to individual patients, and so on.

Other researchers, as I mentioned, have their own promising algorithms, but nobody has fully solved the problem of epilepsy seizure prediction yet. But I think we're on the right track, and if we are, maybe Iasemidis's dream will come true. He dreams that some day an early warning system can be programmed on an implantable computer chip, planted in the brain, and then when the beast appears, the chip will trigger a kind of "brain defibrillator" to fire a preventive electrical volley or maybe release an anti-seizure drug.

Now speaking of defibrillators, that brings us to our second example of how chaos theory is being applied to medicine. The topic is cardiac arrhythmias.

Arrhythmias are irregularities of the heartbeat, specifically of the heart's electrical activity, the electricity that triggers contractions in the heart, the muscular contractions that pump the blood. In the worst kind of arrhythmia, called *ventricular fibrillation*, the electrical activity becomes so disorganized and so uncoordinated that the heart just quivers uselessly and fails to pump blood, causing death within a few minutes if untreated. Doctors often describe a fibrillating heart as feeling like a bag of wriggling worms. I once participated in an experiment in a physiology class at a medical school, and a dog's heart was made to fibrillate in the course of teaching medical students about fibrillation. It's true. It has a very slippery, slimy, wormy feeling during fibrillation.

Sudden cardiac arrest, which is an attack of ventricular fibrillation in an otherwise healthy person with no history, necessarily, of heart disease, causes sudden death in over 300,000 Americans each year. In other words, ventricular fibrillation is a killer, and we really don't see it coming in many cases.

There are other less deadly forms of arrhythmia. Tachycardia, where the heart beats much too fast, is dangerous, not as dangerous as fibrillation, but dangerous because it often degenerates into fibrillation. It's particularly dangerous when tachycardia occurs in the ventricles, the main powerful lower chambers of the heart that pump blood to the rest of the body and the brain.

When arrhythmias occur in the upper chambers of the heart, the auricles or the atria—those are responsible for squeezing blood into the pumping chambers in the ventricles—when you get an arrhythmia in your atrium, that is also a serious cause for concern, but it's not typically as deadly as a ventricular arrhythmia.

Medical textbooks often describe arrhythmias, and particularly fibrillation, as a chaotic state, which raises the question: Should we take that literally? Is fibrillation a form of chaos in the sense that we now understand the term—deterministic chaos ruled by deterministic laws and yet seemingly unpredictable? What about other kinds of arrhythmias? How much can chaos theory tell us about arrhythmias? For example, if these states are truly deterministic chaos, can they be controlled and brought back into a much more benign periodicity using the methods that we talked about in Lecture Nineteen for controlling chaos, for nudging chaos into periodicity? That's an exciting thought. Could we do that?

An elegant experiment along these lines was reported in 1992 by a team led by Alan Garfinkel at UCLA. Alan was actually trained in philosophy and has gone on to become both a chaos theorist and a physiologist.

Garfinkel's team used as its model for arrhythmia a piece of tissue from the heart of a rabbit and induced arrhythmia in it [by] applying certain drugs. The action of the drugs was gradual. The tissue beat periodically at first, as I'll show you in this next slide. What you're seeing here is the spontaneous periodic beating of the tissue from the rabbit heart. Spontaneous beating is not the normal thing that would happen in heart tissue. Maybe you assumed that all heart tissue just beats periodically, but that's not true. The heart beats because of naturally occurring pacemakers in the heart that trigger the beat. Isolated pieces of tissue typically don't beat. They just sit there waiting for a signal from the pacemaker before they contract. But under this drug-induced condition, on this tissue, it was sent into spontaneous rhythmicity, so it was just beating on its own.

That's what you're seeing in the diagram. This looks like some kind of very simple electrical depolarization, a beat, in other words, very rhythmic as a function of time at first. But then as the drugs started to take effect, the beating changed character. Look at the next slide. Now it's in a period-2 cycle. We talked about period-2 rhythms back when we were discussing the logistic map around Lecture Ten or so, and what we're seeing in period-2 is that there is a conspicuous low-level beat followed by high, low, high. And also the durations are different. Here is a long interval followed by a short interval. You have a long beat and a short beat. This is reminiscent of exactly what we saw in the logistic map going from periodicity to period-2, if you remember the scenario; that went to period-4 and then period-8, period-doubling all the way to infinity and eventually becoming chaotic.

So here, in the experiment, not nearly as clean an environment as a computer experiment, the researchers saw the period-2 cycle. Then they saw period-4, but it's starting to look noisy, as you'd expect in a real system. Notice, though, that every fourth beat is a low one, but they're not all exactly the same height. So it's a noisy version of period-4. Finally, it gave way to something that looks like chaos. Here is chaotic beating with the amplitudes changing in no predictable way.

So as I say, this is similar to what we saw in the logistic map, and the question arises: Is this genuine chaos? One test for that would be to look to see if there is there an underlying iterated map. Is there some analogue of the logistic map? So to do that, the researchers plotted the time from one beat to the next, and then the next, to the next after that. They looked at the interbeat interval map. That's what is shown here in this next slide. If you look at the interval between beats measured in seconds against the next interval, the data don't just form a blob. They're pretty scattered, but on the other hand, you do see certain structure in here. There are open regions in the diagram. That suggests, although not completely convincingly, that the arrhythmia was genuine deterministic chaos and not just pure randomness, which would have produced a buckshot pattern all over the place.

Taking it that this is a kind of chaos, next the researchers asked: Could they control it using some variant on the Ott, Grebogi, and Yorke algorithm for controlling chaos that we discussed a few lectures ago? The hope was to harness the order hidden within it. To do that, they nudged the tissue into periodic behavior by stimulating it electrically. Not by manhandling it, not by driving it hard, by pacing it, but instead, by just tweaking it from time to time with electrical impulses carefully timed according to what chaos theory indicated.

Here is the result. If they looked at the time interval between beats as a function of the beat number, basically time into the record, here you see a very erratic beating pattern. They turned on the control scheme right here, and look. It has stabilized onto something that has period-3. It has become periodic. The intervals between beats now repeat every 3. It's not exactly a healthy rhythm, but it's much healthier than the arrhythmia that it had. When they turned the control off, it suddenly falls apart again.

So they were able to control the system onto periodicity. It was very encouraging. On the other hand, if you just try the naive thing of imposing a rhythm on the tissue by just driving it periodically with some electrical driver, that doesn't work. Simple periodic pacing was much less effective at controlling the chaos. Look at how it works. The time interval between beats, just like we measured before, when the control is on, it puts the system in an approximate period-2. It's still working. But then when they put on periodic pacing, right here, you don't see any real change in the record. It basically did nothing.

What are the larger implications here? I would like to say that this offers us a lot of hope towards controlling arrhythmias, but I'm honestly not so sure about that. I think the work is intriguing, but unfortunately, many real arrhythmias are far more complex than this drug-induced one in the rabbit tissue.

In particular, what happens during fibrillation—the deadliest one—on the heart, you get spiral waves of electricity circulating around, and they're abnormal excitation that causes the heart to quiver, and as I say, it can sometimes degenerate into this very bad form where it's like Hydra splitting off heads, where the spirals can degenerate into multiple tinier spirals, really disrupting the heart's normal function. But the point is that real arrhythmias, in particular fibrillation, are complex in space as well as time; that is, all these spirals zooming around on the heart, lashing against each other—that's a serious problem. When there is complexity in time and space together, these are problems that, unfortunately, lie beyond the edge of the chaos theory that has been developed so far, as we discussed in Lecture Twelve, when I gave the analogy of the thermometer of complexity. We're not really there yet to understand spatial-temporal complexity at the same time.

Many mathematicians are working on this, with cardiologists. Teams of collaborators are working to try to decipher these extremely complicated dynamics of arrhythmias. So I'm hopeful that we'll make progress, and with the insights that may emerge, we may be able to develop gentler defibrillators, gentler than what you see on those TV shows when they tell everyone "Clear!" and they put the paddles on and shock the patient. That is very painful. Basically, you have to burn the heart in order to save it. We would like something much gentler than that, and with chaos theory, we might be able to develop something more rational than just blasting the heart with the paddles. Or we might also be able to develop more effective anti-arrhythmia drugs if we knew exactly what the causes were.

Turning to our third and final example of chaos in medicine, I think you may find this one to be somewhat surprising, possibly more surprising than what you've heard so far. The question is: When chaos occurs in the human body, in particular in the heart, could it ever be a sign of health rather than sickness? Could chaos be good for you? With what I've just been talking about with arrhythmias, chaos seemed to be the villain. The doctors refer to arrhythmias as chaos. There has always been this thinking that chaos is terrible. In medical textbooks, there is this revered concept of *homeostasis*, keeping the body flat and even-keeled. That is, the body has all kinds of control mechanisms for evening out disturbances, sort of like a thermostat. If the temperature goes too high, it turns on the air conditioner, and if it's too low, it turns on the heat, trying to regulate the body. Doctors have been taught for a very long time that homeostasis is the body's desired condition.

The idea that chaos could ever be good for the body is the total opposite of this way of thinking, and so that's the question: Could it be good for you? We saw one example already. When I talked about epilepsy, I mentioned that the normal state of the brain is closer to chaos than homeostasis, so that already suggests that in the brain, chaos might actually be better for you. Not utter chaos but a certain low level of chaos.

What about in the heart, which is a simpler organ than the brain? Its job is a pump. It's just supposed to beat periodically, right? So how could chaos be good for your heart? That paradoxical idea, that it might sometimes be good for your heart—or maybe even normally it's good for your heart—was proposed by a leading cardiologist, Ary Goldberger, at Beth Israel in Boston. This is someone who has to be taken seriously. This is not a mathematician talking about the heart. This is a practicing cardiologist who sees patients every day at one of the top hospitals in the world.

Goldberger is an interesting case in that he is one of the rare medical people who has also mastered the ideas and language of chaos and fractals. We've seen in this course instances of mathematicians and physicists gently stepping into fields they don't know much about, like communications or biology, trying their hand to make a helpful contribution. We haven't seen so much in the other direction, where biologists are starting to learn about chaos theory, but Goldberger is a sterling example of that.

I've seen [Goldberger] give very nice lectures about his ideas, and in his lectures, he typically likes to show a graph of four different patients, showing their heart rate variability, how much their heart rate is changing from moment to moment. And then he asks the audience with a dramatic flourish: Which of these four patients is the sickest? Taking a page out of his book I'm going to try to play the same game with you. I think you'll find this interesting. Here are the four charts that Goldberger likes to show.

Heart rate variability, measured as beats per minute. When the doctor checks you, you sort of expect your pulse to be something like 70 beats per minute or thereabouts, maybe 60 if you're an athlete. Look at these four patients. You're seeing their heart rate versus time. They have very different patterns. This one is rock steady. If you like homeostasis, this person is your man or woman. Here is something that looks complicated and jittery, and it has a big bump in it, noisy-looking. Here's one that is close to homeostatic, close to flat, but it has regular oscillations in it, up and down cycles of about one minute. Here's one that has a lot of high-frequency, jittery, jumpy variations in heart rate. They're four very different-looking examples. Now, take your guess. Which one of these patients is the sickest? Here is the answer: That one is. The one with the rock-steady heart rate is suffering from a pretty serious disease called congestive heart failure. When you have heart failure, it means that your heart is failing to pump sufficient blood to the body's other organs. Because of the reduced outflow from the heart, the blood that returns to the heart from other organs starts to back up, causing congestion in the tissues: hence the name, congestive heart failure. Also, fluid can collect in the lungs, causing shortness of breath.

This person in Panel A is very sick. What about this person here in C, whose heart rate looks pretty rhythmic? It's basically flat, but with these very regular cycles to it. They're predictable cycles with a time of about one minute. Sick or healthy? That person also has heart failure and a pathological form of breathing that leads to periodicities. It's a pathological condition known as *Cheyne-Stokes breathing*, in which the patient alternates between breathing very rapidly and then not breathing at all. It's a very serious problem for this patient.

Maybe by now you're getting the idea that homeostasis as measured here is not such a great thing. Maybe it would be better to have a heart rate that fluctuates a lot, like the one in the bottom. Not quite, because that person has atrial fibrillation—excessive randomness in heart rate, as if it were uncorrelated from beat to beat, just independent, random beats. That is also bad for you.

Atrial fibrillation you may have heard about. Bill Bradley has it; the elder President Bush has it. It's a serious problem, too. It requires careful medical attention. It's not as deadly as ventricular fibrillation. You can live with it, but it is a problem.

So who is left? This person with the erratic heart rate that's not quite as erratic as the one on the bottom, that person is normal. That's a healthy person. He or she has a surprisingly variable, almost chaotic and dance-like rhythm, with variations from beat to beat. If we look at it, though, we can see that there is a kind of order to it. If we look closer, that kind of rhythm is a fractal. It's self-similar in time, at least in a statistical sense. It's a fractal process in time, much like we discussed in Lecture Sixteen, when we were talking about stock prices or Internet traffic. We've seen what spatial self-similarity is, where small pieces of some structure, when blown up, look exactly like the original. Here, we're seeing statistical self-similarity in heart rate, and it's a self-similarity in time, not in space. So the panel is showing heart rate as a function of time but now over three different time windows—long, 300 minutes. Take a 30-minute piece of that record and blow it up to full scale, and you'll see you still have this dance-like rhythm in heart rate. Blow up a 3-minute segment of that, and it's still dancing. So whether we look at any of those time scales, we see the same kind of variability.

Goldberger believes this reflects the normal workings of a healthy involuntary nervous system. To put it this way, he believes a bit of chaos is good for you. It allows you to adapt, to avoid rigidity in your physiology. He has also shown that as we age, we lose this kind of flexibility; our heart rate variability decreases as we age, a kind of rigidity that is not good for you. So my advice, to the extent that you can do it: Stay young, stay chaotic.

Thanks, and I'll see you next time.

Quantum Chaos
Lecture 22

You can push one uncertainty down, but then the other one goes up. And if you push that one down, then the other one goes up. There's no way around it. It's built into the structure of the universe. There's no way to measure position and velocity simultaneously with unlimited precision. Why is that so terrible for chaos? Because it destroys our whole concept of state space.

This lecture discusses the blending of two great theories—chaos and quantum mechanics. They don't mix as easily as you might imagine. But once we find the right recipe, they're terrific together. Quantum chaos opens up all sorts of possibilities for future electronic devices, for making abstract art, and maybe for solving the greatest mystery in mathematics. Why aren't chaos and quantum theory kindred spirits? You might think they have a lot in common. They both highlight the unpredictability of nature and the limits of human knowledge. The frenzied motion of subatomic particles certainly seems "chaotic." But at a deeper level, the theories conflict.

First, their worldviews are diametrically opposed. Chaos is founded on determinism: The present uniquely determines the future. Whereas quantum theory speaks only of probabilities: Nothing is ever certain to happen. Second, chaos is mathematically forbidden in quantum mechanics. Quantum theory describes everything as a changing blur of probability waves. The differential equation that governs how these waves evolve (the *Schrödinger equation*) is linear. As such, it has no chaos in it. Only nonlinear equations can support chaos.

The third conflict is the worst. It involves Heisenberg's uncertainty principle. Heisenberg says we can't know the position and velocity of a particle simultaneously. But that's precisely the information we need to predict the motion of a pendulum or other classical mechanical system. So states and trajectories become blurry and ill-defined at the atomic scale. The fine structure required for chaos is smoothed out by quantum effects. So when we speak of quantum chaos, what we really mean are quantum signatures of deterministic chaos. What are the vestiges of chaos when we shrink a classically chaotic system down to atomic size? This question was first examined theoretically.

One model system was billiards. A particle (a billiard ball) moves in straight lines on a frictionless pool table, bouncing elastically whenever it hits the walls, and continuing like this forever. The question is: What are the trajectories like in the long run, regular or chaotic? The character of the motion turns out to depend on the shape of the table. If the table is rectangular or circular, the motion is regular. Two balls that start close together with similar velocities will stay together for a long time, diverging only slowly. But the motion becomes chaotic for a table shaped like a stadium. The trajectories diverge rapidly after a carom or two. This is the billiard version of the butterfly effect.

So far we've described billiards in a classical, deterministic way. What is the quantum counterpart? The trajectories of the billiard ball smear out to probability waves. A picture of these waves shows where the ball is most likely to be found. If the classical motion was regular, the waves look like the vibration patterns of a drumhead. The surprise comes when the classical motion is chaotic. Everyone expected the quantum picture would look speckled, like random waves colliding. But the actual picture has prominent "scars," as discovered by Eric Heller. Although such perfectly repeating orbits are very rare and unstable, they leave their mark by concentrating waves along them.

More recently, Heller has taken up a second career—as a quantum artist. His images of quantum waves and chaos are as beautiful as they are scientifically revealing. Some depict the paths of thousands of electrons, treated as classical particles, as they ski over a landscape with random bumps on it. This problem arises in nanoelectronics, in trying to understand what a 2-dimensional gas of electrons will do as it moves through a tiny semiconductor device called a *quantum dot*.

Quantum chaos may prove of practical value in the design of the nano-scale devices that will revolutionize the electronics of the future. At this scale, and at sufficiently low temperatures, electrons have to be treated as waves, not particles, and the principles of traditional electronics go out the window. Researchers have begun exploring this new realm by building quantum dots in the shapes of stadiums and circles. Now the billiard balls are electrons. At temperatures close to absolute zero, the electrons travel freely. They only suffer electrical resistance when they bounce off the walls of the device. The experiments show the telltale signs of quantum chaos. The electrical resistance of the device changes dramatically, depending on whether the classical motion is chaotic or regular.

The real shocker about quantum chaos, however, is that it links atoms to prime numbers, thus connecting the bedrock of reality to the most ethereal realm of human thought. For thousands of years, mathematicians have sought patterns in the prime numbers. Carl Friedrich Gauss discovered an approximate formula for number of primes less than a given number n. In the mid-1800s, Bernhard Riemann improved this formula by adding certain waves to it. The frequencies of those waves were based on a brilliant guess, now called the *Riemann hypothesis*, sometimes described poetically as "the music of the primes." Just as real music is made of sound waves, Riemann's waves are the building blocks of the primes. Proving Riemann's hypothesis is considered the greatest unsolved problem in math.

Thus, it seems there's a chaotic system—as yet undiscovered—whose quantum counterpart would unlock the secret of the primes.

What does this have to do with quantum chaos? Quantum systems have discrete energy levels, corresponding to waves vibrating at certain frequencies. Likewise, the primes are built from waves vibrating at a discrete set of frequencies, called *the zeros of the Riemann zeta function*. The shocker is the frequencies of the Riemann waves look uncannily like those for a quantum *chaotic* system, as pointed out by physicist Michael Berry. Thus, it seems there's a chaotic system—as yet undiscovered—whose quantum counterpart would unlock the secret of the primes. ∎

Essential Reading

Rockmore, *Stalking the Riemann Hypothesis*, chap. 12.

Supplementary Reading

Berry, "Quantum Physics on the Edge of Chaos."

Du Sautoy, *The Music of the Primes*, chap. 11.

Gutzwiller, "Quantum Chaos."

Internet Resource

Explore the dynamics of billiards on tables of various shapes here: http://serendip.brynmawr.edu/chaos/home.html.

Eric Heller's online gallery of quantum art: www.ericjhellergallery.com.

Questions to Consider

1. Could chaos be the fundamental source of the randomness inherent in quantum mechanics?

2. Why do you think mathematicians are so obsessed with prime numbers? Explain why it makes sense to call these numbers the "atoms of arithmetic."

Quantum Chaos
Lecture 22—Transcript

Welcome back to chaos. It occurs to me only now, for some reason, that there has been a certain arc to this course that we haven't talked about yet. Remember, we began with the creation of the universe itself out of chaos in the creation myths of ancient peoples around the world? Then, before long, we were talking about the solar system and the orbits of the planets, Poincaré's work and so on. So, from the universe to the solar system. Then we spent most of our time in the course at the everyday scale of the world around us. There is one place left to go. It's natural for us to now think about what happens in the microscopic realm of atoms, where quantum mechanics reigns supreme.

I hope you'll find this lecture on quantum chaos to be really thought-provoking, and maybe even delicious. Why delicious? Because it's about a stew. It's the story of an amazing scientific stew in which we blend two of the greatest theories in all of science—chaos and quantum mechanics. What happens when we try to blend them? You will see that they don't mix as easily as you might imagine. But once we find the right recipe, they're terrific together.

Quantum chaos opens up all sorts of possibilities for future electronic devices, for making beautiful abstract art, and maybe even for solving the greatest mystery in mathematics.

Let me begin by telling you why it is that chaos and quantum theory aren't kindred spirits. You might have thought that they were. You would think they have a lot in common, really, because they both highlight the unpredictability of nature and the limits of human knowledge. In that way, they seem like bedfellows. Certainly, the frenzied motion of atoms and subatomic particles sounds like chaos. But at a deeper level, there are profound conflicts between the two theories. Let me focus on three of them in ascending order of seriousness.

The first one has to do with their worldview. They have divergent worldviews. Chaos is founded on a deterministic worldview, in which the future is determined by the present. Only one thing can happen. Given current conditions of the universe, there is one unique future.

Quantum theory, on the other hand, doesn't see the world like that at all. Quantum theory sees the world in terms of probability. That is the only meaningful thing. There is no certainty; there is only probability. Nothing is certain to happen. You can only talk about the probability or likelihood of something, say, finding an electron at a certain place at a certain time.

The second conflict is more severe. Strictly speaking, chaos is forbidden in quantum mechanics. It's outlawed. There is no quantum chaos, so we could stop this lecture right now, you might think. But we're not going to. Let me try to explain why chaos is outlawed. What is so impermissible about it? To make sense of this, we have to know a little more about the math behind quantum theory. Specifically, we need to know about the differential equation at the heart of the theory.

Quantum theory, as you probably heard somewhere, describes everything as a blur of probability waves. As I say, you can only describe the probability of something being somewhere, or moving at a certain speed, or whatever. So there are waves that control those probabilities, and the waves themselves are not fixed. They change and evolve over time, deterministically, it turns out. That's an interesting paradox. Although they have an interpretation in terms of probability, [in] their own evolution, the waves evolve deterministically. They change from moment to moment according to a certain differential equation, a certain law of motion.

Actually, that's what we have come to expect in this course, ever since Lecture Two, where we first encountered Newton's great secret, "It's useful to solve differential equations." Remember that point? The laws of nature are written in the language of differential equations, and that is still true even in quantum theory. But here's the rub: The differential equation that happens to govern the evolution of quantum waves, called the *Schrödinger equation*, turns out to be linear. Linear, the key idea that we've talked about so many times now—linear versus nonlinear. Linear equations cannot support chaos. They have no chaos. It can be proven mathematically. There is no chaos in linear systems. You need nonlinear equations to see chaos. That means there can be no chaos in any system governed by the Schrödinger equation. There is no chaos in quantum theory—period.

As if that weren't bad enough, there is a third conflict that I would say is the worst of all. It has to do with the famous Heisenberg's uncertainty principle. Let's remember what that says. It has to do with making simultaneous measurements of a particular type. Suppose I were trying to measure some particle moving, like a ball moving through the air or an electron moving through a microscopic device. I might want to know its position right now and its velocity, how fast it's moving and in what direction. Heisenberg tells us we can't do that. We can't measure both position and velocity simultaneously. You could say, "Why not? What will happen if I try?" You can try, and you can measure one of them as well as you like. You can pin the position down with the best instruments, you can make it as precise as you wish, but if you do that, by narrowing the uncertainty in position, you automatically raise the uncertainty in velocity. It's like a see-saw. You can push one uncertainty down, but then the other one goes up. And if you push that one down, then the other one goes up. There's no way around it. It's built into the structure of the universe. There's no way to measure position and velocity simultaneously with unlimited precision.

Why is that so terrible for chaos? Because it destroys our whole concept of state space. From the beginning of this course, we've been talking about state space. What is a state? A state is the amount of information you need to predict the future of a deterministic system. For example, we talked about a simple pendulum swinging back and forth. Early on, I pointed out that the state of that pendulum is determined by two numbers: its initial angle and its initial velocity. It's not enough to know how far I deflect the pendulum. I might also give it a push, or I might release it at rest. I need to know initial position and initial velocity to determine the future of its swing.

But that's what Heisenberg will not permit. You cannot measure or know position and velocity simultaneously, and that's exactly what we need to describe the state of a pendulum. It's not just a problem with a pendulum. It's true of all classical mechanics that we need the two things that we're not allowed to have by quantum theory. So the upshot is that states and trajectories become blurry and ill-defined once we get to the atomic scale, where quantum mechanics takes over. The fine structure required for chaos to persist gets smoothed out and washed away by quantum effects.

So for all these reasons, quantum chaos is a very problematic subject. You might even say it doesn't exist, but it does. Here is what people mean when they speak of quantum chaos. It's sort of imprecise language. This is the better way to say it: What we really mean are quantum signatures of deterministic chaos. In other words, suppose we have some classical system out here in the macroscopic world that we live in that's chaotic. That can happen. We have such things. We've seen them throughout the course.

So we have a chaotic system, and now we imagine shrinking it down to microscopic size, to atomic size. Is there any vestige, any remnant, of its chaos when we look at the system at that small scale? That's the question. What are the vestiges or the signatures of classical chaos in the quantum realm? That's our subject for today.

This question was first examined theoretically with math and in computer simulations. One model system that I want to discuss for a few minutes is something you'd encounter if you go to a pool hall: billiards. Just a ball bouncing around on a pool table. It might seem like a strange thing for a lofty academic scientist to be studying, but it's not. It's a beautiful mechanical system, and it has taught us a lot about quantum theory, as well as classical mechanics.

So here is the setup: Imagine a particle, which if you like, think of as a billiard ball, and it's moving in straight lines on a frictionless pool table. It's not so realistic, but it makes life simpler to imagine no friction. So this particle is moving on a frictionless pool table, and then it bounces off the cushions whenever it hits one of the walls. This just goes on forever. We're assuming there's no loss of energy; there's no friction. Also, the collisions at the walls are perfectly elastic. So this particle is just going to bounce around in the pool table, bouncing off the cushions forever. The question is: What are the trajectories like in the long run? Will they be regular or will they be chaotic?

The character of the motion turns out to depend on the shape of the table. If the table has a simple shape like a rectangle, like the usual pool table, or say, a circular pool table, it turns out the motion is regular. I'll show you that in a computer simulation in a minute. *Regular* in this case means that two balls that start close together—imagine two shots that start very close together and move in practically the same direction at the same speed—they will stay together for a long time after subsequent bounces, diverging but only very slowly, not exponentially fast. There's no chaos in a circular pool table.

But the motion does become chaotic if the table has a shape like a stadium. Think of a football stadium with two straight lines capped off by two semicircles. If you watch the motion of a ball bouncing around in such a stadium, it turns out that the trajectories of a pair of such balls will rapidly exponentially diverge after just a few caroms. That's the billiard version of the butterfly effect, sensitive dependence on initial conditions, the exponential amplification of small uncertainties or differences, the trademark of a chaotic system.

To give you a feel for what this looks like, let me show you these simulations of billiards: first, the regular case of the circle, and then, the stadium case where we have chaos. Let's see if I can get this simulation to work. Here, you see the initial setting is that we're using a circular pool table. What I can do is choose the number of bounces that I want to watch. Just to keep it really simple, suppose I do 1 bounce. There, I've hit the wall. That's not too interesting. Let's watch 3 bounces. That is also not terribly interesting. That's going to just bounce back and forth, so let's try starting somewhere else with 3 bounces. That looks better. You see that, as you would expect with any bouncing off of the wall, as they say, the angle of incidence equals the angle of reflection. That keeps happening as it bounces. You get the idea of what's going to happen, but let's watch it a little bit longer; go for 10 bounces. You start to see a little pattern emerging, and it's probably obvious to you what's going to happen now in the long run. If we do something like 100 bounces, it fills out a very nice pattern. That is regular motion, as expected for the circular pool table.

Let's change now to a stadium. To do that, I need to change what's called the ratio here, so I'm going to make this number 2, and I will start. There's a stadium, and I now pick a point somewhere in the stadium. Let me only make 2 or 3 bounces initially to get the hang of things. I've chosen my initial point, click on start, and so far, it doesn't look like anything very dramatic. Let's try 10 iterations. You might see that motion looks a little more complicated, but to really appreciate the full magnitude of the unpredictability of this motion, watch the long-term pattern after 100 bounces and compare it to what you just saw in the case of the circle. I'll increase this number to 100, click on start, and watch. That's what chaos looks like, a very irregular trajectory in the case of the stadium pool table.

So far, we've been focusing on billiards in just an ordinary, classical sense—deterministic motion of a particle moving in a pool table. That's the classical world. We haven't done anything quantum yet. Now we're going to go to the quantum counterpart of this classical chaotic system. How will things be different? We can't really talk about trajectories any more. In fact, I can't even talk about a state. I can't say what the initial position and velocity are, so the best I can do is think of the trajectory as a kind of smeared-out, blurry probability wave, whatever that means. There's a kind of wave of fuzz going around what used to be the classical trajectory. A picture of these waves, if we can make one, will show where the ball is most likely to be found; that is, where the wave is highest, that means high probability of finding the particle there. But you can find it anywhere, just with lower probability at the places where the wave is low.

I really do mean to speak about waves because you can think about these in a classical way. If the classical motion were regular, like in the case of the circular stadium, the waves turn out to look like the vibration patterns of a drum. You may have seen vibration patterns of a drumhead. It's a standard exhibit that they do in science museums. You take a plate—it could be a circle or a square—and sprinkle a little bit of sand on it. Have you ever done this? You have sand on the metal plate, and then they give you a violin bow, and you start bowing on the edge of the plate to cause it to vibrate at different frequencies. The sand gets shaken, and in places where the plate is not vibrating at all would be the nodes of the vibration. Then the sand collects and you get a nice pattern of sand, which shows you the vibration pattern of the plate. You may be able to visualize those. They're not very complicated patterns. They're quite regular looking, as you would expect for regular motion. The surprise comes if we do the same experiment with a chaotic plate, so to speak. If the classical motion were chaotic, like it is in the stadium, what would the wave pattern be? That's the question.

People thought about this question, and before computer simulations were done, there was intuition about what would happen. The intuition was, "Look, chaos is basically the same thing as randomness, and so what we will expect to see [is] a random collision of waves. It will look like whatever a random wave pattern will look like." We know that from other branches of physics, that random wave patterns with waves colliding from all different directions look speckled. People thought you would expect a speckled pattern.

You probably noticed that I said something pretty specious a minute ago, that chaos is basically randomness. I hope by now, here in Lecture Twenty-Two, you know that is dead wrong. Chaos is not the same as randomness. Chaos has lots of latent order, hidden order in it. It's quite different from randomness. It's poised between randomness and regularity. In fact, when the first simulations were done by Eric Heller, a physicist, he found something that didn't look random at all. It didn't look like speckles. The actual picture he found has prominent "scars" in it. I'll show you the scars right now in an image that Heller has made.

Here is Eric Heller's image that he calls *Scar*. You see the stadium, and what he is showing is the picture of the quantum waves inside the stadium, corresponding to the billiard problem we were just talking about. As you see, this doesn't look like a speckled pattern at all. There are very prominent tracks through here. Look at this one that I'm tracing with my cursor. It looks like there is a kind of X-shaped region capped off by two lines on the top and bottom. What is this thing? You can see the way I'm moving the mouse here that this is a possible trajectory for a ball. I could bounce happily around and around in a periodic orbit—bounce over to here, bounce down to here, over to there, and back to where I started. There's a periodic orbit that repeats perfectly, a trajectory that is conceivable in this chaotic stadium.

Such perfectly repeating trajectories are very rare. I happened to pick one, and actually the *Scar* is highlighting that one. They're rare and unstable. Remember, this system is chaotic, so if I move just a little bit off that trajectory, I won't repeat it; I'm going to go diverging exponentially away from it and do something else. So these perfectly repeating trajectories are rare and very unstable. Nevertheless, they manage to leave their mark because they concentrate quantum waves along them. That's why you see scars.

In the years since his discovery of scars, Heller has continued to do pioneering work in quantum chaos. Perhaps more surprisingly, he's taken up a kind of second career. Maybe it's more like a hobby, but he's quite serious about it. He's a quantum artist. He has made images of these quantum waves and also classical chaos, and they're as beautiful as they are scientifically revealing. In a minute I'll show you two images of classical chaos that depict the paths of hundreds of thousands of electrons, treated as classical particles. Think of them as skiing over a bumpy landscape with randomly placed moguls on it. These random moguls are a natural thing to think about in connection with a problem in nanoelectronics, the electronics of the future. Heller and his colleagues were trying to understand what a 2-dimensional gas of electrons would do as it moves through a tiny semiconductor device called a *quantum dot*. The random moguls are produced by interactions between these electrons and the donor atoms that have contributed the electrons. They are now positively charged ions sitting in two layers that are confining the electrons to this 2-dimensional sheet, the gas.

Given the randomness of these moguls, just sort of skiing your way over the bumpy landscape—you, the electron—Heller and his colleagues expected to find a mess of crisscrossing paths as all these electrons—the 100,000 of them launched in—would be veering past each other. But their simulations showed something surprising and important for understanding these future electronic devices.

Here's what he found you would get if you launch the electrons in from the side as a thin tube of 100,000 trajectories, all close together. He calls this picture *Exponential*, and it's magnificent. Here's the thin tube of 100,000 trajectories, but they all start separating, diverging, because of chaos. That's not so surprising, but what is surprising is that they later get concentrated again along these wispy structures. That was not expected. The concentration on certain pathways was unexpected and beautiful.

Similarly, here is what would happen if you launch the electrons from the center. The same wispy structures, not expected. So you can see why people like these as art, and you can actually buy them from—by the way, he goes by Rick Heller—Eric Heller's gallery that's called Resonance Art. Buy a poster if you like. They're gorgeous images.

One image that I don't think is his most gorgeous but that I like very much is this next one. It shows something very deep and satisfying: how to reconcile the classical and quantum views of reality. If you think about a traditional classical trajectory, there is a way to give it a quantum flavor. What you need to do is imbue that trajectory with a color. Think of a color like a color wheel; red is close to orange, yellow, green, blue, indigo, violet, which is close to red again. There's a circle of phases of different colors. If you attach a changing phase to the trajectory as it moves, that is, a changing color, that's the right recipe for going between the classical particle description and the quantum wave description. That's what this next picture shows. Here, you see a trajectory with the changing colors along it. At this point we're still thinking classically, but Heller wondered: How do we go over into the quantum version of this? As this trajectory continues to move, threading its way almost like crochet here, it makes a structure which, by the end, looks exactly like wave patterns interfering. This is the quantum version of the same phenomenon that's being shown up here classically with these changing phases, the changing color. So this is a way of visualizing how quantum mechanics and classical mechanics can blend seamlessly into each other. That's why he calls it *Rosetta Stone*. It's the key to decoding the correspondence between the classical and quantum worlds.

Quantum chaos may also have practical applications. I've already hinted that it may play a role in the design of nano-scale devices that will revolutionize electronics in the years ahead.

At this scale of tens of atoms or hundreds of atoms and at sufficiently low temperatures, electrons have to be treated as waves. You can't really think of them as particles any more, and the principles of traditional electronics go out the window. Researchers have begun exploring this new realm by building miniature quantum dots in the shapes of stadiums and circles, except now the billiard balls are electrons. At temperatures close to absolute zero, they find that the electrons travel freely. They only suffer electrical resistance when they bang off the walls. These experiments show the telltale, predicted signs of quantum chaos. The electrical resistance of the device changes dramatically depending on whether you used a stadium, where the classical motion would be chaotic, or regular, like in a circle. So quantum chaos leaves its mark on electrical conduction, and it's going to be important in understanding these future electronic devices.

I want to conclude, though, with what I take to be the greatest shocker of all about quantum chaos. It uncovers a mysterious pattern linking two realms that you might think have nothing to do with each other. One is the realm of reality—atoms. We're going to be talking about atoms, the bedrock of reality. The other is the realm of the imagination, the most ethereal realm of human thought, pure number theory.

Here is the question. Let's speak in the language of number theory for a few minutes and then come back to quantum mechanics for the astonishing connection. So in the world of number theory, an ancient question is: Can we find a formula for prime numbers? Remember prime numbers? Whole numbers are divisible—6 is divisible by 2 and by 3. That's not a prime. Seven is not divisible by anything, just 1 and itself, 7. So we're interested now in the prime numbers, and the question is: Can we find a formula for the next prime number or for any given prime number? No one has ever been able to find a formula that produces just prime numbers. Short of that, mathematicians have wondered: Can we at least understand the statistics of prime numbers? They have random-like qualities. They don't come at a regular clip. Sometimes there's a big space between two prime numbers. Sometimes they're just next to each other separated by two [numbers], like 17 and 19. What are the rules governing the spacing of the prime numbers? That's one question. But what are their statistical properties viewed in aggregate?

For example, suppose I ask you this: Look at all the numbers between 1 and 100. How many prime numbers are there in that region? If you count—just looking for prime numbers like little jewels or beautiful little shells on the beach—you would find that there are 25 numbers between 1 and 100 that are prime. Suppose I ask how many between 1 and 1000, to try to get better statistics? It turns out the percentage thins out. It's not 25% any more. It thins out as we go to a higher n than 100. The primes become rarer.

Here is what the pattern looks like, showing the distribution of prime numbers. How many prime numbers are there less than a given number? Here is a graph. On the vertical axis I'm showing the number of primes less than a specified number, n, and then here's n going from 0 to 100. I already mentioned that if I look for how many prime numbers are less than 100, it should be 25, and you see that this graph goes up to a point at a height of 25. What's interesting is the staircase structure to it. Of course, there has to be a staircase because every time n crosses a new prime number, the count goes up by 1. That's why we have these steps of height 1.

I want you to notice the trend. There is a staircase structure, but there's a very clear trend. You could draw a smooth line through here. That's the first thing that caught mathematicians' attention. What is the trend line?

Well, the great mathematician Carl Friedrich Gauss figured that out. He found a formula for the trend, and it's called the *prime number theorem*. It says that the number of prime numbers less than n is given by this formula: $\frac{n}{\log n}$. Logarithms again. We've talked about logarithms a few times. They mysteriously reappear here, but that's not what I want to focus on. The trend line is not really our story. It's interesting, but that's not the important point for right now. The question is: Can we improve over the trend line? Can we capture the wiggles in the staircase?

In the mid-1800s another great mathematician, Bernhard Riemann, improved Gauss's formula. Here's what his trend line looks like. It has the right general shape, what you would get from eyeballing the picture. But Riemann went farther and found a stunning formula for the missing staircase part. What he needed to do was add certain waves to the trend to make it look more staircase-like. He did that and got something that looks like this. You see the trend line, but there are wiggles on it. These are certain magic waves which, if Riemann added enough of them, would exactly reproduce the distribution of prime numbers.

This is where the plot thickens because these magic waves were based on a guess. [Riemann's] improved formula, the trend line plus the waves, was based on a hypothesis that is now called the *Riemann hypothesis*, about the specific frequencies of the waves that he needed to add. So his guess is sometimes called "the music of the primes." Why music? Of course, because music is made of sound waves. Riemann's waves are the secret to the mystery of the primes. Proving Riemann's hypothesis is regarded as the greatest unsolved problem in all of math.

Back to quantum chaos. What does this have to do with quantum chaos? Waves. I've been talking about waves. Quantum systems have discrete energy levels, corresponding to waves vibrating at certain frequencies. Likewise, the secrets of the primes are encoded in a discrete set of wave frequencies, the magic frequencies that Riemann found that he needed, which are called by this foreboding name: *the zeros of the Riemann zeta function*. Just think of them as the magic frequencies.

Here's the shocker that I hinted at earlier: The frequencies of the Riemann waves look uncannily like the frequencies of a quantum *chaotic* system. That's the big point. In a nutshell, it seems that there is some chaotic system—no one has discovered it yet—whose quantum counterpart would hold the secret to the music of the primes. The connections here—chaos, atoms, prime numbers—are as spooky as they are far-reaching. The atoms of arithmetic, the prime numbers, are connected to the atoms of reality, and the link between them [is] chaos. I get chills just telling you about this. I hope this problem will be solved in my lifetime. It is magnificent.

I will see you next time. We will talk about nature's incredible ability to organize itself.

Synchronization
Lecture 23

You may have thought about this already. Women who are good friends who suddenly find that they are having their period around the same time as their best friend, month after month; their menstrual cycles have become synchronized. We now understand how that works. It's through some silent, chemical communication between women, mediated through pheromones.

It's surprising to learn how complicated the dynamics are of simple systems. In particular, all of these simple systems displayed chaos. At the frontiers of chaos research today, scientists have begun taking the natural next step. They've started exploring much larger dynamical systems with huge numbers of variables. We've devoted much of this course to dynamical systems having just a few variables. Examples include:

- The three-body problem (see Lecture 4).

- Lorenz's convection model (see Lectures 7 and 8).

- The one-variable logistic map (see Lectures 9 and 10).

You might expect the chaos to be exacerbated, and sometimes it is. On the other hand, sometimes these complex systems can behave very simply. Patterns emerge unexpectedly, and the system organizes itself. The individual parts of the system spontaneously cooperate.

In this lecture we'll discuss the simplest kind of self-organization, in which different parts of a system begin oscillating in unison. In other words, they *synchronize*. We've already discussed synchronized chaos in the context of encryption (see Lecture 20) and epilepsy (see Lecture 21). We're going to be talking about the synchronization of periodic things called *oscillators* rather than synchronization of chaos.

Rhythmic synchronization is important and pervasive. We tend to overlook it or think it trivial, perhaps because it comes so easily to us. Groups of people can march in step or keep a beat, even without a leader. Consequently, we have little intuition about what synchronization requires. Does it depend on human consciousness? No—animals can synchronize. Think of the graceful coordinated movements of flocks of birds, schools of fish, or fireflies that flash on and off in silent, hypnotic unison. These organisms may not be fully conscious, but still they can perceive what's happening around them. Maybe perception is necessary for synchronization? No. Thousands of mindless pacemaker cells in your heart synchronize their electrical rhythms automatically, triggering your heart to beat. Menstrual synchrony sometimes occurs among women who are close friends or roommates. The precise mechanism is unknown, but it seems to rely on unconscious chemical communication through odorless pheromones contained in sweat.

Even inanimate things can synchronize spontaneously. Christiaan Huygens serendipitously discovered this remarkable phenomenon in 1665, while observing a pair of pendulum clocks hanging from the same beam. Their swinging pendulums shook the beam imperceptibly, but enough to bring the clocks into lockstep. Any collection of oscillating systems—fireflies, metronomes, heart cells—can synchronize, if they are similar enough and if they interact in the right way. It doesn't matter whether the coupling occurs by chemicals, gravity, light, sound, vibrations, or electrical currents.

Rhythmic synchronization is important and pervasive. We tend to overlook it or think it trivial, perhaps because it comes so easily to us.

Working out the mathematics of collective synchronization has been a major thrust of nonlinear dynamics, running parallel to chaos theory. The pioneer was theoretical biologist Art Winfree. As an undergraduate at Cornell in 1965, he used a computer to simulate the dynamics of large populations of biological oscillators such as heart cells or fireflies. He assumed the oscillators were diverse, as expected for any biological population. In a thought experiment, Winfree imagined making the oscillators progressively more similar. He discovered that synchronization broke out suddenly, not gradually, as he made the population more uniform. A phase transition occurred at a certain threshold, somewhat like the sudden freezing of water into ice at a critical temperature.

Forty years later, Winfree's ideas shed light on a seemingly unrelated phenomenon: the spontaneous crowd synchronization that caused London's Millennium Bridge to wobble on its opening day. The Millennium Bridge is an elegant, minimalist footbridge across the Thames River. When thousands of Londoners streamed onto the bridge, it began to sway unexpectedly from side to side. Its designers were flabbergasted. The bridge was closed 2 days later and was repaired over the next 18 months at a cost of 5 million pounds and great embarrassment.

Investigations revealed that the swaying was caused by a strange cooperative effect. The pedestrians were unconsciously synchronizing their footfalls to the slight sideways movements of the bridge to keep their balance. In so doing, they inadvertently pumped energy into the bridge, making the vibrations worse. Video footage from the BBC showed the whole crowd rocking from side to side in unison.

This is not merely the old chestnut about soldiers having to break step while crossing a bridge. Soldiers arrive at a bridge already in sync. The pedestrians were not. Marching soldiers drive a bridge up and down. The pedestrians drove the Millennium Bridge sideways. Diagnostic experiments uncovered a threshold effect akin to Winfree's. The swaying occurred only if there were more than a critical number of people on the bridge. Below that number, the bridge was essentially motionless and the crowd remained desynchronized. The point is that civil engineers (and the bridge building code) were in the dark about the possible outbreak of synchronized dynamics that this complex system could display.

Fortunately, no one got hurt, and the bridge has now been stabilized. In 2005, my colleagues and I published a mathematical explanation of what happened on opening day. Our model combined ideas from chaos theory, Winfree's work on biological rhythms, and bridge mechanics. The analysis revealed that the wobbling and crowd synchronization were two aspects of a single instability process. The model also predicts the critical number of people for bridges with different properties, which may help in the design of future structures.

In the final lecture, we'll discuss why large dynamical systems are so important, not just for chaos theory but for the future of all of science. ∎

Essential Reading

Strogatz, *Sync*, chaps. 1, 2, 4, 6.

Supplementary Reading

Pikovsky, Rosenblum, and Kurths, *Synchronization*, chaps. 1–5.

Strogatz et al., "Crowd Synchrony on the Millennium Bridge."

Winfree, *The Timing of Biological Clocks*.

Internet Resource

Play with a model for the synchronization of fireflies, using this applet: http://ccl.northwestern.edu/netlogo/models/Fireflies.

Questions to Consider

1. Describe some real-world situations in which synchronization (of people, companies, stock markets, and the like) is undesirable or even dangerous.

2. What are some possible adaptive explanations for synchronization among animals? For instance, why do birds flock? Why do crickets sometimes chirp in unison?

Synchronization
Lecture 23—Transcript

Welcome back to chaos. We've devoted much of this course to the smallest dynamical systems, to systems involving just a few variables. For example, when we talked about the three-body problem or Lorenz's convection model, there were really just three variables that we needed to be thinking about. The convection problem was in Lectures Seven and Eight; the three-body problem was in Lecture Four. The logistic map, our model for studying the transition to chaos, that just had a single variable, that was in Lectures Nine and Ten. In the last lecture, we talked about a billiard ball bouncing around in a stadium—two variables: its location and its velocity. All of these were very simple systems.

With systems as simple as these, the surprise was how complicated their dynamics could be. In particular, all of them displayed chaos. At the frontiers of chaos research today, scientists have begun taking the natural next step. They've started exploring much larger dynamical systems with huge numbers of variables. You might expect the chaos to become more severe. Sometimes it does. But what's perhaps more surprising is that these complex systems can also behave very simply in certain cases. Patterns emerge unexpectedly. The system seems to organize itself, with no outside intervention. Maybe most unexpected of all, the individual parts of the system spontaneously cooperate.

In this lecture, I'll be talking about the simplest kind of self-organization, in which different parts of a system begin oscillating in unison. In other words, they *synchronize*. We've discussed synchronization before. You might remember that when we were talking about encryption, we discussed the possibility of synchronizing chaos for encryption, to use the chaos as a cloak, as a mask over a secret message. Also, two lectures ago, when I was talking about epilepsy, that was an example of unwanted synchronization among brain cells, pathological synchronization. Both of the instances used in encryption and in epilepsy were cases of synchronization of chaos.

Going to larger systems, which is what we aim to do in this lecture, it's really much easier if we consider individual units that are rhythmic rather than chaotic. In other words, we're going to be talking about the synchronization of periodic things called *oscillators* rather than synchronization of chaos.

Another reason we're doing it besides simplicity is that this is a very common phenomenon in the world around us. Nature has built many synchronizing systems out of inherently rhythmic devices or rhythmic entities. You might find some of the examples that we'll be discussing pretty easy to relate to because they have a human dimension to them. In some cases, the individual parts of the complex system were, literally, people.

In the first part of the lecture, I'll begin by mentioning a large number of examples of synchronization in the natural world to help you see just how pervasive and important it is. If you find this topic interesting, you might like to take a look at a book I wrote called *Sync* about this very theme, synchronization in nature and the math that we use to understand it.

In the next part of the lecture, I'll be describing some groundbreaking work done in the 1960s on modeling these huge populations, in this case, huge populations of biological oscillators. Think of the pacemaker cells in your heart. Remember, your heart doesn't beat on its own. It has a pacemaker region of about 10,000 cells, in fact, in what's called the *sinoatrial node*. Each of them is a competent little oscillator. It has its own voltage rhythm. But to work properly to trigger the rest of the heart to beat, those 10,000 cells have to be electrically synchronized to send a coherent signal. How does the pacemaker do it? Who is the pacemaker for the pacemaker? Are they following a leader? No, they're organizing themselves. That's what this is all about: self-organization, emerging spontaneously.

We'll also discuss other examples in nature, like fireflies that flash in unison. Then, at the end of the lecture, I'll close by applying these ideas that have been developed to understand self-synchronizing systems to a fascinating case study, the unanticipated synchronization of pedestrians on London's Millennium Bridge that caused it to wobble on its opening day and led to the closure of the bridge for about a year and a half while it was being repaired. This was an unprecedented thing in the engineering literature. It was quite mysterious, but we can understand it now by what we have learned about the mathematics of synchronization.

You may be tempted to think, "Wait a second; is he serious? Synchronization is a subject? This seems so trivial. People are easily able to synchronize." That's true. It comes easily to us. We can march together; we can sing in a chorus; we can dance together. All these are kinds of synchronization, and there's really nothing so sophisticated about it. We can all do it. We can even do it without a leader. A symphony orchestra doesn't need a conductor to stay in step. Trained musicians can do that. Even people marching or clapping their hands after a concert can clap in sync. It's just something people are good at. But because we're so good at it, we have very little intuition about exactly what's required.

What does it take to get in sync? Does it need the full measure of human consciousness and sensitivity? That can't be right because we know that animals can synchronize. Think about flocks of birds, or schools of fish, or fireflies in Southeast Asia that flash on and off in perfect time together. They don't need a leader. They self-organize, too. You might start to think, "All right, it's true. They're not quite as sensitive as human beings, but nevertheless these different creatures—fish or birds or fireflies—can sense what's happening in their environment, and they can sense what their comrades are doing and make adjustments accordingly, so maybe what we need is something that has some minimal level of perception." Is that true? Is that what's needed for synchrony?

What about cells? I just mentioned the pacemaker cells in your heart. They're not conscious. They can't perceive anything. Still, they can synchronize, and in fact you need them to keep you alive. Or the brain cells that inadvertently synchronize during epilepsy. They're also not particularly conscious, and yet for electrical reasons, they do get in step.

What about the case of women? You may have thought about this already. Women who are good friends who suddenly find that they are having their period around the same time as their best friend, month after month; their menstrual cycles have become synchronized. We now understand how that works. It's through some silent, chemical communication between women, mediated through pheromones. There is some substance in sweat that conveys information about the phase of a woman's menstrual cycle, and even though it's weak and it's not something you can smell or detect, it's there and it's powerful enough to synchronize their hormone rhythms.

All the examples I've mentioned so far are living things, from cells to people. You might think, "Maybe it's being alive. That's the essential thing for synchrony." But no, of course, that can't be right, either, because inanimate things can synchronize. We've already heard in Lecture Twenty about chaotic circuits that were able to synchronize their electrical rhythms.

Let's talk for a minute more about the case of inanimate synchronization because there is something mind-boggling about it. When you see an instance of it, it's hard to believe. It's like order emerging out of chaos. It's like nature running the wrong way. This was noticed very long ago, back at the beginning of our subject, in 1665, almost 350 years ago, by the second-greatest scientist alive at the time, Christiaan Huygens. Remember, at that point, Newton was a young man; 1666 was his big year. Huygens had maybe the misfortune of being alive at the same time; otherwise, he would have been the greatest.

Huygens was a tremendous scientist, and rather than go into all his accomplishments, let me focus on what he did that sort of founded the subject of synchrony. It was an accidental discovery. He seemed to have had a weak constitution. It's not exactly clear what was wrong with him, but he was writing to his father, and he talks about how he was confined to his bedroom, feeling ill. It seems like something vaguely intestinal, but the letter isn't clear. Anyway, something is wrong. He's in his room, and he's kind of bored, and he happens to look over at two pendulum clocks that he recently built. Huygens, among other things, invented the modern version of the pendulum clock. He had two of them, and they were made to be nearly identical. The reason he had built them in the first place was that they were supposed to be put on ships that would be taken out to sea, and they would help the ship keep time and thereby solve the longitude problem, figuring out where the ships were, east or west, along the globe.

There he was lying in bed sick, looking at his identical clocks, when he noticed something strange. They were ticking in perfect anti-phase. That is, when one pendulum would swing this way, the other pendulum would swing that way, and they stayed in perfect time like this for as long as he watched them, for half an hour. He thought this was pretty peculiar because he knew that although they were nearly identical, they weren't that identical. They couldn't possibly be that good. He thought that they were communicating somehow. They're interacting. How?

First, he thought it was air currents, and he put a big plank between them to block any air currents between them, but they still stayed synchronized. So how are they talking to each other? Then he realized, they're hanging from the same wooden beam. Maybe they're sending subtle vibrations through the beam and shaking each other, and that's enough. So he separated them and moved one to the opposite side of the room, and then they fell right out of sync. So this serendipitous discovery of inanimate synchronization of pendulum clocks kind of got the subject going, although for Huygens, he felt it was a nuisance and he ignored it, surprisingly, because it defeated his purpose. He wanted these clocks to keep very good time out at sea, and the Royal Society, where he submitted his results—[he] told the Royal Society about this great discovery that he thought would solve the longitude problem. He was told no, in fact, he was showing just the opposite: If your clocks are so sensitive to tiny vibrations, they will never work on a ship that's heaving under the ocean waves.

So I want to show you a demonstration, not quite of Huygens's pendulum clocks, but of a modern-day analogue, which is metronomes. This will convey what inanimate synchronization looks like. What I've got here are two metronomes. These are said to be the world's smallest metronomes. I'm going to place them on a platform that can roll. The rolling platform is going to provide the same kind of interaction that the wooden beam did in the case of Huygens's accidental observation.

This is a live demonstration, so anything could happen, but let's see if it works. I'll set this metronome going, and let me set this one going. Now this may take a while to get itself in sync. [Sound of metronomes ticking.] Do you hear it? But now they're losing it again. That's it; they're synchronized, and you can see as the platform moves, it provides a little jiggle that keeps them in sync. This is not specific to metronomes. Any oscillating system, whether it be fireflies flashing, metronomes, heart cells having electrical rhythms, just about any collection of oscillators, can spontaneously synchronize, like I just showed you, if they are similar enough and if they interact in the right way. It doesn't matter how the interaction is mediated. If it's mediated by vibrations—by light in the case of fireflies, by sound for crickets chirping in unison, by gravity for planets that are orbiting in sync, electrical currents in the heart—any kind of communication between oscillators can be enough to synchronize them. Working out the mathematics of synchronization, especially when huge numbers of oscillators are involved, as in the case of heart cells or in the brain, has been a big part of nonlinear dynamics, running parallel to the development of chaos theory.

A pioneer in this respect was a man named Art Winfree. Art was interested in large populations of biological oscillators, for instance, this case of the fireflies that I've been mentioning, coming to synchronize their flashes. To study problems like this, Art did computer simulations. He was, at that time, an undergraduate at Cornell, around 1965, majoring in engineering physics, an interesting major for someone who was curious about biology. He thought he would acquire different tools from the usual biologist, and maybe by knowing something about differential equations and computers and circuits, he would have different insights than the usual kind of person who studied biology.

This work was being done around the same time, remember, as Ed Lorenz was looking at the unpredictability of the weather, simulating his convection model, using computers there, too, and discovering strange attractors. Winfree, who as far as we know had no contact with Lorenz, is doing a parallel thing but for oscillators, not strange attractors. He needed to know some nonlinear dynamics, the sorts of things that we've been talking about in this course, to model each oscillator. Why did he need nonlinear? Because biological oscillators always have attractors. That is, they're not just periodic, but they settle into a periodic rhythm at a certain amplitude that is well regulated. If they're disturbed away from that amplitude, they come back. Think about your own heart rate. If you're disturbed, you may transiently speed up your heart rate, but then you'll be brought back into normal rhythm soon by normal processes in your nervous system. So whenever you have attractors—in this case, an attracting cycle for the periodic behavior of an oscillator—you must have nonlinear interactions going on or nonlinear dynamics. Linear systems only have very boring attractors that are just pure equilibrium. We saw attracting cycles earlier in the course when we talked about Lorenz's waterwheel rocking back and forth periodically or when the logistic map settled into period-2 or period-4 cycles.

Winfree was looking at these nonlinear oscillators, and his problem was very hard, unprecedented in its difficulty, the sort of thing that only a young person would tackle. He was looking at thousands of interacting nonlinear oscillators. Even two would be very difficult to study, but thousands? It was unheard of. Worse than that, his were diverse. They were not identical oscillators. In any real biological system, there is always diversity because of genetic differences among, say, the fireflies. So he tried to think about how he could make progress on this problem of a huge collection of attracting-cycle oscillators, diverse in their properties. He had a metaphor in his mind, which we'll think of this way: Imagine a club of runners, like a jogging club, and they're friends. They would like to run together, to synchronize as they run around a circular track, but it's hard for them to synchronize because some are inherently faster than others. Some people are better runners than others. So here is a slide to indicate a little bit about what Winfree found in his computer simulations.

What you're seeing first in this little cartoon at the bottom is a diverse population of runners. These are the analogue of the oscillators that we're studying in the biological case. They have different shapes. Inherently, some are fast and slow. This bell curve that I'm showing is supposed to show the diversity in their natural frequencies, their natural running speeds, let's say.

Winfree then did a thought experiment where he imagined, "What if I make the population more uniform? Suppose I narrow the bell curve so that the people become less diverse, more uniform, more homogenous, all the way to the limit to where the people are almost clones of each other, almost perfectly identical, with a very narrow bell curve of their properties." The question he asked himself was: "As I make them more similar, what happens to the amount of synchronization in the system?" That's what's shown in the top graph. If I look at the amount of synchronization versus how homogenous the system is, what you might naively expect is that the more uniform you make it, the more they're able to synchronize. But that's not what he found.

It's interesting what he found. He found that as he made the system more and more similar, more and more uniform, there was nothing happing, no synchronization at all. It stayed completely incoherent. It would be as if the runners didn't care about each other. They're just running around at their own rate on the circle, ignoring each other, desynchronized.

But then at a critical point, at a critical amount of homogeneity, suddenly, part of the system started to synchronize. It was as if you lowered the temperature on water, it stays water; keep lowering it, it's still water; but at a critical temperature, it freezes. What Winfree had found was the analogue of a phase transition, a kind of freezing. What he thought of was very creative here because—in the case of water, molecules are lining up in space—the phase transition that Winfree discovered in his simulations was a kind of lining up in time. It was the temporal counterpart of a phase transition. That's what I'm showing in the graph here, that the amount of synchrony builds up sharply at this critical point, a tipping point or phase transition, and then as you make the population more and more homogenous, it just keeps getting more and more synchronized. So synchrony breaks out suddenly, not gradually, as you might expect. It's a kind of bifurcation. You should recognize that by now at this point in the course. We've seen these sudden changes in qualitative behavior as we change a parameter. Here, we're changing the width of the distribution of their properties, and we're seeing a sudden bifurcation to synchrony.

The bifurcation in this problem is more difficult to analyze than the ones we saw in the case of period-doubling in the logistic map or in the case of Lorenz's waterwheel. But Winfree was able to do it, and it was a monumental breakthrough in our understanding of the collective behavior of oscillators.

Something like this bifurcation was seen a few years ago on London's Millennium Bridge. Maybe you've been there. It's gorgeous. It's a beautiful, elegant, minimalist structure. It's as if you took rubber bands and pulled them tight across the Thames River and then attached a bridge to it. It's not a suspension bridge of any type you've ever seen. It connects to the Tate Modern Museum on one side of the Thames to St. Paul's Cathedral on the side that you can't see. This is a foot bridge for pedestrians, no cars going over it. These are the rubber bands. They're actually steel tension cables on the side of the bridge pulled taut across the river. You might think these piers are holding up the bridge, but they're not. The piers are, as in any suspension bridge, holding up the cables. They're holding the cables, and the bridge is hanging suspended from the cables by these elements, and those elements are holding up the deck. So it's a true suspension bridge, but it's like walking on something that's held up by piano wire or by rubber bands, the minimum amount of structure needed to make a bridge.

The point was to give London a beautiful "blade of light," as one of its designers, Lord Norman Foster, maybe England's greatest architect, called it. He said, "I want to give England a blade of light across the Thames," a sort of thin, shiny ribbon.

Other architects thought this was an arrogant design, and they expected trouble. It just looked like something that might have trouble, and there was trouble. On opening day, thousands of Londoners showed up, excited by their new bridge. It was built to commemorate the millennium. The queen was there, a big fanfare. Everyone showed up, and when people streamed onto the bridge, the bridge started to sway unexpectedly. This was not seen in computer simulations. It wasn't seen in the wind tunnel tests. The engineers looked at each other; they looked at the police. What is going on? The bridge is swaying sideways, side to side. The designers were flabbergasted.

It turns out the bridge had to be closed. These oscillations, the side-to-side swaying, that was so severe, the bridge had to be closed, and it stayed closed for about a year and a half and had to be repaired at the cost of several million pounds and great embarrassment to everyone involved.

What was causing the problem? Investigations done while the bridge was closed revealed that the swaying was caused by a strange cooperative self-synchronization effect. The pedestrians were unconsciously synchronizing their footfalls to the slight sideways swaying of the bridge, to keep their balance. In so doing—picture me as a pedestrian. I'm walking on the bridge. It's swaying sideways. It's like you're standing up in a train that's moving too fast and it's wobbling, or maybe you're in a rowboat and as you try to keep your balance, you may change your gait a little bit, and in so doing, you might inadvertently pump energy into the bridge.

The BBC was there filming opening day, and their footage showed the way people walked. It showed that they adopted a strange gait like a novice ice skater, with feet out wide, pushing sideways. By walking in this peculiar way to keep their balance, the pedestrians inadvertently pumped energy into the bridge, into its sideways vibration, making it worse, which made more people adopt the weird gait, which made the vibrations worse, and so on. It was a runaway effect.

Video footage from the BBC shows, dramatically, the whole crowd rocking side to side in unison. Let me show you that footage. What you will see first is just a big crowd on the bridge milling around. You won't see much happening, but a few seconds into the footage, keep your eye on the crowd and you'll see. It looks almost like they're deliberately pushing on the bridge, but they're absolutely not. They didn't even know they were doing it, and they're just trying to keep their balance. Also, you may want to look at the tension cables on the side of the bridge, and you'll see them moving significantly from side to side. Let me go to that footage.

Just to get yourself oriented, look at the wobbling cables on the side. Some people are holding. They've given up even trying to walk. Here is the dramatic footage. Look at these people. Do you see whole sections of the crowd rocking in perfect time, sideways? Unbelievable. It's sort of like the Ministry of Silly Walks on the Millennium Bridge.

Let me be clear about one thing. People sometimes get confused on this point. They say, "Wait a second. I've heard about bridges vibrating before. What about soldiers who are told to break step when they cross a bridge?" Absolutely. Soldiers are taught that, and that's real. There have been bridges that have collapsed when soldiers were marching in step, causing the bridge to resonate with their footfalls, causing the bridge to vibrate and potentially undergo such large oscillations that it could break. That has happened. That has nothing to do with the Millennium Bridge for a few reasons.

First of all, those soldiers were told to march in step by a commander. Leaving the commander aside, they arrive at the bridge marching in step. That is nothing like the Millennium Bridge. These are just ordinary pedestrians. They're not marching in step. They're just walking in both directions across the bridge, grandmothers, little kids, people with different walking speeds. There's no organization initially. That's the first thing.

The second thing is when soldiers are marching in step on a bridge, they make very severe vertical oscillations, up-and-down oscillations. That is not what happened on the Millennium Bridge. That bridge swayed sideways. That's the strange thing. Every bridge engineer knows you don't build a bridge with a resonant frequency of 2 cycles per second in the vertical direction because that's the frequency of normal human walking. That's the soldier effect. This bridge doesn't have a vertical resonance frequency of 2 cycles per second. But what happened on the Millennium Bridge was a sideways sway at 1 cycle a second, half the frequency of normal walking. It's the frequency that you make with your left foot—that is, in the normal 2 strides per second, one of them is with your left foot, one is with your right foot. So you put sideways force on the bridge at half the normal walking frequency, 1 cycle per second, to the left and right, and it happened that the bridge had a sideways resonant frequency of 1 cycle per second, and so it became resonated.

But that's not all. Normally, these people are walking incoherently. They're walking out of step from each other, and engineers don't usually think about anything happening sideways. First of all, we put very little force sideways, much less than you put vertically when you take a step. Secondly, some people are putting their left foot down when some are putting their right foot down, and so those forces tend to cancel out. That's true. That's what normally happens. But if for some reason all the people start walking in sync, the forces won't cancel out, and then you might have a problem.

Diagnostic experiments done on the bridge after it was closed uncovered an unexpected effect related to the nonlinear dynamics of the walkers interacting with the bridge. What they found was another kind of bifurcation very, very similar to what Winfree had discovered in his biological oscillator simulations about 40 years earlier. So here are the results of some experiments done by the engineering firm that built the bridge when they were trying to figure out what was wrong.

What you're seeing is a graph of two things as a function of time. On the one hand, you see the number of people on the bridge rising in a staircase pattern.

What was done in this experiment, the engineering firm had its own employees get on the bridge—first 50 of them, then 60, then 70, then 80—and had them walk around in circles to see if it was the number of people that was causing the problem on the bridge. They also recorded the bridge's movements, as shown here, using acceleration to see how much the bridge is shaking sideways. What you see is, as the number of people, these poor volunteers— actually not really volunteers; it's their own employees. They said, "You're going to walk on that bridge." When they got between somewhere between 150 and 160 people, shown on this axis, look at that. The bridge suddenly exponentially started moving and vibrating.

There was a kind of bifurcation to sudden oscillations that occurred at a critical number of people, much like Winfree's critical amount of uniformity that led to the phase transition in his studies. Below that number, the bridge was essentially motionless and the crowd desynchronized. The point is that civil engineers and the bridge building code before this event were completely in the dark about the spontaneous outbreak of synchronization that a complex system like this could display.

Fortunately, nobody got hurt, and the bridge has now been stabilized. It won't do this if you go there. I don't know if I actually believe it's fortunate because if the Leaning Tower of Pisa weren't leaning, would it be so great? I think this Millennium Bridge would have been a terrific tourist attraction when it was wobbly. Still, it could have been dangerous. Anyway, it has been repaired and stabilized. It doesn't sway any more. They put viscous dampers like shock absorbers to stabilize it.

In 2005, my colleagues and I developed a mathematical model of the interactions between the pedestrians and the bridge, and our model combined ideas from chaos theory, from Winfree's work on biological oscillators, and mechanics about the bridge. Our analysis was able to explain what I just described, as I show in this next slide, when we did a simulated crowd test. We could show that the wobbling of the bridge and the amount of synchronization in the crowd both took off suddenly at the same point with respect to the number of pedestrians. In other words, these two effects, the wobbling and the synchronization, are two sides of the same coin.

The model also predicted the critical number of people needed for this to happen, which may help bridge engineers in the future avoid it by figuring out how damped bridges should be.

In my final lecture, I'll have more to say about why this kind of phenomenon, the cooperative behavior of huge dynamical systems, is so important, not just for chaos theory but for the future of all of science.

See you next time.

The Future of Science
Lecture 24

We've come to the end of our whirlwind tour of chaos theory, and it's time to reflect on the subject as a whole.

Let's begin with what should be the easiest question: What's the science of chaos all about? Our view of the subject has changed as we learned more and more. The same evolution happened historically, and it's still happening. In ordinary language, chaos is synonymous with disorder and utter confusion. But that's too broad for us. Such a definition would subsume chaos under probability theory, the study of randomness. That's wrong, because chaos theory deals exclusively with non-random, deterministic systems (where the present completely determines the future).

On the other hand, chaotic systems do have a disorderly aspect to them. They can impersonate random behavior because of their long-term unpredictability. That unpredictability is caused by the butterfly effect, the extreme sensitivity of a chaotic system's behavior to small changes in the initial conditions. So a second try would be: Chaos is the science of disorder in deterministic systems. And for a long time, it *was* that. That unpredictable side of chaos was dominant in Poincaré's work on the three-body problem (see Lecture 4). And also in Lorenz's work on his artificial weather model, through which he discovered the butterfly effect (see Lecture 5 and 6).

But soon, it came to be realized that there was also a tremendous amount of *order* in chaos. It was just order in many unfamiliar guises. Of course there was short-term order and predictability provided by the underlying differential equation. This would be true of any deterministic system, not just chaotic ones. But there was also unexpected long-term order in the way the system moved around in state space. This is the order embodied by strange attractors (see Lecture 7). And there was medium-term order in the way the system hopped along its attractor, when viewed under a strobe light. This is the order embodied by iterated maps (see Lecture 8).

Then we focused on iterated maps, instead of differential equations, as the dynamical systems of interest. This allowed us to uncover still more kinds of order (see Lectures 9 through 11). First, there was order as *parameters* were changed. (Recall, parameters are like knobs that can be manually controlled, like the growth rate of the insects modeled by the logistic map, or the setting of the brake on the waterwheel.) This order with respect to parameters was displayed vividly by the orbit diagram (see Lecture 10) and even more stunningly by the Mandelbrot set (see Lecture 18). Those diagrams were like Rosetta stones. By consulting them, we could see all the possible behavior of the system, in one icon. These diagrams showed where bifurcations occur— where the system dramatically changes its long-term behavior, from stable equilibrium, to oscillations, to chaos. Both of these Rosetta stones had incredible fractal structure, as did strange attractors themselves. And one particular aspect of this fractal order— the Feigenbaum numbers for the spacing between bifurcations leading to chaos— turned out to be *universal* for many chaotic systems, both real and mathematical (see Lectures 11 and 12). This led to the most stunning experimental confirmations of chaos theory, in experiments on fluids, circuits, chemical reactions, heart cells, lasers, and so on.

And the order in chaos soon got harnessed to practical advantage for all kinds of purposes, from encryption to space travel, and from quantum electronics to forecasting epileptic seizures.

Meanwhile, fractals took on a life of their own, as models of self-similar shapes or processes in the real world (see Lectures 13 to 18), not just in the abstract worlds of state space and orbit diagrams. And the order in chaos soon got harnessed to practical advantage for all kinds of purposes, from encryption to space travel, and from quantum electronics to forecasting epileptic seizures (see Lectures 19 to 22).

Behind all of this was a hidden puppeteer pulling the strings. The puppeteer was *nonlinearity*. None of our story would have been possible with linear systems—the ones studied by traditional science, in which causes are strictly proportional to effects and the whole equals the sum of the parts. So we've come to this definition: The science of chaos is about order and disorder in nonlinear, deterministic systems.

The real point of this definitional exercise, however, is that it shows us why chaos theory is going to be crucial to the science of the 21st century. Why? Because virtually all the major unsolved problems of science are fundamentally nonlinear. Consider the following:

- The orchestration of thousands of biochemical reactions in a single cell and their disruption when the cell turns cancerous.

- The battle between the immune system and the AIDS virus.

- The emergence of consciousness from the interplay of billions of neurons in the brain.

- The origin of life from a meshwork of chemical reactions in the primordial soup.

What makes these problems so vexing is their decentralized, dynamic character, in which enormous numbers of components keep changing their state from moment to moment, looping back on each other in ways that can't be studied by examining any one part in isolation. In such cases, the whole is surely not equal to the sum of the parts. These phenomena are nonlinear.

Unfortunately, our minds are bad at grasping nonlinear problems. We're accustomed to thinking in terms of centralized control, clear chains of command, and the straightforward logic of cause and effect. But in huge, interconnected systems, our standard ways of thinking fall apart. Verbal arguments are too feeble, too myopic. The good news is that, thanks to the development of chaos theory, we have now mastered the simplest nonlinear systems, the ones studied in this course. What awaits is a similar mastery of complex nonlinear systems.

By complex, I mean something with an enormous number of parts. These parts are typically diverse and interact with one another in bewildering networks. For example, consider the vast network of genes, proteins, and enzyme reactions that controls how every cell in your body grows and divides. One node in the network is p53, the most important tumor suppressor gene. It is known to be mutated (broken, basically) in about 50% of all cancers.

Mathematical models of the cell division cycle (for yeast, much simpler than the mammalian cell cycle, but still very complicated) are being developed by such pioneers as John Tyson and his colleagues at Virginia Tech. Their models are based on methods of chaos theory. Although such models are still avant-garde from the point of view of traditional cell biology, I believe they foreshadow the kinds of theoretical work that will prove indispensable in solving the riddle of cancer. After all, genes are nonlinear devices; stimulate them twice as strongly and they don't respond twice as much. The same is true of neurons. So all the mysteries of gene regulation and brain function—and all of biology for that matter—will ultimately require us to decipher complex nonlinear systems.

There are grounds for optimism. Even the most hard-boiled mainstream scientists now acknowledge that reductionism alone won't solve all the great mysteries we're facing. At every major research university, institutes are springing up with names like "functional genomics" and "systems biology." Biologists are teaming up with computer scientists and mathematicians to try to make sense of the dance of life at the molecular level. Chaos theory is playing a vital role in the new biology. Here's why:

- Although sequencing the human genome gave us an enormous list of parts (23,000 individual genes and the proteins they encode), we still have almost no clue how the interlocking activities of those genes and proteins are choreographed in the living cell, or how they go awry when cells turn cancerous.

- These are problems of dynamics.

- Traditionally, the genome has been viewed in more static terms, as a blueprint for the construction of proteins (the building blocks and molecular machines essential to life).

- But today we see that metaphor as too linear, a vestige of the assembly-line mentality of an earlier era.

- Some of the most important genes (the so-called regulatory genes) code for proteins that alter the activity of other genes, turning them on or off, forming circuits and feedback loops.

- The genome starts to seem less like a blueprint and more like a computer.

- The functioning of this computer—and its malfunctioning when cells turn cancerous—will not be deciphered until we understand the dynamics of gene networks.

- Our era awaits its own Kepler, Galileo, and Newton. They found the laws of inanimate dynamics; we're still looking for the laws of life.

Fortunately, biologists at the cutting edge are now being trained in chaos theory. As part of their graduate education, systems biologists learn nonlinear dynamics to help them reverse-engineer the networks of life. Synthetic biologists are using chaos theory to design their own biological circuits, building the living versions of toggle switches, oscillators, and amplifiers out of genes and proteins instead of wires and transistors. Likewise, chaos theory is helping to shape other newfangled branches of modern science, from computational neuroscience to econophysics.

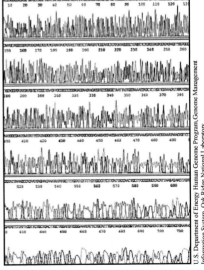

Genome Sequence Trace.

U.S. Department of Energy Human Genome Program, Genome Management Information System, Oak Ridge National Laboratory.

Chaos theory is also shedding light on perhaps the most puzzling question of all: consciousness. How can 100 billion neurons, each mindless on its own, fall in love or remember a favorite song or fret about its own mortality? Neuroscientists have found that when we recognize a face, or decide to pay attention to a conversation, far-flung parts of the cortex suddenly fire a volley of electrical signals, at 40 cycles a second, in synchrony. The synchronized oscillations emerge spontaneously out of chaos, the brain's baseline state. These studies paint a fascinating, though also disconcerting, picture of human existence. As we go about our lives, all that we feel and perceive and think may be just a reflection of the fleeting processes of synchrony in our brains. In every branch of science today, the tools, ideas, and interdisciplinary spirit of chaos theory are proving indispensable as scientists confront the greatest challenges of our times.

Let me end by saying how much fun it's been to teach this course. Thanks to The Teaching Company for giving me this wonderful opportunity. I hope you've enjoyed it too. And although trajectories are not allowed to cross in a deterministic system, I do hope our trajectories will cross again soon. Until then, be well and stay chaotic! ∎

Essential Reading

Strogatz, *Sync*, chaps. 9–10, epilogue.

Supplementary Reading

Alon, *An Introduction to Systems Biology*.

Barabasi, *Linked*.

Buchanan, *Nexus*.

Questions to Consider

1. Having finished the course, what aspects of chaos theory did you find most surprising? Most appealing?

2. Select a story from the newspaper that is somehow related to the ideas in this course. In what ways do you see the issues differently now as a result of what you've learned?

The Future of Science
Lecture 24—Transcript

Welcome back to chaos theory. We've come to the end of our whirlwind tour, and now it's time to reflect on the subject as a whole: what it's about, what it has accomplished, and where it's headed.

Let's begin with what should be the easiest question of all, especially at this stage: What's the science of chaos all about? Our view of the subject has changed throughout the course, evolving as we learned more and more. That's what happened historically, too, and it's still happening.

In ordinary language, chaos is synonymous with disorder, utter confusion, randomness. But that's too broad for us because if that were our definition of chaos, that would put it inside of probability theory, inside the study of randomness, which it really isn't. Chaos theory deals exclusively with deterministic systems. Remember, we've emphasized that throughout. Deterministic systems are those in which the present state completely and uniquely determines the future. There's nothing random about it.

On the other hand, chaotic systems do have a certain aspect of disorder to them. They can impersonate random behavior, masquerade as if they were random because of their long-term unpredictability. Remember where that comes from. That's the hallmark of chaos due to the butterfly effect, the extreme sensitivity of a chaotic system to imperceptible, tiny changes in its initial conditions.

So a second try would be: Chaos is the science of disorder in deterministic systems. That's better. Actually, for a long time, it *was* that. That unpredictable side of chaos was what we emphasized early in the course, when we were looking at the work of Poincaré on the three-body problem in astronomy and also in Lorenz's work in his artificial weather model, where he was getting to the essence of what makes the weather unpredictable and where he discovered the butterfly effect. So that's a good start. But that's not quite the definition we want either because it soon came to be realized that there is a tremendous amount of *order* in chaos. It's not just disorder. It's not just the negative side of chaos. There is also magnificent order, structure in chaos. It was just order in an unfamiliar guise. For a long time, we couldn't see it. We couldn't recognize it.

Of course, there was short-term order, the kind that comes from the pure determinism of these systems. That meant that we could predict in the short term. We could predict a few moments ahead. You can do that in any deterministic system. That's what it means to be deterministic, after all. So that would be true of any deterministic system, not just chaotic ones. So this short-term predictability, that kind of order, was no surprise.

But there was unexpected long-term predictability in a certain sense. It was a kind of order, not of what an individual system would do as far as its behavior moment by moment, but in an average sense. Long-term order in the way that the system moved through this abstract space that we created called state space, showing its motion as an imaginary point moving from state to state, telling what the system is doing throughout its history, throughout its time course. When we traced out the motion in state space, we found that chaotic systems don't just make a random blob. They don't just fill up their state space with a mess, a snarl of spaghetti. They form very delicate, beautiful structures [called] strange attractors. We came to call them "strange" because their geometry seemed strange to the early pioneers.

There was long-term order in the sense of the strange attractor. And there was a kind of medium-term order, also unforeseen, which was in the way that the system hopped around on its strange attractor in state space when viewed under a strobe light. Do you remember we had this analogy where we said a differential equation is like a movie, a cinematic picture of reality, unfolding instant by instant? That was Newton's great contribution. Then, we said another way to look at things was like through time-lapse photography, like an overnight exposure, seeing the cumulative shape of chaos in its state space. That gave us the strange attractor. Then there was a third way of doing things where we flashed a strobe light on a system in its state space. It's in the dark, and we can't see it, and then—flash!—we look at where it is. Then it's in the dark again—flash!—now we see where it is next. By tracking its motion from flash to flash, we created a mathematical device called an *iterated map*. That tracked the motion under the strobe light. That embodied the medium-term order that we discovered in chaotic systems. I'm being silly here. I'm saying "we" discovered. "We" didn't discover anything. I'm just using that shorthand.

Then, having realized that iterated maps were a shortcut through the order that we were interested in, we focused on them as the dynamical systems of interest for the next few lectures in the course because we could go farther with iterated maps than with differential equations. They allowed us to uncover still more kinds of incredible order in chaos.

First, there was order with respect to *parameters*. Remember what parameters are. That word is used commonly in colloquial speech, and I get the feeling most people have no idea what they're talking about when they speak of parameters. When we speak of parameters, what we mean are knobs. They're the sort of thing that you can turn to change the characteristics of a system. Think of the intrinsic growth rate of the insects when we were making a model of population growth, the logistic map. That was a knob that could be manually adjusted, the growth rate. We literally had a knob that was the brake on the waterwheel when we were looking at the waterwheel analogue of the Lorenz convection model.

So by changing parameters of these types, we could send the system into different long-term behavior. It could, under some conditions, be very steady in equilibrium. Sometimes it would oscillate periodically. Sometimes it would show the kind of wild oscillations that we associate with chaos. Those transitions, which we came to call *bifurcations*—splitting in the behavior of a system—those bifurcations and all the order that was contained in them were captured in two magnificent diagrams, icons of chaos. One of them showed the order in the logistic map as we changed the growth rate parameter, and we called that the *orbit diagram*. It was gorgeous. That was the scroll that we unrolled.

And then the order in chaos intermingled with periodicity was illustrated even more vividly in the Mandelbrot set, with its fantastic, infinite richness of structure all the way down to the infinitesimal level.

Those diagrams were like Rosetta stones for us in that they helped us decode chaos. By consulting them, we could see all the possible behavior a system could exhibit: chaos intermingled with rhythmicity, with periodicity, and then back to chaos. It was all there, stretched out for us in these beautiful charts, allowing us to see everything at once, in just one icon.

The diagrams were about bifurcations; where do the transitions occur? Where do the bifurcations between different types of behavior occur as we change the parameters to send the system from equilibrium, to oscillation, to chaos, and back again?

Both of those Rosetta stones themselves had incredible fractal structure. This is where fractals started to enter the course. We saw that within the orbit diagram, there were tiny copies of the orbit diagram at the end of each periodic window. They were unexpected bonuses just sitting there. Inside of each of those mini-diagrams, the whole structure repeats, and so there would be mini-mini-orbit diagrams in every window. The fractal structure was even more spectacular in the Mandelbrot set, as we saw when we zoomed in all the way down to a resolution 1000 times greater than what we started [with].

The fractals that we saw in parameter space in the Mandelbrot set and the orbit diagram were mimicked, you might say, by different kinds of fractals that we saw in state space, the space where we watch a system's behavior unfolding. We saw that strange attractors had this kind of mica-like structure, infinitely many sheets stuck together so closely we couldn't see that they were separate. But they were. We knew they had to be. So we saw fractals in state space as well.

Finally, one aspect of all this fractal order—the Feigenbaum numbers that we discussed when we were looking at the orbit diagram. These were the numbers that controlled the ratio of successive spacings in what we called the fig tree, showing where equilibrium split to period-2, repeating every 2 years in the case of a population biology problem, and then split again to 4-year cycles, and then 8-year, approaching an impenetrable wall that's an accumulation point. That fig tree that we looked at had a fractal branching structure to it controlled by these Feigenbaum numbers. This was, I consider, the greatest shock in the whole subject, that the Feigenbaum numbers turned out to be *universal*.

They were the same for our little iterated map—the logistic map that we could play with by punching a calculator—they were the same numbers for that seemingly numerological system, meaningless potentially. We found those same numbers in nature. We saw them in the chaos that was appearing in fluids undergoing convection. We found them in electronic circuits, in heart cells making their way from standard rhythmicity all the way into irregular beating associated with chaos. The same numbers appeared in lasers and on and on, in chemical reactions. It is unbelievable the universal predictions that Feigenbaum was able to make, and in case after case, they were right. This was the great triumph of chaos theory in the mid-1980s. It was a very, very heady time for the subject. I remember it well. There was champagne being uncorked everywhere.

Meanwhile, fractals took on a life of their own. They weren't just the handmaidens of chaos theory. They were a vigorous subject in their own right as models of self-similarity wherever it occurred in nature. Remember, self-similarity means that a small part of a structure looks like the original, like the whole. We could see self-similarity in certain shapes, as well as in processes, things that unfold in time. So not just in the abstract world of state space or parameter space for dynamical systems, but out there, really out there in nature, in the world: in the shape of broccoli, in coastlines, in mountains, in the ups and down of the stock market, in the patterns of earthquakes, in the fluctuations of human heart rate variability—fractals in all of them.

Then, toward the end of the course, we saw how the order embodied in fractals and chaos could be used to good advantage, could be harnessed, tamed, and used for purposes like encryption of secret messages or, potentially, in the future of quantum electronics, using the quantum chaos inherent in tiny semiconductor devices, quantum dots. They haven't made it to the market yet, but they will for radical new types of electronic circuits and devices. For space travel, we saw the use of chaos in getting to the Moon with practically no fuel, which will revolutionize space travel in the future, in the sense that we can now take freight to any planet through the solar system at practically no cost in fuel by harnessing chaos, by surfing the gravitational chaos of the solar system.

We also saw how the chaos inherent in the brain's normal functioning is leading researchers toward having a hope of being able to predict the breakdown of that chaos into epilepsy. So chaos is helping us finally come to a way of being able to predict when an epileptic seizure might occur, with enough time, enough warning that a patient can do something about it. So it's a thrilling story. I hope you've enjoyed it.

I want to leave you with a few thoughts here. Behind all of this incredible order, there was a hidden puppeteer pulling the strings. It has an ugly name but is the most beautiful idea in the whole subject. That puppeteer was *nonlinearity*. None of our story would have been possible with linear systems. Everything I just told you—poof! Gone! As soon as we insist on linearity—a simple proportionality between cause and effect, the idea that the whole is nothing more than the sum of the parts; they're exactly equal—if we lived in a linear world, all of that structure would be gone. In fact, all of the richness of the world would be gone. Life and everything beautiful around us depends on nonlinearity, where the whole can be more than the sum. The parts are interacting, cooperating, sometimes interfering. But the whole is not just the sum of the parts in all the things that matter to us, and to make sense of that scientifically, we have to come to grips with nonlinear dynamics, with nonlinear phenomena We, finally, in all that I've been saying, come to a sensible definition now. You're supposed to give a definition at the beginning, but of course, you can't do that. You can only give a definition at the end after you understand something.

So here is our definition of what the science of chaos is all about: It's about order and disorder in deterministic systems that happen to be nonlinear. That's what this subject is about.

It's not just an exercise of definitions that we've been going through. It may seem like that, but there's a much bigger point to this exercise, which is that it shows us why chaos theory is going to be crucial to the science of the 21st century. Why? Because virtually all the major unsolved problems in every branch of science today are fundamentally nonlinear.

Consider the orchestration of thousands of biochemical reactions in a single living cell and the disruption of that orchestration when the cell turns cancerous. Consider the battle between the immune system, with its thousands of different components—antibodies, T cells—and the AIDS virus. Consider the emergence of consciousness from the interplay of billions of neurons in the brain. Consider the origin of life from a meshwork of chemical reactions in the primordial soup. What makes these problems so vexing is their decentralized, dynamic character, in which enormous numbers of components keep changing their state from moment to moment, looping back on each other in ways that can't be studied in the usual reductionist way, by examining one part in isolation. That misses the essence of all of these problems, and it's why reductionism is not strong enough. It's not strong enough to help us tackle these mysteries.

In cases like I've just described, it seem obvious that the whole is surely not equal to the sum of the parts. These phenomena are intrinsically nonlinear.

Unfortunately, our minds are very bad at grasping nonlinear problems. We're just not wired that way for some reason. We're accustomed to thinking in terms of centralized control, simple top-down hierarchies, clear chains of command, straightforward logic of cause and effect and each subsequent cause: just chains, linear thinking in that sense, but in huge, interconnected systems, where every part affects every other part, our standard ways of thinking fall apart. Verbal arguments are certainly too feeble, too myopic. You can't get anywhere with just pure verbal reasoning. But the good news is that thanks to the development of chaos theory, we have now mastered the simplest nonlinear systems, the ones we have been studying in this course.

What awaits deep in the jungle is a similar mastery of complex nonlinear systems. That's the future. That is the great frontier. By complex, I mean something with an enormous number of parts. These parts are typically not identical. They're diverse, and they interact with one another in bewildering networks.

Let me give you a concrete example. It's just a tiny piece of a much bigger network, a vast network of genes, proteins, and enzyme reactions that control how every cell in your body grows and divides. Here's what it looks like. Biologists have spent decades mapping this out, and I'm just showing you a tiny corner of the map. Take a look. This is a picture of a fraction of what we know about the mammalian cell cycle, the cycle that controls cell division. You're not supposed to see much here. You're supposed to see a lot of lines and little circles, and you're supposed to be intimidated. If I've succeeded in doing that, I've succeeded. This looks very complicated, and what's being shown are different genes with complicated names, and then there are different colors for arrows showing what things bind to what, or which things inhibit each other, or which stimulate.

Here's an important one, very tiny, called the p53 BOX. Here's p53. Here's something interacting with it, Mdm2. That may not mean much to you, though if you are a biologist, or especially if you're an oncologist, it means very much to you because p53 refers to the most important tumor suppressor gene. It's known to be mutated—basically, broken—in about 50% of all cancers. The function of p53 is to act as a brake on cell division, to suppress the growth of tumors. When it's not working right, that's something that predisposes that cell to becoming cancerous. It's like driving without a brake.

Mathematical models of this unbelievably intricate cell cycle—actually, not of this one because we're not even close to understanding this one yet—a much simpler cell cycle of one of the most fundamental eukaryotic organisms, yeast. You might not think we have much to do with yeast, but we're in the same family. We're eukaryotes. We have nuclei like they do. We're not bacteria. We're not talking about them. The yeast are sort of like us. They're in the eukaryotic family, and they can teach us a lot about the cell cycle. Mathematical models of the yeast cell cycle are being developed right now by pioneers like John Tyson and his colleagues at Virginia Tech. Their models are based on the methods of chaos theory. Tyson was trained in chemistry as a chemical physicist. He's very comfortable with math and computers but also good in the lab and totally tuned in to biology. He's a modern scientist who can do it all and is not afraid of advanced math, like the type we've been talking about. In his work, he combines biology with bifurcations. He's equally at home in both worlds—chaos theory and cell biology.

The models that are coming out of Tyson's group are still avant-garde from the point of view of traditional cell biology. But I'm certain that they're the forerunners of the kinds of models that will prove indispensable in solving the riddle of cancer. It has to be like that. After all, genes are nonlinear devices; stimulate them twice as strongly and they don't produce twice the output. It doesn't work like that in biology. It's much more complicated.

You're probably familiar with that in the case of neurons, the famous all-or-none response. You can stimulate a neuron electrically and it may not fire at all until you cross a threshold, and then it fires. It doesn't fire in proportion to the stimulus. So neurons are nonlinear devices, too, which means that whether we're talking about enormous collections of neurons, the brain; enormous collections of genes, in the case of the control of the cell cycle; and all of biology for that matter—ultimately, all of these things will require us to master complex nonlinear systems.

It's a daunting task, no question. But there are grounds for optimism. I think it's a sign of the times that mainstream science is now coming onboard with what I'm saying. This used to be fringe, wacko stuff I'm telling you, and may be a little bit still, but it's not regarded that way by the leading mainstream scientists.

Here is why I can say that. Here's the proof: Even the most hard-boiled reductionists now acknowledge that reductionism won't solve all the great mysteries we're facing. For example, if you look at every major research university, you can follow the money. See where the money is being spent, and you'll find that new institutes are springing up with names like Institute for Functional Genomics and Institute for Computational Neuroscience. Harvard Medical School has created a new program in systems biology, their first new department in years. Harvard University, not to be outdone by its medical school, has its own center for systems biology. What is systems biology? It's where chaos theorists and computer scientists are teaming up with molecular biologists to try to make sense of the dance of life at the molecular level.

Chaos theory is playing a vital role in the new biology. Here's why: You've heard about the sequencing of the human genome. It's a great story. But what did it do? It gave us an enormous list of parts: 23,000 individual genes and the proteins that they encode. But we still have almost no clue about how the interlocking activities of all those genes and proteins are choreographed in the living cell and how they go awry when the cell turns cancerous. These are problems of dynamics, our subject, how things unfold in time.

That's not traditionally the way we thought of the genome. You've probably heard it called the "blueprint of life." That's a very static metaphor. That's the old-fashioned metaphor of an assembly line or if you're building a house or something. It's true, to some extent, that the genome is a blueprint. It does give instructions for building proteins, which are the molecular machines that carry out all the functions of life. But that's not all it is. It's not just a blueprint. That is much too linear. It's a vestige of an earlier mentality. Here's why: Think about what some of those proteins do. Some of them act back on other genes. Regulatory genes produce proteins that then act on other genes, turning them on or off, forming feedback loops.

Once you have a picture like that in your head, of genes making proteins that affect the activity of other genes, now it's starting to seem less like a blueprint and more like a computer. It's more like a very complicated circuit. The functioning of this computer—and its malfunctioning when cells turn cancerous—that will not be deciphered until we understand the dynamics of gene networks of the type I just showed you.

Our era awaits its own Kepler, its own Galileo and Newton, especially Newton. We need a Newton. They found the laws of inanimate dynamics, and we are still looking for the laws of life.

Fortunately, biologists at the cutting edge are now being trained along the lines I've just described, in chaos theory, along with their work at the bench in the wet lab. As part of their graduate education, systems biologists are learning nonlinear dynamics and chaos theory to help them reverse-engineer the basic networks of life.

There's a field you may not even have heard of yet called *synthetic biology*. The name is pretty descriptive. We're going to create biology ourselves. The people being trained in synthetic biology are using chaos theory to design their own biological circuits, building living versions of things that electrical engineers have built, like toggle switches, oscillators, and amplifiers, except they're not being built out of transistors. They're being built out of genes and proteins. It's incredible. It's really a very, very exciting development.

Likewise, chaos theory is helping to shape other newfangled branches of science, from computational neuroscience, where computer people collaborate with neuroscientists on the problems of the brain and nervous system, to econophysics, where physicists, with their mastery of these systems with an enormous number of components, are starting to look at the economy itself—the world economy, the financial markets—and using the data analysis and analytical techniques of physics in what may be one of the most perplexing systems of all, the world of money.

Chaos theory is also shedding light on perhaps the greatest question of all: consciousness. It's spooky to think about consciousness. I don't know if you've ever felt this, but have you ever looked in the mirror and had a creepy feeling suddenly, like "Who is in there?" as you look at yourself? Who is this pile of atoms that's looking at me and thinking about itself?

How can 100 billion neurons, each mindless on its own, fall in love, or remember a favorite song, or fret about its own mortality? Neuroscientists are starting to uncover clues about how this might work. They found that when we recognize a face or decide to pay attention to someone who's talking to us, far-flung parts of the cortex suddenly fire a volley of electrical signals at a specific frequency, about 40 cycles a second, in sync. It's not just electrical chatter any more, the idle state of the brain. It's a chorus. Parts of the brain start singing in unison, and this chorus arises spontaneously out of chaos. It's a self-organizing thing when we have consciousness.

These studies paint a fascinating but somewhat unsettling [or even] truly disconcerting picture of human existence. It's starting to look this way: As we go about our lives, all that we feel and perceive and think may be just a reflection of fleeting processes of synchronization among neurons in our brain.

In every branch of science today, the tools, ideas, and interdisciplinary spirit of chaos theory are proving indispensable as scientists confront the greatest challenges of our times.

Let me end by saying how much fun it has been to teach this course. Thanks to The Teaching Company for giving me the opportunity. It has been wonderful, and I really appreciate it. Thanks to my family for their patience. And I hope you've enjoyed this course, too. Although, as we've learned, trajectories are not allowed to cross in a deterministic system, I hope our trajectories will cross again soon. Until then, be well and stay chaotic!

Timeline

B.C.E.

c. 1000 .... Ancient peoples worldwide develop
creation myths in which the universe
arises from "chaos," a primeval state
of emptiness or utter disorder.

c. 900–500 .... Chaos figures prominently in the
majestic opening chapter of Genesis:
"In the beginning God created the
heaven and the earth. And the earth
was without form, and void ..."

c. 600–500 .... Lao Tzu, the founder of Taoism,
writes, "There was something
chaotic yet complete" before the
universe was born, "I do not know its
proper name but will call it Tao."

585 .... Thales correctly predicts a solar
eclipse. Generally regarded as
the founder of Greek science, he
teaches that the world is orderly and
ruled by comprehensible laws.

c. 520 .... Pythagoras discovers the laws of musical
harmony and proves the Pythagorean
theorem. He leads a mystical cult
that holds that "all is number."

c. 300 .... Euclid writes the *Elements*, perhaps
the greatest mathematical textbook

of all time, and establishes a style
of logical reasoning that dominates
Western science for 2000 years.

C.E.

1621...Johannes Kepler publishes his
three laws of planetary motion,
thus solving an ancient mystery.

1638...Galileo publishes *Discourses and
Mathematical Demonstrations
Concerning the Two New Sciences*,
which summarizes his discoveries
about the laws of motion on Earth.

1666...Isaac Newton invents calculus
and differential equations and
discovers the law of universal
gravitation. He is 24 years old.

1686 ....Newton finishes his masterpiece *Principia
Mathematica*, which synthesizes and
explains the laws found by Kepler and
Galileo. This book launches modern
science and gives rise to the concept
of the universe as clockwork.

c. 1700–1900...................................Elaboration of Newton's clockwork
universe. High points include
the prediction and discovery of
Neptune (1846) and the discovery
of the laws of electricity and
magnetism, thermodynamics, solid
mechanics, and fluid mechanics.

1889 .... While investigating the three-body problem of astronomy, Henri Poincaré discovers "chaos" (in the modern sense). Specifically, he finds that a deterministic system obeying Newton's laws can be unpredictable in the long run because of its extreme sensitivity to tiny changes in its initial conditions. The clockwork begins to creak, but few people notice.

1896 .... Economist Vilfredo Pareto finds that the number of people whose personal incomes exceed a large value obeys a power-law distribution. Power laws later prove important in analyzing chaos and fractals.

1904 .... Mathematician Helge von Koch constructs an example of a curve that is continuous but has no tangent anywhere. Once viewed as monstrosities, such examples later seem natural when fractal geometry is developed.

1900–1960 .... Two gigantic upheavals in physics—relativity and quantum mechanics—overshadow Poincaré's discovery of chaos. A few mathematicians keep the flame of Poincaré alive. They refine his geometric approach to differential equations, leading them to insights about chaos in iterated maps and classical mechanics.

1920–1960 .... Applied mathematicians and electrical engineers use Poincaré's methods to analyze the nonlinear oscillators at the heart of radio, radar, transistors, and lasers.

1932... Veterinary scientist Max Kleiber reports his 3/4-power scaling law of metabolism, later explained as a consequence of fractal branching networks in living things.

1943–1952...................................... Artist Jackson Pollock invents a new style of painting, deliberately channeling chaos and mirroring the fractals of nature.

c. 1950.. Invention of the modern computer.

1954... Geologists Beno Gutenberg and Charles Richter report a power law relating the frequency of earthquakes to the amount of destructive energy they release.

1960... The meteorologist Edward Lorenz creates an idealized computer model of weather and discovers its sensitivity to initial conditions, now recognized as a signature of chaos. This suggests an explanation for why real weather is so hard to predict.

1960... The mathematician Stephen Smale constructs the horseshoe map. It reveals stretching and folding to be the basic geometric mechanism behind chaos. Smale's rigorous math builds on earlier studies of chaotic differential equations by the mathematicians Mary

Cartwright, J. E. Littlewood, and Norman Levinson. Yet like their earlier work, Smale's contributions make no dent on the wider scientific community.

1963..Lorenz uncovers order in chaos. He publishes the first image of a strange attractor, infers its infinitely layered structure, and extracts an iterated map that underlies the chaos.

1965..Arthur Winfree, then a senior in engineering physics at Cornell, uses computer simulations to investigate synchronization in large systems of biological oscillators. He discovers that the onset of synchronization is analogous to a phase transition.

1971..Mathematicians David Ruelle and Floris Takens publish an important article that gets scientists thinking about the longstanding mystery of turbulence in new ways. They also coin two buzzwords that soon catch on, when they suggest that "strange attractors" may underlie the "very complicated, irregular and chaotic" behavior that turbulence represents.

1972..The "butterfly effect" enters the lexicon soon after Lorenz gives a lecture titled "Predictability: Does the Flap of a Butterfly's Wings in Brazil Set Off a Tornado in Texas?"

1975...Physicists Harry Swinney and Jerry
Gollub test the predictions of Ruelle
and Takens. They report experiments
on turbulence in rotating fluids that
contradict the conventional theory but
seem consistent with the new approach.

1975...Mathematicians T. Y. Li and
James Yorke publish "Period
Three Implies Chaos." "Chaos"
now becomes the nickname of an
entire intellectual movement.

1975...Benoit Mandelbrot publishes a
book in French, *Les objets fractals:
Forme, hazard et dimension*,
in which he introduces fractal
geometry, aimed at quantifying all
that is fragmented and irregular.

1976...Theoretical biologist Robert May
publishes a review article in *Nature* in
which he makes an "evangelical plea"
for educators to start teaching students
about "the wild things that simple
nonlinear equations can do." This article
introduces chaos and the logistic map
to a broad community of scientists.

1976...Astronomer Michel Hénon reveals, in
dramatically clear computer graphics,
the infinitely layered, self-similar
microstructure of a strange attractor.
In so doing, he helps others to grasp
what Lorenz had meant back in 1963,
when he described the attractor as
an "infinite complex of surfaces."

1978.. After several rejections, physicist
Mitchell Feigenbaum manages to
publish his ideas about universality in
the period-doubling route to chaos.

1980–1982...................................... Experiments on fluids, electronic
circuits, lasers, chemical
reactions, and heart cells confirm
Feigenbaum's predictions.

1984.. Physicist Eric Heller predicts
"scars" in the wave functions of
certain quantum systems whose
classical counterparts are chaotic.

1985.. The Mandelbrot set appears on
the cover of *Scientific American*.
The lay public becomes enchanted
by fractals, which soon appear on
T-shirts, posters, and screensavers.

1986.. Physicist Michael Berry links
quantum chaos to the Riemann zeta
function, which controls the enigmatic
distribution of prime numbers.

1986.. Sir James Lighthill publishes a
remarkable collective apology on
behalf of all scientists for misleading
the public about the unpredictability
lurking in Newton's laws.

1987.. James Gleick's book *Chaos*
becomes a bestseller.

1990.. Chaos theory stars in Michael
Crichton's novel *Jurassic Park*.

Timeline

1990.. A new trend begins: Having understood chaos, scientists try to use it for practical purposes.

1990.. NASA scientists Edward Belbruno and Jim Miller devise a way to get an unmanned Japanese space probe to the Moon using hardly any fuel, by exploiting chaotic trajectories through the solar system.

1990.. Ed Ott, Celso Grebogi and Jim Yorke publish an influential paper, "Controlling Chaos."

1990.. Physicists Louis Pecora and Tom Carroll show that two identical chaotic systems can be synchronized, opening the door to practical applications in communications.

1992.. Using chaos-control techniques, physiologist Alan Garfinkel and collaborators show that a form of cardiac arrhythmia in lab animals can be stopped by gentle, precisely timed electrical stimuli.

1993.. Electrical engineers Kevin Cuomo and Alan Oppenheim experimentally demonstrate the use of chaos for enhancing the privacy of electronic communications.

1993.. Steven Spielberg brings chaos theory to the masses with his blockbuster movie version of *Jurassic Park*.

1993 ....Tom Stoppard brings chaos theory to the theatre-going elite with his play *Arcadia*.

1994...Mathematician Kevin Short decodes communications masked by chaos, raising doubts about the security of this encryption method.

1996...Bioengineer Leon Iasemidis and neurologist Chris Sackellares propose methods based on chaos theory to predict epileptic seizures.

1997...Fractal theory proposed by Geoffrey West, Jim Brown, and Brian Enquist explains Kleiber's law of metabolism (1932) and scaling laws for other biological processes.

1998...Applied mathematicians Duncan Watts and Steven Strogatz report that the small-world phenomenon is a generic feature of networks in nature, society, and technology. Networks become the next big topic in chaos and complexity research.

1998 ....Laser physicists Gregory VanWiggeren and Rajarshi Roy experimentally demonstrate the use of synchronized chaotic lasers for encryption of high-speed data transmission in optical fibers.

1999...Fractal analysis of Jackson Pollock's drip paintings reported by physicist Richard Taylor and colleagues.

1999.. Power laws proposed as another
organizing principle for complex
networks, by physicists Albert-
Laszlo Barabasi and Reka Albert.

2000.. Human genome sequenced,
in "first draft" form.

2000 ....London's Millennium Bridge wobbles
unexpectedly on opening day when
pedestrians spontaneously synchronize
their footfalls, exemplifying the surprises
lurking in complex nonlinear systems.

2001.. The NASA Genesis mission uses
the interplanetary superhighway, a
freeway through the solar system
uncovered by chaos theory.

2005.. First successful implementation
of chaotic encryption outside the
lab, demonstrated in the fiber-optic
network of Athens, Greece.

Glossary

accumulation point: The limit of a converging sequence of numbers. For example, the sequence 1, 1/2, 1/4, 1/8, 1/16, … has an accumulation point at 0, because the sequence approaches (or "accumulates" at) 0 as a limit. In the context of the period-doubling route to chaos, the accumulation point refers to the parameter value where the fig tree becomes infinitely bifurcated and chaos ensues. (For the logistic map, this occurs when the parameter is equal to 3.5699... .) Hence the accumulation point also marks the onset of chaos.

aperiodic: Changing without ever repeating or settling down to equilibrium.

attractor: A state, or collection of states, that represents a system's natural, long-term behavior. If a small disturbance nudges the system away from these states, the system quickly relaxes back onto them, as if "attracted" by them. There are three main kinds of attractor: (1) an attracting fixed point, which represents a stable equilibrium state; (2) an attracting cycle, which represents a stable periodic behavior; and (3) a strange attractor, which represents a permanently self-sustaining form of chaos.

bifurcation: A sudden, qualitative change in a system's long-term behavior caused by changing one of its parameters.

butterfly effect: The phenomenon in which a small change in a chaotic system's initial state (metaphorically caused by the flap of a butterfly's wings) leads to a huge change in what follows, compared to what would have happened otherwise.

chaos: (1) Colloquially: utter confusion. (2) Theologically: the primordial state before creation. (3) Scientifically: seemingly random, unpredictable behavior in a system that is nevertheless governed by deterministic laws. Chaos in this last sense is a subtle mix of order and randomness; it is predictable in the short run (because of determinism) but unpredictable in the long run (because of sensitivity to initial conditions).

chaos theory: The interdisciplinary study of nonlinear dynamical systems and their applications. Broader than the study of chaos per se, in the sense that it includes such non-chaotic phenomena as oscillations, waves, and patterns. Broader also in that it is not confined to mathematics; chaos theory encompasses all the scientific and technological areas in which nonlinear phenomena arise.

clockwork universe: A worldview in which everything in the universe unfolds according to Newton's laws, with no room for chance or free will.

complex system: A nonlinear dynamical system composed of an enormous number of interacting parts.

convection: Rotating flow of a fluid, such as a large mass of air or water, driven by differences in its temperature from place to place.

cycle: Anything that repeats. Represented by closed orbits in state space (for a system governed by a differential equation) or by a discrete set of points (for a system governed by an iterated map).

deterministic system: A system in which the present uniquely determines the future. Later states evolve from earlier ones according to a fixed law; only one thing can happen next.

differential equation: An equation that dictates how the state of a system changes in the next instant, given its current state. More generally, any equation that relates a system's variables to its derivatives (its rates of change with respect to time or space). Traditionally, one tried to solve such equations by using formula-based, analytical methods that rely on calculus. This approach requires great ingenuity and works best when the equation is linear or can be recast as linear. But for nonlinear differential equations, formulas are typically impossible to obtain, so one needs to resort to qualitative methods (as pioneered by Poincaré) or computational methods (as illustrated by the work of Lorenz).

dimension (of a fractal): Any of a set of related numbers that quantifies the degree of irregularity, jaggedness, or roughness of a fractal and gives information about its scaling and self-similarity properties. Reduces to the commonsense notion of dimension for Euclidean shapes: A line or curve is one-dimensional, a plane or surface is 2-dimensional, etc. Fractals have higher dimensions than their Euclidean counterparts, e.g., a fractal curve has dimension somewhere between 1 and 2; higher dimension means more convolution.

dimension (of a state space): The number of axes, one for each state variable. For example, the state of a pendulum is defined by two variables: its current angle and velocity. Hence its state space is 2-dimensional.

double pendulum: A toy used to demonstrate chaos. A lower pendulum hangs from an upper one. If given enough initial energy, the system behaves crazily, with the lower pendulum sometimes swinging and sometimes whirling, all unpredictably.

dynamical system: Any system that changes over time according to a fixed rule. There are two main types: differential equations (in which time is regarded as flowing continuously) and iterated maps (in which time is regarded as advancing in discrete steps).

equilibrium: An unchanging state. Note that nothing is implied about stability. An equilibrium could either be stable (a ball resting at the bottom of a bowl) or unstable (a pencil balancing on its point).

Euclidean: Relating to Euclid's classical geometry of lines, circles, planes, cones, etc. Smooth, as opposed to fractal.

exponent: The power p in an expression like x^p .

exponential (growth or decay): A rapid form of growth or decay, in which a variable changes by a constant *factor* over equal intervals of time. Much faster than linear growth or decay, in which a variable changes by a constant *amount* over equal intervals of time.

fig tree: The tree-like portion of the orbit diagram for any one-humped map (such as the logistic map or the sine map) that displays an infinite sequence of period-doubling bifurcations as a parameter is varied. Whimsically named in honor of Mitchell Feigenbaum (*Feigenbaum* = "fig tree" in German), who elucidated its universal properties. He showed that each wishbone in the tree has a width (i.e., distance along the parameter direction) that is always about 4.7 times narrower than the wishbone to the left of it and a height (i.e., distance along the state of the system) that is always about 2.5 times shorter. Close to the onset of chaos, these shrinkage factors universally converge to very specific, very peculiar numbers: 4.6692... and 2.5029... . These are fundamental constants of chaos, as basic to period doubling as pi is to circles.

fixed point: The state-space representation of equilibrium; an unchanging state.

fractal: An object, shape, or temporal process in which small parts resemble the whole. The resemblance may be exact in certain mathematical cases, but in most real-world cases it is only approximate or statistical. Furthermore, it may hold only over some limited range of length or time scales.

fuzzy boundary: A complicated region between the Earth and the Moon (or between two other celestial bodies) in which a spacecraft would be chaotically balanced, like a surfer riding the crest of a wave, pulled by the competing gravity of both bodies and pushed by centrifugal forces created by its own orbit.

growth rate parameter: The parameter r in the logistic map that controls how fast the population grows and hence regulates its dynamics. Sometimes called the **steepness parameter**.

heavy tail: The tail of a power-law probability distribution. "Heavy" in the sense that extremely large events occur much more often than they would in a normal distribution (standard bell curve). Mathematically, heavy tails decay slowly (in inverse proportion to the size of the event, raised to some positive power), rather than exponentially fast.

Hénon map: A pedagogical example, devised by the astronomer Michel Hénon, for exploring chaos and the fractal structure of strange attractors. Defined mathematically as a map involving two variables x_n, y_n, updated according to the rule $x_{n+1} = y_n + 1 - ax_n^2$, $y_{n+1} = bx_n$. Following Hénon, the parameters are usually taken to be $a = 1.4$, $b = 0.3$, because this produces a strange attractor whose internal structure is plainly visible.

horizon of predictability: Roughly speaking, the time up to which a chaotic system's behavior can be successfully predicted despite some uncertainty in its initial state. For times much longer than this, accurate prediction becomes impossible. Mathematically, the horizon of predictability is defined as the reciprocal of the system's Lyapunov exponent.

hump-shaped: Resembling an arch or upside-down parabola.

infinite complex of surfaces: Lorenz's description of the multi-sheeted structure of the strange attractor he discovered.

initial conditions: The state of a system at the beginning of an experiment, observation, or any other stretch of time that may be of interest to an investigator.

iterated map: A mathematical operation in which a number is fed into a function (a "map" or "mapping") to generate a new number. This new number is then fed back into the map to generate another new number, and the process is repeated, or "iterated," indefinitely.

Kepler's laws: The three laws of planetary motion discovered by the astronomer Johannes Kepler: (1) The planets move in elliptical orbits around the sun, with the sun at one focus of the ellipse. (2) An imaginary line joining the planet to the sun sweeps out equal areas in equal times. (3) The ratio T^2 / a^3 is the same for all planets, where T is the time required for one orbit, and a is the average of the planet's longest and shortest distances from the Sun.

Kleiber's law: An empirical regularity linking the average basal metabolic rates of all mammals—namely, a mammal of mass M (in kilograms) needs to eat about $70M^{3/4}$ calories per day. First reported by the veterinary scientist Max Kleiber in 1932, in a study spanning rats to steers. The law is for basal metabolism, assuming no special exertion (you can change your law by running or swimming vigorously each day).

Koch curve: A mathematical fractal formed by starting with a line segment, then repeatedly replacing its middle third with the other two legs of an equilateral triangle. Constructed by Swedish mathematician Helge von Koch in 1904.

linear system: A system in which alterations in the initial state will result in proportional alterations in any subsequent state. More loosely, a system in which causes are proportional to effects, and the whole is equal to the sum of the parts.

logistic map: A canonical model of the transition from order to chaos, originally inspired by studies of population growth. Mathematically defined as the iterated map $x_{n+1} = rx_n(1-x_n)$. Here x_n is a real number between 0 and 1, representing the state of the system at discrete times $n = 0,1,2, \ldots$, and r is a parameter, usually taken to lie between 0 and 4. In its original biological context, x_n is proportional to the population size of generation n, and r is the per capita growth rate.

long tail: See **heavy tail**.

long-term behavior: The behavior of a dynamical system after it has settled onto its attractor and "forgotten" the artifacts introduced by the particular way it was started.

Lorenz attractor: The butterfly-shaped strange attractor of the Lorenz system.

Lorenz map: An iterated map defined by the consecutive maximum values of the variable z in the Lorenz system. The map takes one maximum to the next.

Lorenz system: A system of three nonlinear differential equations studied by meteorologist Edward Lorenz, in connection with a simplified model of fluid convection. Can be visualized intuitively using its mechanical analogue, a chaotic waterwheel that erratically changes its direction of rotation. Famous for being the first system in which strange attractors were discovered. (For those with a math background: The Lorenz equations are given by the differential equations $x' = \sigma(y-x)$, $y' = rx - y - xz$, and $z' = xy - bz$, where prime denotes time derivative. Here $\sigma, r, b > 0$ are parameters, usually taken as $\sigma = 10$, $r = 28$, and $b = 8/3$, as in Lorenz's 1963 paper.)

Lyapunov exponent: The rate at which nearby trajectories diverge in a chaotic system. Defined mathematically as λ, the rate of separation, in the formula $d(t) \approx d_0 \exp(\lambda t)$, where $d(t)$ is the exponentially growing distance between the trajectories, and t is the time for which they've been separating.

Mandelbrot set: An amazingly complex and beautiful fractal set associated with a certain family of 2-dimensional maps of the plane to itself. (For those with a math background: The Mandelbrot set is defined as follows. Start with a complex number c, and generate the sequence c, $c^2 + c$, $(c^2 + c)^2 + c$, and so on; the next number is always the previous number squared plus the original c. Iterate indefinitely. The question is: Does this sequence head off to infinity or not? If any member of the sequence ever lies outside a certain square [namely, the square region of complex numbers having both real and imaginary parts between 2 and -2], the sequence is provably heading off to infinity. In this case, the point c is considered outside the Mandelbrot set, and c is color-coded by how many iterates it requires to escape from the square. Otherwise, the sequence remains in the square forever, and the point c is colored black, indicating it is in the Mandelbrot set.)

map: A function; a mapping; a transformation (all these terms are synonymous). Roughly speaking, a map eats states and spits them out, transformed. Think of a map as a machine that takes a state as input and "maps" it to a new state as output, using a given mathematical rule.

nonlinear: A system that is not linear; alterations in the initial state need not produce proportional alterations in subsequent states. More loosely, a system where the whole can be more or less than the sum of the parts, and causes can generate surprisingly disproportionate effects. All chaotic systems are nonlinear, but nonlinear systems comprise a much larger universe than just the chaotic ones. In that sense, "chaotic" is a subset of "nonlinear."

non-periodic: See **aperiodic**.

onset of chaos: Value of a parameter at which periodic or otherwise regular behavior ceases and chaos begins.

orbit: A geometric representation of the successive states of a dynamical system as it changes from moment to moment. Appears as a smooth curve in state space (for a differential equation), or a discrete set of points, each one hopping to the next (for an iterated map). Called an orbit by analogy with the orbit of a planet as it glides through outer space; the difference is that here the gliding occurs in an abstract mathematical state space. Synonymous with **trajectory**.

orbit diagram: A diagram showing how the attractor of a system changes as a parameter is varied. Called an orbit diagram because it depicts the long-term orbit (the attractor) of a system as a function of one of its parameters.

oscillator: Any dynamical system that cycles repetitively through a set of states.

parameter: A feature of a system that stays constant as time progresses. The natural evolution of the system does not change any of its parameters; only its variables will change. (For instance, when a pendulum swings back and forth, its angle changes, but its mass and length do not. Thus, by this definition, the mass and length of a pendulum are parameters, whereas the angle is a variable.) However, it is often possible for an experimenter to intervene and manually change a parameter. The most useful parameters are those that an experimenter can adjust, as if by turning a knob—especially if those adjustments cause the system to behave in new, interesting ways.

period: The time required for one complete cycle of an oscillation.

period doubling: A bifurcation seen in certain systems, in which a periodic attractor abruptly changes to another form and now takes twice as long to complete a cycle.

period-doubling route to chaos: An infinite sequence of periodic-doubling bifurcations as a parameter is varied, culminating in chaos. Seen in the logistic map and in many real systems investigated experimentally. The documentation of its predicted universal features provided one of the great triumphs of chaos theory.

periodic: Repeating exactly; cyclic.

periodic window: A continuous set of parameter values within which a system shows stable periodic behavior. Typically bracketed between parameter values for which the behavior is chaotic.

phase transition: A bifurcation in the state of a complex system, in which the overall character of the system changes radically, altering enormous numbers of components at once. A classical example is trillions of water molecules freezing into a solid block of ice when the temperature falls below the freezing point. Spontaneous synchronization of flashing fireflies or other oscillators is an analogous phenomenon, except it represents an alignment of rhythms in time, not molecules in space.

power law: An algebraic relationship in which one variable is proportional (or inversely proportional) to another variable raised to some power. For example, Newton's inverse square law of gravity is a power law because it relates the gravitational force between two bodies to the inverse second power of the distance between them. Kleiber's law relates basal metabolism to the 3/4 power of a mammal's body mass. Power laws are often associated with fractals, but not always.

probability distribution: A mathematical function (often graphed as a bell curve or the like) that shows the relative likelihood of a random variable taking on a certain value. For example, human heights lie on a bell-shaped probability distribution centered somewhere between 5 and 6 feet tall. Probability distributions are also useful for showing the likelihood of possible states for a chaotic system because some features of chaotic systems resemble those of truly random ones.

quantum chaos: A misnomer; not chaos in quantum systems. Rather, the telltale sign in a quantum system that chaos lurks in its classical Newtonian counterpart. Sometimes called "quantum signatures of classical chaos" for this reason.

random system: A system in which the progression from one state to another is not uniquely determined by any law; a system that is not deterministic. Loosely speaking, anything that can happen, can happen *next*.

reductionism: A philosophy of science that holds that all complex phenomena can (and should) be understood by "reducing" or breaking them into simpler pieces.

route to chaos: A series of bifurcations, starting from equilibrium and periodic states and ending in chaos, as a parameter is varied.

scale: The typical size of the main features of an object, shape, or process. Somewhat subjective; depends on which features are of interest in a given setting. For example, the scale of vortices in the atmosphere is hundreds of miles if we're interested in hurricanes, or a few feet if we're interested in piles of swirling leaves.

scale-free: A shape or process that displays similar behavior over such a wide range of scales that it is effectively independent or "free" of any one scale. Often used in reference to fractals or other systems showing power-law behavior.

scaling law: An algebraic relationship, often a power law, that describes how the measured value of some property depends on the resolution or scale (the length of the "yardstick") used to make the measurement.

scar: A region of high probability in a quantum system. Caused by probability waves adding up constructively along a closed orbit of the classical version of the system. The name is intended to convey the idea of a remnant: Just as a real scar is the remnant of a wound, a quantum scar is the remnant of a classical closed orbit, still visible even after descent into the quantum regime.

self-similar: Composed of parts that, when magnified, resemble the whole. In idealized mathematical fractals, the resemblance is exact and extends down to arbitrarily small parts; in real fractals, the resemblance is approximate and holds over a limited range.

sensitivity to initial conditions: The defining property of chaos; initially small disturbances to the state of a system grow exponentially fast as time progresses.

stable: A fixed point or cycle is stable if trajectories that start close to it remain close to it forever. In physical terms, a state is stable if small disturbances to it stay small. In many cases, these small disturbances eventually die out. Not synonymous with equilibrium; an equilibrium may be stable or unstable.

state: The condition of a system at one instant; the totality of information that (along with the governing equation of motion) is needed to predict what the system will do next.

state space: A geometric representation of all possible states of a system, it has as many dimensions as the number of variables needed to specify the system's state. The state space of a pendulum has two axes, one for its angle and one for its velocity. The state space for the most general three-body problem of astronomy has 18 dimensions, since we must specify the x, y, and z coordinates of each particle as well as its velocities along those axes. Thus each particle requires 6 numbers, and there are 3 particles, hence a state space of $6 \times 3 = 18$ dimensions.

steepness parameter: The parameter in the logistic map that controls how steep it is and hence regulates its dynamics. Also called the **growth rate parameter**.

strange attractor: An attractor with a fractal structure. Called "strange" because of its fractal geometry; all attractors studied previously were points, circles, or other smooth shapes in state space.

synchronization: A coordinated form of motion in which parts of a system move in unison or at the same frequency.

system: Any entity that can change over time.

tail: The far end of a probability distribution, corresponding to outliers or extreme events (large earthquakes, enormous incomes, etc.).

thermometer of complexity: A metaphor to illustrate what is known and unknown about dynamical systems.

three-body problem: The problem of determining the motion of three bodies pulling on each other by gravity. Now known to be unsolvable, except by brute-force computer simulation. Analysis of it led Poincaré to the discovery of chaos.

trajectory: The sequence of states followed by a dynamical system as time progresses. Synonymous with **orbit**.

transient behavior: Short-term behavior seen in a dynamical system before it reaches its attractor.

turbulence: Irregular motion of a fluid, such as air or water, in which the velocity at a given point varies erratically in magnitude and direction as time progresses.

two-body problem: The problem of determining the motion of two bodies pulling on each other by gravity. Solved by Newton, and used by him to explain Kepler's laws of planetary motion.

universal: Independent of the specific details of a mathematical or physical system.

unstable: Not stable. Small deviations from an unstable state grow over time.

U-sequence: The sequence in which stable cycles occur for the logistic map, or any other hump-shaped iterated map, as its steepness parameter is increased. Called "U" for "universal"; the same sequence occurs for any map with the same overall shape, independent of its precise algebraic form. The U-sequence, up to period 6, is 1, 2, 4, 6, 5, 3, 6, 5, 6, 4, 6, 5, 6. The U-sequence up to period 5 means the 6s drop out, leaving 1, 2, 4, 5, 3, 5, 4, 5. Notice that any definition of the U-sequence also requires (as part of the definition) an arbitrary upper limit on the cycle periods being considered. Otherwise, the period-doubling sequence at the beginning causes trouble; accounting for it alone requires listing infinitely many numbers 1, 2, 4, 8, 16,

32, ... , and you would never get to list the other, subsequent periods like 6, 5, 3, etc., that occur farther to the right in the orbit diagram. Also, once you allow any higher period as the upper limit, that new period will insert itself in places throughout the sequence in various complicated ways.

variable: A feature of a system that changes as time progresses.

vector field (on state space): A rule that assigns a vector to each point in state space, indicating the direction and speed with which one state changes to the next; the geometric counterpart of a differential equation.

waterwheel, chaotic: A mechanical analogue of the Lorenz system, used in lab demonstrations of chaos. Devised by applied mathematicians Willem Malkus and Lou Howard, then at MIT. Shows irregular reversals, turning clockwise and then counterclockwise after an unpredictable number of rotations each time.

window, periodic: See **periodic window**.

Biographical Notes

Belbruno, Edward (b. 1951). American mathematician with expertise in celestial mechanics and dynamical systems. An artistic soul and a bit of a maverick. While working for NASA's Jet Propulsion Laboratory, he used chaos theory to devise a new strategy for getting a spacecraft to the Moon at extremely low cost in fuel. Rather than fighting the gravitational tugs of the Earth and Moon, the spacecraft goes with the flow. It glides along chaotic trajectories created by the delicately balanced pulls of the Earth, Moon, Sun, and centrifugal forces. The savings in fuel are tremendous, but the journey is slow and so is suitable only for unmanned spacecraft. This method was used to rescue the Japanese space probe *Hiten* in 1990.

Feigenbaum, Mitchell (b. 1944). American theoretical physicist. Looks the part of a genius, with the brow and swept-back hair of a Romantic composer. While working at Los Alamos in the mid-1970s, he used a pocket calculator to uncover a stunning kind of universality in chaos. Roughly speaking, he found that diverse systems go chaotic in the same way. In particular, systems that follow a period-doubling route to chaos obey certain universal scaling laws, and always with the same scaling exponents. He explained how such universality arises by borrowing a Nobel Prize–winning technique (the "renormalization group") from statistical physics. His predictions were confirmed in controlled experiments on chick heart cells, electronic circuits, chemical reactions, and convection in liquid mercury.

Heller, Eric (b. 1946). American computational physicist. Predicted "scars" (regions where quantum waves add up coherently) as a signature of quantum chaos. These are produced by unstable periodic orbits in the chaotic classical counterpart of the quantum system. Heller has generated gorgeous computer graphics of quantum and chaotic systems, leading him to a second life as a quantum artist.

Hénon, Michel (b. 1931). French mathematician and dynamical astronomer. Invented a 2-dimensional chaotic mapping, now known as the Hénon map, that rendered the infinitely layered microstructure of strange attractors plainly visible for the first time.

Lorenz, Edward (1917–2008). American mathematician and dynamical meteorologist. Modest and prone to speaking in a soft monotone, he gave the impression of a laconic Yankee farmer, not that of one of the world's greatest scientists (which he was). Around 1960, he concocted a simple computer model of artificial weather, and in so doing discovered the butterfly effect in chaotic systems. He later simplified this model to a system of three innocent-looking differential equations, known now as the Lorenz system. This "little model" (as he called it) launched the chaos revolution. Unlike the chaos found by Poincaré in the three-body problem, which was transient and therefore hard to visualize, Lorenz's system showed never-ending, self-sustained chaos. While studying it, Lorenz uncovered two kinds of order within the chaos, now called strange attractors and iterated maps. Lorenz died, shortly after the lectures for this course had been recorded, at the age of 90.

Mandelbrot, Benoit (b. 1924). French-American mathematician, born in Poland. A polymath, he has analyzed the firing of nerve cells, the fluctuations of cotton prices, the shapes of galaxies and coastlines, the turbulence of fluids, and the dynamics of iterated maps. Much of his work has centered on the themes of irregularity, self-similarity, and heavy-tailed distributions, all of which are conspicuous in the natural world, yet all of which had been shunted off to the fringes of science as isolated curiosities. Synthesizing and extending work from many fields, he pulled these curiosities into a coherent framework that he called fractal geometry. Fractals also arise in the analysis of chaotic systems. In a sense they are the footprints of chaos, the geometric stamp of chaos.

May, Robert (b. 1936). Theoretical biologist. Born in Australia and trained as a theoretical physicist, his interests shifted early on to the dynamics of animal populations. His work in mathematical ecology is among the most highly regarded in the field. Later he turned his attention to the dynamics of infectious diseases. An elegant, witty speaker and writer, he published a review article in 1976 that brought the incredible nonlinear dynamics of the logistic map to a much wider scientific audience and helped to trigger the chaos boom that followed. He has been chief science adviser to the British government and is now Baron May of Oxford, a crossbencher in the House of Lords.

Newton, Isaac (1642–1727). Perhaps the greatest scientific genius of all time. Made unparalleled contributions to physics and mathematics. Invented calculus and differential equations. Postulated the law of universal gravitation and the three fundamental laws of motion. From them, he deduced that planets move in elliptical orbits and that they obey two other laws of planetary motion that Kepler had found empirically. As a person, he was difficult, secretive, and sexless. But the astonishing synthesis he created changed the world forever and ushered in the Scientific and Industrial Revolutions and the Enlightenment. Philosophically, his work led later thinkers to a supreme sense of confidence in the power of human reason.

Pecora, Louis (b. 1947). American physicist. Light-hearted and self-effacing, he was one of the first to see chaos as potentially useful, not just as an interesting curiosity. Working with his postdoctoral fellow Tom Carroll at the Naval Research Laboratory in Washington, DC, he used computer simulations and electrical circuits to demonstrate what others had thought impossible: Two identical chaotic systems could become perfectly synchronized while nevertheless fluctuating erratically. This work opened the door to using chaos for novel communication schemes.

Poincaré, Jules Henri (1854–1912). French mathematician. From a prominent family (cousin of Raymond, later president of the French Republic during World War I). The supreme mathematical mind of his era, comparable in sheer brainpower to Einstein, though less iconoclastic in spirit. He worked across the entire range of mathematics and physics, from geometry and topology to fluid mechanics and relativity theory. Intuitive and intensely geometric, yet unable to draw a decent sketch, he invented a new approach to differential equations, emphasizing the global behavior of all their solutions rather than focusing on special cases as others had done before him. While applying this approach to the greatest unsolved problem in astronomy—the three-body problem—he was shocked to discover the phenomenon we now call chaos.

Pollock, Jackson (1912–1956). American artist. Often classified as an Abstract Expressionist, he was inspired by the chaotic forces of nature and channeled them in his own work. In the 1940s he created a new technique, called drip painting. Pollock laid the canvas flat across the floor of his windswept barn and then walked around it, leaning in a state of controlled off-balance, while dripping, flinging, or pouring paint continuously onto the canvas. Though he was reviled as a drunk and a fraud by some critics at the time, his paintings are now valued at tens of millions of dollars.

Swinney, Harry (b. 1939). American experimental physicist. Interested in patterns, instabilities, and chaos in a wide range of nonlinear systems, from the Great Red Spot of Jupiter, to oscillating chemical reactions, to the flow of sand, ball bearings, and other granular materials. His 1975 experiments (with Jerry Gollub) on the onset of turbulence in fluid flow between two rotating cylinders lent early support to Ruelle and Takens's theory of turbulence based on strange attractors and decisively contradicted an older theory.

Taylor, Richard (b. 1963). Australian physicist, also trained in art theory. Works on applications of fractals to art and the visual sciences, as well as to electronic and optical nano-devices. He and his colleagues have applied fractal analysis to Jackson Pollock's drip paintings. They controversially claim that his paintings qualify as fractals and that their fractal dimension systematically increased over the decade while Pollock was refining his technique.

West, Geoffrey (b. 1940). Theoretical physicist. Born in a rural town in western England, he worked at Los Alamos as a particle physicist and then switched in midcareer to working on complex systems in biology and society. Currently president of the Santa Fe Institute, which is devoted to complexity studies. In 1997, he and his biological colleagues Jim Brown and Brian Enquist proposed a far-reaching explanation for Kleiber's law of metabolism and other scaling laws in biology, based on the pervasiveness of fractal branching networks in all living things.

Winfree, Arthur (1942–2002). American theoretical biologist. He worked on the dynamics of biological rhythms, from sleep-wake cycles to arrhythmias of the heart. Creative and exceptionally visual, his ideas and discoveries repeatedly opened new branches of study in mathematics, physics, and biology, for which he was awarded a MacArthur genius award (as his son Erik later was), along with top prizes in cardiology and applied mathematics. His first discovery, made as an undergraduate, concerned the collective synchronization of biological oscillators. He showed that as the coupling between the oscillators is increased, synchronization erupts spontaneously beyond a certain threshold, in a manner reminiscent of a phase transition.

Bibliography

Essential Reading:

Belbruno, E. *Fly Me to the Moon: An Insider's Guide to the New Science of Space Travel*. Princeton, NJ: Princeton University Press, 2007. A fun, fast-paced memoir by the creator of a new approach to space travel: surfing the gravitational chaos of the solar system.

Gleick, J. *Chaos: Making a New Science*. New York: Viking, 1987. The best book on chaos. It has everything—wonderful storytelling, memorable characters, an exhilarating sense of intellectual adventure, and exceptionally good explanations of the main ideas and why they matter. Read this book!

Lorenz, E. N. *The Essence of Chaos*. Seattle: University of Washington Press, 1993. Part memoir and part tutorial on the basics of chaos, this popular book by one of the giants of the field is characteristically understated, occasionally wry, and always illuminating.

Mandelbrot, B. *The Fractal Geometry of Nature*. San Francisco: W. H. Freeman, 1982. Idiosyncratic masterpiece by the genius who put fractals on the map. Hard to follow, but amazingly wide-ranging and original.

Peterson, I. *Newton's Clock: Chaos in the Solar System*. New York: W. H. Freeman, 1993. Excellent popular account of the development of celestial mechanics, written by a superb science journalist.

Rockmore, D. *Stalking the Riemann Hypothesis: The Quest to Find the Hidden Law of Prime Numbers*. New York: Pantheon, 2005. A terrific introduction to the Holy Grail of mathematics—the Riemann hypothesis—and its tantalizing connection to quantum chaos.

Schroeder, M. *Fractals, Chaos, Power Laws: Minutes from an Infinite Paradise*. New York: W. H. Freeman, 1991. Witty and erudite, this dazzling survey contains insights you won't find anywhere else. Aimed at readers comfortable with college-level mathematics.

Stewart, I. *Does God Play Dice? The Mathematics of Chaos.* Oxford: Blackwell, 1989. An outstanding popular account of chaos theory. Covers many of the same topics as Gleick's book, but in more mathematical depth and with less colorful stories.

Strogatz, S. *Sync: The Emerging Science of Spontaneous Order.* New York: Hyperion, 2003. Aimed at the general reader, this book explores nature's amazing ability to synchronize itself, from traffic patterns to brain waves.

Taylor, R. P. "Order in Pollock's Chaos." *Scientific American,* December 2002, 116–21. Taylor's excellent introduction to his fractal analysis of Jackson Pollock's drip paintings.

Whitfield, J. *In the Beat of a Heart: Life, Energy, and the Unity of Nature.* Washington, DC: Joseph Henry Press, 2006. Very nice book about the scaling laws of life and their proposed explanation by West, Brown, and Enquist in terms of the fractal networks inside all living things. Balanced, clear, and authoritative.

Recommended Reading:

Alon, U. *An Introduction to Systems Biology: Design Principles of Biological Circuits.* Boca Raton, FL: Chapman & Hall/CRC, 2006. An excellent text on the hot new field of systems biology, by one of its leaders.
Ball, P. *The Self-Made Tapestry: Pattern Formation in Nature.* New York: Oxford University Press, 1999. Offers broad coverage of patterns in nature, from fractal branching networks in plants and animals, to sand dunes, snowflakes, and zebra stripes.

Barabasi, A. L. *Linked: The New Science of Networks.* Cambridge, MA: Perseus, 2002. An engaging discussion of networks in science and society, by one of the pioneers in this emerging field.

Berry, M. "Quantum Physics on the Edge of Chaos." Chap. 15 in *Exploring Chaos: A Guide to the New Science of Disorder*, edited by N. Hall. New York: W. W. Norton, 1992. A fascinating introduction to the riddle of quantum chaos, by one of the world's most inventive theoretical physicists.

Buchanan, M. *Nexus: Small Worlds and the Groundbreaking Science of Networks*. New York: W. W. Norton, 2002. An excellent, wide-ranging introduction to the recent breakthroughs in the science of networks.

———. *Ubiquity: Why Catastrophes Happen*. New York: Three Rivers Press, 2001. A nice introduction to self-organized criticality, an ambitious and controversial theory that grew out of the chaos revolution of the 1970s and '80s. The goal was to explain why fractals and power laws are so ubiquitous and why so many diverse systems seem to organize themselves right to the brink of catastrophe (and sometimes beyond).

Cohen, I. B. *Science and the Founding Fathers: Science in the Political Thought of Jefferson, Franklin, Adams, and Madison*. New York: W. W. Norton, 1995. Especially intriguing for its discussion of Newton's impact on the thinking of Thomas Jefferson, as reflected in the language of the Declaration of Independence.

Crichton, M. *Jurassic Park*. New York: Knopf, 1990. Chaos theory is a central character in this bestselling novel about a theme park where cloned dinosaurs run amok, despite the best laid plans of the genetic engineers who brought them back to life.

Cvitanovic, P. *Universality in Chaos*. 2nd ed. Bristol, UK: Adam Hilger, 1989. A reprint collection of the seminal articles in chaos theory, up until the late 1980s. The famous theoretical papers of Lorenz, May, Feigenbaum, and Hénon are included, as are several experimental papers confirming the theory.

Diacu, F., and P. Holmes. *Celestial Encounters: The Origins of Chaos and Stability*. Princeton, NJ: Princeton University Press, 1996. The first chapter gives a great account of Poincaré's work on the three-body problem and his tortuous path to the discovery of chaos. Later chapters dive more deeply into

Bibliography

the mathematics and discuss the ongoing interplay between chaos theory and celestial mechanics.

Ditto, W. L., and L. M. Pecora. "Mastering Chaos." *Scientific American*, August 1993, 78–84. Nicely explains how chaos can be harnessed for communications and used to stabilize erratically pulsating lasers and arrhythmic animal hearts.

Du Sautoy, M. *The Music of the Primes: Searching to Solve the Greatest Mystery in Mathematics*. New York: Harper Collins, 2003. Like Rockmore's book, this is an excellent popular account of the Riemann hypothesis of prime number theory and its intriguing connection to quantum chaos.

Feder, J. *Fractals*. New York: Plenum, 1988. A good technical monograph about the physics of fractals.

Frank, A. "Gravity's Rim." *Discover*, September 1994. Available online at http://discovermagazine.com/1994/sep/gravitysrim419. Entertaining popular treatment of Belbruno's work on using chaos in space travel.

Galison, P. *Einstein's Clocks, Poincaré's Maps: Empires of Time*. New York: W. W. Norton, 2003. The dramatic story of two scientific giants, Einstein and Poincaré, in a race to understand the nature of time. Gives a unique account of their scientific and technological milieu in the years just before the development of relativity theory. Galison nicely sets Poincaré's work on chaos in this larger context.

Garfinkel, A., M. L. Spano, W. L. Ditto, and J. N. Weiss. "Controlling Cardiac Chaos." *Science*, August 28, 1992, 1230–35. A demonstration that drug-induced arrhythmia in a rabbit heart can be stabilized by intermittent, gentle electrical stimulation applied at specific times dictated by chaos theory.

Girardot, N. J. *Myth and Meaning in Early Taoism: The Theme of Chaos (Hun-Tun)*. Los Angeles: University of California Press, 1983. Explores the Taoist concept of chaos, which foreshadowed many of our contemporary scientific notions.

Glass, L., and M. C. Mackey. *From Clocks to Chaos: The Rhythms of Life.* Princeton, NJ: Princeton University Press, 1988. Very accessible monograph on biological oscillations and chaos.

Goldberger, A. L., D. R. Rigney, and B. J. West. "Chaos and Fractals in Human Physiology." *Scientific American*, February 1990, 42–49. Popular account of the role that chaos and fractals play in human health and disease.

Gutzwiller, M. C. "Quantum Chaos." *Scientific American*, January 1992, 78–84. Enjoyable summary of quantum chaos, by a physicist who made important contributions to this area.

Hall, N., ed. *Exploring Chaos: A Guide to the New Science of Disorder.* New York: W. W. Norton, 1992. A collection of excellent expository articles about chaos, reprinted from the British magazine *New Scientist* and written by such luminaries as Robert May, Benoit Mandelbrot, and Michael Berry.

Iasemidis, L. D., et al. "Adaptive Epileptic Seizure Prediction System." *IEEE Transactions on Biomedical Engineering* 50 (2003): 616–27. Technical article showing that there are recognizable precursors to epileptic seizures, as revealed by analysis of a patient's EEG (brain waves) in the minutes before the seizure. The article goes on to describe a seizure prediction algorithm that is genuinely predictive, meaning that it warns of impending seizures and is not merely based on retrospective data.

Klarreich, E. "Navigating Celestial Currents: Math Leads Spacecraft on Joy Rides Through the Solar System." *Science News*, April 16, 2005, 250–52. Excellent account of how chaos can be used to send unmanned probes through the solar system at very low cost, starting with Belbruno's early work and continuing to the more recent concept of an interplanetary superhighway, now used by NASA for some of its missions.

Kline, M. *Mathematics in Western Culture.* New York: Oxford University Press, 1953. A magnificent book, feisty and learned and full of insight. Kline persuasively documents the importance of math in all aspects of Western culture. The chapters about Newton's influence on science, philosophy, religion, literature, and aesthetics are wonderful, but so is the rest of the book.

Liebovitch, L. S. *Fractals and Chaos Simplified for the Life Sciences*. New York: Oxford University Press, 1998. A pedagogically creative text, organized in an unusual way: The pages come in pairs, with crisp explanations of basic concepts on the left-hand page accompanied by real data or line drawings on the right-hand page. Includes good discussions of fractal statistics, fractal processes in biology and their significance, the medical benefits of chaos, and techniques for controlling chaos.

Lighthill, J. "The Recently Recognized Failure of Predictability in Newtonian Dynamics." *Proceedings of the Royal Society of London, Series A* 407 (1986): 35–48. Sir James Lighthill held the Lucasian Professorship at Cambridge, the same chair that Newton once held and that is now occupied by Stephen Hawking. His article is notable for its lucid discussion of the "predictability horizon" that plagues all chaotic systems.

Mandelbrot, B., and R. L. Hudson. *The (Mis)behavior of Markets: A Fractal View of Risk, Ruin, and Reward*. New York: Basic Books, 2004. An iconoclastic, fractal approach to finance.

May, R. M. "Simple Mathematical Models with Very Complicated Dynamics." *Nature*, June 10, 1976, 459–67. Impassioned and memorable, this review article helped launch the chaos revolution by alerting the wider scientific community to the wonders of even the simplest nonlinear systems.

McMahon, T. A., and J. T. Bonner. *On Size and Life*. New York: Scientific American Library, W. H. Freeman, 1983. A delightfully written and beautifully illustrated book about the scaling laws of life by a pair of eminent scientists—one an engineer, the other a biologist.

Ott, E., T. Sauer, and J. A. Yorke. *Coping with Chaos: Analysis of Chaotic Data and the Exploitation of Chaotic Systems*. New York: Wiley, 1994. Reprint collection of classic articles about controlling, harnessing, and synchronizing chaos.

Peak, D., and M. Frame. *Chaos Under Control: The Art and Science of Complexity*. New York: W. H. Freeman, 1994. Lively text about chaos and fractals. Unusual for its broad coverage, including connections to music, art, and fiction.

Peitgen, H. O., and P. H. Richter. *The Beauty of Fractals: Images of Complex Dynamical Systems*. Berlin: Springer-Verlag, 1986. The stunning multicolor images of the Mandelbrot set, now commonplace, first appeared here. Still great as a coffee-table book, though its mathematics is presented at an advanced level.

Pikovsky, A., M. Rosenblum, and J. Kurths. *Synchronization: A Universal Concept in Nonlinear Sciences*. Cambridge: Cambridge University Press, 2001. Very good technical monograph on synchronization in physical, biological, and chemical systems.

Poincaré, H. *Science and Method*. New York: Dover, 2003. Most famous for Poincaré's essay on the psychology of mathematical discovery, this lucidly written book also contains prescient discussions of chance and chaos.

Rehmeyer, J. J. "Fractal or Fake: Novel Art-authentication Method is Challenged." *Science News*, February 24, 2007, 122–124. Terrific popular article about the controversy surrounding the use of fractal analysis in authenticating Jackson Pollock's drip paintings.

Roux, J. C., R. H. Simoyi, and H. L. Swinney. "Observation of a Strange Attractor." *Physica* D 8 (1983): 257–266. Careful experimental study of chaos in a chemical reaction, confirming the reality of strange attractors, iterated maps, and other predictions of chaos theory.

Stoppard, T. *Arcadia*. New York: Faber and Faber, 1993. Improbable as it may sound, Tom Stoppard managed to weave beautifully clear explications of chaos, fractals, iterated maps, and entropy into this hit play about the mysteries of time, determinism, and love.

Strogatz, S. H. *Nonlinear Dynamics and Chaos: With Applications to Physics, Biology, Chemistry, and Engineering*. Cambridge, MA: Perseus, 1994. The most widely used textbook on chaos theory. Emphasizes applications, concrete examples, and geometric intuition. Assumes knowledge of first-year calculus and physics.

Strogatz, S. H., D. M. Abrams, A. McRobie, B. Eckhardt, and E. Ott. "Crowd Synchrony on the Millennium Bridge." *Nature*, November 3, 2005, 43–44. A mathematical model of what happened to the Millennium Bridge on opening day.

Taleb, N. N. *The Black Swan: The Impact of the Highly Improbable*. New York: Random House, 2007. A lively (and sometimes comically belligerent) book about the way the world really works. Taleb argues that life is far more random than we are willing to admit and that history is dominated by rare, unforeseeable events that change everything.

Turcotte, D. L. *Fractals and Chaos in Geology and Geophysics*. 2nd ed. Cambridge: Cambridge University Press, 1997. Authoritative treatment of fractals and power laws applied to earthquakes, wildfires, landscape topography, and other phenomena in the Earth sciences, by a leader in this field.

Winfree, A. T. *The Timing of Biological Clocks*. New York: Scientific American Library, W. H. Freeman, 1987. A fascinating, sometimes quirky survey of daily rhythms in plants and animals, written by an exceptionally creative scientist. Winfree uses visual mathematics (specifically, mappings between circles) and multicolored pictures to reveal the underlying principles of biological oscillations, and in this way exposes a gorgeous unity among diverse living things.

———. *When Time Breaks Down: The Three-Dimensional Dynamics of Electrochemical Waves and Cardiac Arrhythmias*. Princeton, NJ: Princeton University Press, 1987. Winfree was also a pioneer in the use of mathematics to explain why biological rhythms sometimes stop abruptly. Here he explores the mystery of sudden cardiac death, which kills hundreds of thousands of seemingly healthy people each year, when their hearts lapse into fibrillation, a mysterious kind of electrical turbulence.

Internet Resources:

Eric Heller's online gallery of quantum art: www.ericjhellergallery.com. Frequently asked questions about chaos (somewhat out of date, but still valuable): http://www.faqs.org/faqs/sci/nonlinear-faq/.

A fascinating conversation with the playwright Tom Stoppard about the starring role of chaos theory in his hit play *Arcadia*: http://www.siam.org/news/news.php?id=727.

For all things fractal: http://classes.yale.edu/Fractals/.

Online simulations (Java applets) for chaos and fractals: http://math.bu.edu/DYSYS/applets/.

Excellent compendium of Java applets for chaos and fractals: http://www.student.math.uwaterloo.ca/~pmat370/JavaLinks.html.

Advanced software for dynamical systems (mostly for mathematically sophisticated users): http://www.dynamicalsystems.org/sw/sw/.

Advanced tutorials about dynamical systems, many accompanied by online simulations: http://www.dynamicalsystems.org/tu/tu/.

Laboratory demonstrations of chaos and other nonlinear phenomena—a video hosted by Professor Strogatz, with the help of several colleagues: http://ecommons.library.cornell.edu/handle/1813/97.

Java applets for billiards: http://serendip.brynmawr.edu/chaos/home.html.

More links for chaos, fractals, nonlinear dynamics, etc.: http://mathforum.org/library/topics/dynamical_systems/.

For applets about many fascinating complex systems: http://ccl.northwestern.edu/netlogo/models/.

Website about the Millennium Bridge: http://www.arup.com/MillenniumBridge/.

Explore the statistical distribution of earthquakes of various magnitudes: http://www.data.scec.org/Module/s2act08.html.

Internet Resources by Lecture:

Lecture Four

You can play with a simulation of a pendulum and its motion in state space here: http://www.aw-bc.com/ide/idefiles/media/JavaTools/pndulums.html.

Lecture Five

If you want to examine the butterfly effect for yourself in a simulation of another of Lorenz's models (to be discussed in Lectures Seven and Eight), try this online Java applet: http://www.aw-bc.com/ide/idefiles/media/JavaTools/lrnzdscv.html.

Lecture Seven

Play with a Java applet of Lorenz's attractor here: http://www.aw-bc.com/ide/idefiles/media/JavaTools/lrnzr320.html.

For a simulation that ties the mathematical model to the physical problem of convection, try this one: http://www.aw-bc.com/ide/idefiles/media/JavaTools/lrnzphsp.html.

Lecture Eight

Play with Lorenz's iterated map at this site: http://www.aw-bc.com/ide/idefiles/media/JavaTools/lrnzzmax.html.

Lecture Nine

Explore the dynamics of the logistic map yourself. Go to: http://www.student.math.uwaterloo.ca/~pmat370/JavaLinks.html to find some relevant Java applets.

One of the simplest to use is this one: http://www.geom.uiuc.edu/~math5337/ds/applets/iteration/Iteration.html.

Lecture Ten

Explore the orbit diagram yourself. A very nice Java applet is here: http://math.bu.edu/DYSYS/applets/bif-dgm/Logistic.html. Try to find the mini-orbit diagrams at the end of a big periodic window.

Lecture Thirteen

Play with Hénon's mapping and strange attractor here: http://www.cmp.caltech.edu/~mcc/Chaos_Course/Lesson5/Demo1.html.

Lecture Twenty-Two

Explore the dynamics of billiards on tables of various shapes here: http://serendip.brynmawr.edu/chaos/home.html.

Eric Heller's online gallery of quantum art: www.ericjhellergallery.com.

Lecture Twenty-Three

Play with a model for the synchronization of fireflies, using this applet: http://ccl.northwestern.edu/netlogo/models/Fireflies.

Notes

Notes